U0545599

Deepen Your Mind

Deepen Your Mind

序

讀者你好！歡迎閱讀這本書，如果想學習如何製作一個網頁中的 3D 遊戲，又或者想了解 GPU 繪圖的相關原理，WebGL 讓我們可以在 Web 這樣通用且功能豐富的平台上進行入門，本書將從一個三角形開始出發，一步一步做出範例的方式，並搭配過程中每步原理的說明，與讀者分享 WebGL 的來龍去脈、使用方式以及心得，最後製作出一款在網頁中運作的 3D 小遊戲！

筆者從事網站開發多年，其中對於『使用者互動』、『有畫面』的程式設計特別有興趣，工作上也因此經常使用前端技術進行開發，後來想在網站中撰寫遊戲，而遊戲中會讓人第一個想到、也是能在第一瞬間吸引人的就是畫面了，畢竟本來就對前端的生態系熟悉，也就順勢從 WebGL 開始進行研究，也於 2021 iThome 鐵人賽撰寫一系列 WebGL 的技術文章與網友們分享研究的結果，本書內容即為此鐵人系列文章的改良，並擴寫加入延伸互動引導讀者繼續製作成小遊戲。

如果讀者同樣身為前端或全端開發者，對於 Web 生態系最核心的 Javascript 有一定的掌握，同時也知道一些 Web API、HTML 以及 CSS 的使用，能夠製作一個有前端互動的網頁（無論是否使用工具、框架），那麼你就已經具備閱讀本書所需的基礎知識，讓本書幫助你再獲得一項技能吧！

⌘ WebGL 是啥？為何是 WebGL？

簡單來說，WebGL 是一組在 Web 上操作 GPU 的 Javascript API，而 WebGL 絕大部分的 API 都可以找到 openGL[1] 上對應的版本，且名字幾乎沒有差別，猜測制定 WebGL 標準時只打算做一層薄薄的包裝，這樣一方面瀏覽器

1　openGL 是用於渲染 2D、3D 向量圖形的 API，這層 C 語言的 API 實做通常由 GPU 硬體廠商提供，以當今 Web 的角度來說，可說是非常底層的 API。其維基百科：https://zh.wikipedia.org/wiki/OpenGL

可能比較好實做，但是也因此 WebGL 直接使用時是非常底層的，甚至偶爾會需要去算線性代數、矩陣的東西。

看到這邊讀者們可能會想說：『哇，我要來洗洗睡了。』老實說，如果對於基礎原理沒有興趣，想要『快速』做出東西，這邊確實可以左轉 three.js 或是 babylon.js 這種高階的 3D 渲染引擎，筆者是基於下面這個因素決定研究 WebGL 的：

當你了解其原理時，比較不容易受到框架、潮流演進的影響

為什麼？在了解原理的狀況下，比較能知道框架幫你做了什麼，遇到變化比較大的需求的時候可能比較容易想到方法應對；前端技術更迭速度大家都知道，但是基礎原理是不會有太大的變化的，現今渲染技術除了 openGL 之外還有 Apple 的 Metal、微軟的 DirectX 甚至更新更潮的 Vulkan，相信透過 WebGL 學到的原理多多少少也能應用在其他平台上吧。

⌘ WebGL2 以及其相容性

WebGL 有兩個版本：WebGL 以及 WebGL2，WebGL 第一版（本書中接下來會以 WebGL1 來強調是第一版）的標準中有一些綁手綁腳的限制，例如在 WebGL 中渲染時啟用自動縮放的圖樣（Texture）長寬必須是 2 的次方；甚至缺少一些實務上需要的功能，雖然有些可以透過非標準的 WebGL extension 彌補（此 extension 並非指瀏覽器擴充套件，在後續的章節中將會提到），但是就是會多出一些不方便的限制，因此 WebGL2 標準的出現就是將第一版的缺失補起來；在筆者撰文當下，WebGL2 在 Chrome、Firefox、Safari、Edge 等瀏覽器的最新版都有支援，行動裝置的部份，Android Chrome 最新版以及 iOS 15 以後的 Safari（iPhone 6s 以上能安裝 iOS 15）也都能使用 WebGL2，詳細的支援情況可以參考此網頁：https://caniuse.com/webgl2

因為主流瀏覽器都已經支援，本書接下來的說明以及範例都將以 WebGL2 為主，不過遇到使用之功能在 WebGL 第一版有差異時也會稍微提一下讓讀者們知道。

有需要的讀者也可以參考 WebGL 第一版的相容表：
https://caniuse.com/webgl，例如 IE 11 就只支援到 WebGL 第一版，如果讀者需要支援更多更舊的瀏覽器，那麼可能就得使用 WebGL 第一版了。

⌘ 如何使用本書 — 範例程式碼

本書透過實做範例，並在過程中講解來學習，為了方便讀者取得範例程式碼，筆者將範例程式碼放置在 Github 上：
https://github.com/pastleo/webgl-book-examples

讀者可以在 Github repository 頁面點選 "Code" 並 "Download ZIP"（如圖 0-1 所示）下載『範例程式碼包』到電腦中並解壓縮，在壓縮檔的頂層資料夾 webgl-book-examples-main/ 將包含所有範例以及各個小章節的進度快照，接下來在本書中如果看到以下程式碼的圖示以及檔名：

</>	ch1/01/index.html
</>	ch1/01/main.js

即表示其中之 ch1/01/index.html、ch1/01/main.js 兩個檔案，接下來的章節可能會使用這些檔案作為起始點進行修改，或是該章節的進度快照，讓讀者有完整的專案可以參考。

▲ 圖 0-1 電腦下載『範例程式碼包』

如果讀者純粹閱讀本書，不是跟著進度實做範例的方式進行，那麼建議用瀏覽器（電腦或行動裝置皆可）打開上述放置於 Github 網站上的範例程式碼，以便在文章中提到相關程式碼時得以參考對照，下圖 0-2 可能為行動裝置打開 Github 網站範例程式碼所看到的樣子，需要點選『View code』展開範例程式內容，如圖 0-3：

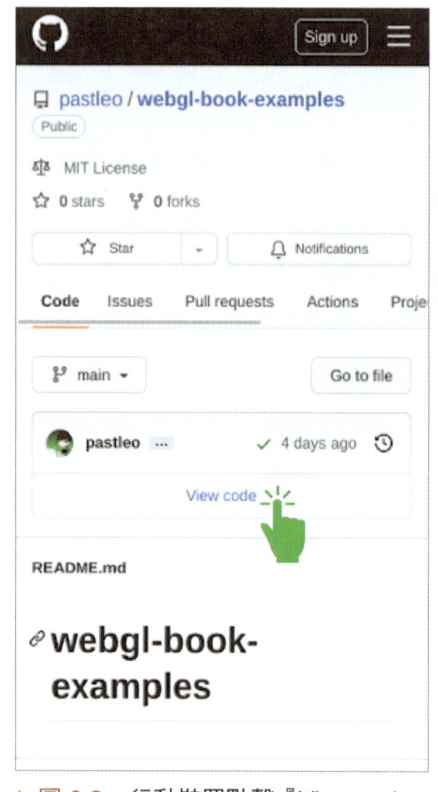

▲ 圖 0-2　行動裝置點擊『View code』　　　▲ 圖 0-3　行動裝置開啟樣子

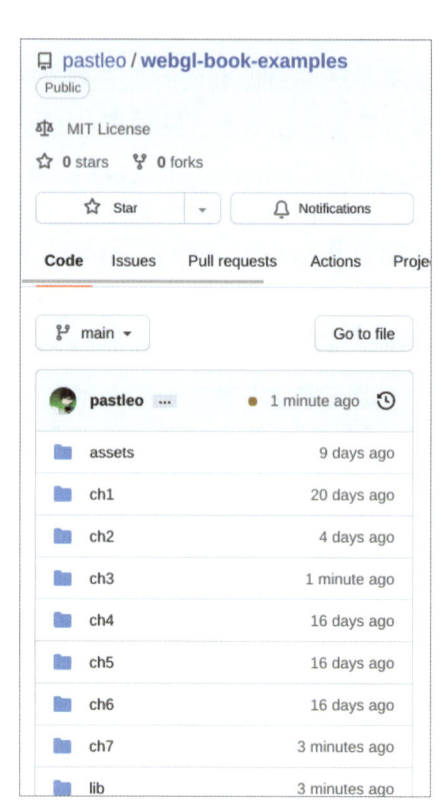

vi

⌘ 如何使用本書 — 程式碼區塊之底色標注

在本書章節段落中，會穿插許多程式碼區塊，這些程式碼區塊可能是當下進度所實做或是修改的範例程式，例如：

```glsl
#version 300 es
in vec2 a_position;
in vec2 a_texcoord;

uniform vec2 u_resolution;
uniform vec2 u_offset;
uniform mat3 u_matrix;
out vec2 v_texcoord;

void main() {
  vec2 position = a_position + u_offset;
  gl_Position = vec4(
    position / u_resolution * vec2(2, -2) + vec2(-1, 1),
    0, 1
  );
  vec3 position = u_matrix * vec3(a_position.xy, 1);
  gl_Position = vec4(position.xy, 0, 1);
  v_texcoord = a_texcoord;
}
```

在這段程式碼區塊中可以看到有部份片段有紅色、綠色或是黃色的底色標注，這些標注分別表示：

紅色底色標注：表示要刪除的程式，常常在取代原本實做或是重構時會有

綠色底色標注：表示要新增的程式，看到此底色通常表示整個程式碼區塊其他程式是現有不用更改的程式

黃色底色標注：表示修改取代後的程式，也有可能是表示值得注意的地方

vii

⌘ 在開始之前

筆者要來推薦一個 WebGL 學習網站：WebGL2 Fundamentals，網址為 https://webgl2fundamentals.org/，這個網站上的教學寫的非常完整，從基礎一路到各個功能的實做，有興趣的讀者不妨交互參考看看。

如果讀者在閱讀的過程遇到問題、想要與筆者聯繫，歡迎參考筆者網站下方的聯絡方式：https://pastleo.me/about

本書在撰寫過程中難免有錯誤，筆者將把出版後發現的勘誤內容放置在範例程式 Github page 的首頁上：https://webgl-book.pastleo.me/#勘誤表

目錄

序

1 Hello WebGL

1-1　準備開發環境 1-2
1-2　畫一個三角形（上）... 1-12
1-3　畫一個三角形（下）... 1-21
1-4　Uniform – shader 之參數 1-28
1-5　畫多個三角形 1-37
1-6　Varying – fragment shader 之資料........................... 1-39

2 Texture & 2D

2-1　在 WebGL 取用、顯示圖片 – Textures................... 2-2
2-2　Texture 使用上的細節 2-12
2-3　互動 & 動畫 2-25
2-4　2D Transform 2-44
2-5　2D transform Continued 2-54

3 3D & 物件

3-1　Orthogonal 3D 投影 3-2
3-2　Perspective 3D 成像 3-17
3-3　視角 Transform 3-34
3-4　使相機看著目標 3-41
3-5　渲染多個物件 3-52

ix

4　Lighting

4-1　法向量（Normals）與散射光（Diffuse） 4-4
4-2　Indexed Element 4-25
4-3　請 TWGL 替程式碼減肥 4-32
4-4　Specular 反射光 4-44
4-5　點光源與自發光 4-57
4-6　Normal Map 4-67

5　Framebuffer & 陰影

5-1　Framebuffer 是什麼？ 5-4
5-2　鏡面效果 5-14
5-3　陰影─拍攝深度資訊 5-26
5-4　陰影─深度 Framebuffer 與 Texture 5-37
5-5　陰影─計算是否產生陰影 5-41
5-6　毛玻璃效果─使用 Normal Map 的鏡面 5-52

6　帆船與海

6-1　主角『帆船』─ obj 3D 模型檔案的讀取與繪製 6-3
6-2　Skybox 6-23
6-3　半透明的文字看板 6-43
6-4　使用 Shader 即時渲染波光粼粼的海面 6-62

7　Catch The Wind 小遊戲

7-1　地形高度圖的產生 7-4
7-2　依照地形高度圖繪製島嶼 7-19
7-3　Set Sail! 航行帆船 7-36
7-4　遊戲標題與 UI 7-60
7-5　碰撞島嶼判定、結束遊戲 7-79
7-6　結語 7-104

Hello WebGL

開始搭建 WebGL 渲染環境，我們將建立一個基本的網頁，讓 WebGL 在網頁中的『畫布』繪製基本三角形；在此章節中可以學習到 WebGL 基本渲染流程、機制，於此過程中撰寫簡單的 shader，並了解傳送資料以及參數到 shader 中的方法。

1　Hello WebGL

第一次接觸什麼工具或是框架時，第一個要做的事情就是設立開發環境，當然 WebGL 也不例外，不過幸好 WebGL 只需要瀏覽器就可以運作，我們需要做的事情就是撰寫 HTML、Javascript 檔案，並且透過一個開發用的 HTTP 伺服器讓瀏覽器讀取網頁，執行 Javascript 去呼叫 WebGL API，便可在畫面上看到 WebGL 繪製的畫面。

1-1　準備開發環境

首先會需要一個文字編輯器，就跟開發一個純 HTML、Javascript、CSS 的網站一樣，事實上任何純文字編輯器皆可，由於有許多的網站開發者都使用 Visual Studio Code 進行開發，本書接下來會使用 Visual Studio Code 進行示範。

在進行本書範例的製作時建議設定編輯器使用序章『範例程式碼包』解壓縮之 `webgl-book-examples-main/` 資料夾作為專案頂層，以 Visual Studio Code 為例，使用『檔案（Files）』拉下式選單選擇『開啟資料夾 ...（Open Folder…）』（如圖 1-1 所示），並選擇 `webgl-book-examples-main/`：

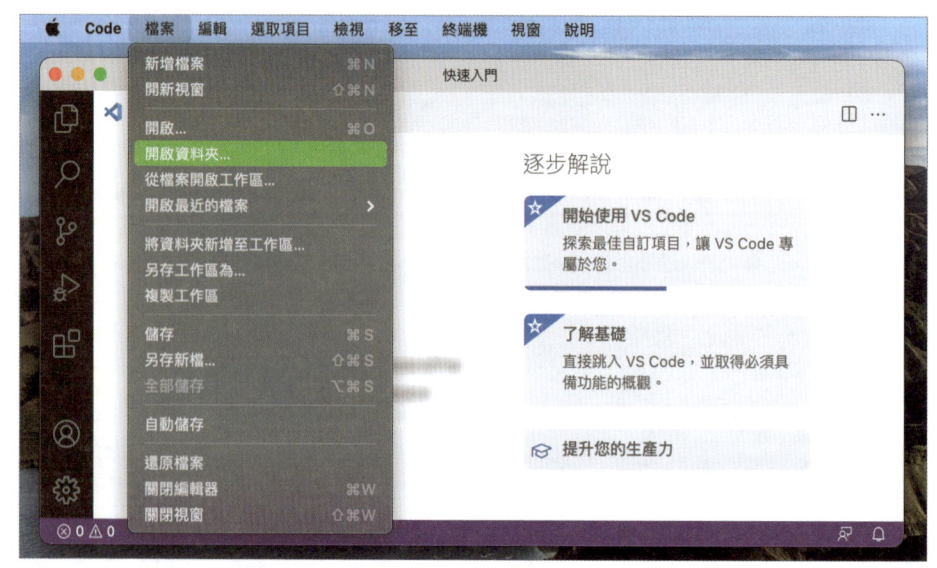

▲ 圖 1-1　Visual Studio Code 裡開啟資料夾

除此之外比較需要準備的就是上面說到的『開發用 HTTP 伺服器』，這邊可以使用任何靜態網頁伺服器，例如以下（擇一即可）：

1. Visual Studio Code 之延伸模組：Live Server

如果讀者使用的編輯器是 Visual Studio Code，這會是最方便的選擇，在 Visual Studio Code 左方點選延伸模組（Extensions）按鈕（如圖 1-2 內之方框 1），搜尋『live server』（如圖 1-2 內之方框 2），選擇並安裝『Live Server』（如圖 1-2 內之方框 3 以及 4）。

安裝完成之後，應該可以在 Visual Studio Code 編輯器視窗的右下角看到『Go Live』的按鈕（如圖 1-3），接下來撰寫好 HTML、Javascript 要測試時點一下這個按鈕即可啟動開發伺服器於 `http://127.0.0.1:5500`，並自帶 live reload 功能。

▲ 圖 1-2　安裝『Live Server』示意圖

▲ 圖 1-3　『Go Live』的按鈕

1-3

2. Node.js 之套件：`http-server`[1]

這是一個 Node.js 的套件，如果有安裝好 Node.js 以及 npm，可以在終端機上執行指令進行安裝：`npm install -g http-server`（有可能因為 Node.js 安裝於 macOS 或是 Linux 上的系統目錄，需要使用最高權限進行安裝：`sudo npm install -g http-server`），安裝完成後會有 http-server 的指令可供呼叫。

接下來開啟伺服器來提供專案網頁檔案給瀏覽器，打開終端機 cd 至專案頂層並執行 http-server，套件將以當前的工作目錄當成網站根目錄，預設把 port 開在 8080，可以透過 http://127.0.0.1:8080 打開網站，並具有檢視資料夾的功能。

3. Ruby 的簡易網頁伺服器：`webrick`

webrick 會內建於版本大於等於 1.9.2 且小於 3.0 的 Ruby 程式語言，如果有安裝好 Ruby 程式語言，那麼就可以使用 Ruby 內建的 `httpd`[2] 作為伺服器，如果版本在 3.0 以後也可以透過此指令進行安裝：`gem install webrick`

webrick 有內建或是安裝到 Ruby 程式語言之後，打開終端機 cd 至專案頂層並執行此指令：`ruby -run -e httpd`

Ruby 的 `httpd` 模組會去使用 `webrick` 來啟動網頁伺服器，與 `http-server` 類似，使用當前的工作目錄當成網站根目錄，而預設把 port 開在 `8080`，可以透過 `http://127.0.0.1:8080` 打開網站。

---

[1] Node.js 套件 `http-server` 詳細使用說明之 npm 頁面：https://www.npmjs.com/package/http-server

[2] Ruby `httpd` 詳細使用說明請見：https://apidock.com/ruby/Object/httpd

4. Python 內建的網頁伺服器

如果有安裝好 Python 3，程式語言也內建了網頁伺服器 `http.server`[3]，使用方式與 `http-server`、`webrick` 很像，同樣打開終端機 `cd` 至專案頂層作為網站根目錄，使用此指令啟動內建的網頁伺服器：`python -m http.server`

Python 的 `http.server` 模組預設把 port 開在 `8000`，可以透過 `http://localhost:8000` 打開網站；如果是安裝的版本是 Python 2，那麼模組名稱則是 `SimpleHTTPServer`，此時的指令應為：`python -m SimpleHTTPServer`

開發用網頁伺服器準備好之後，最後一個是負責執行並顯示結果的瀏覽器，讀者可以使用任何一個有支援 WebGL2 的瀏覽器，撰文當下最新版的桌面版 Chrome、Firefox、Edge 以及 Safari 皆有支援，詳細情況在序章有做過說明，本書接下來使用 Chrome 來進行示範。

⌘ `index.html`

一個網頁最少還是要有 HTML 檔案作為起點，使用 HTML5 建立第一個範例的網頁：

```html
<!DOCTYPE html>
<html lang="en">
<head>
  <meta charset="UTF-8">
  <title>ch1-hello-webgl</title>
</head>
<body>
  <canvas id="canvas"></canvas>
  <script type="module" src="main.js"></script>
</body>
</html>
```

3　Python 3 `http.server` 文件說明：https://docs.python.org/3/library/http.server.html

1 Hello WebGL

可以看到 `<body></body>` 下有這個元素：`<canvas id="canvas"></canvas>`，這就是 WebGL 待會會操作的『畫布』，WebGL 繪製的東西將透過這個元素呈現，另外 `<script type="module" src="main.js"></script>` 以 ES Module[4] 的形式引入接下來主要要寫的 Javascript 主程式：`ch1/main.js`

建立此 `main.js` 並與 `index.html` 一同放置在同一個資料夾下，先寫這簡單的一行：

```
console.log('hello');
```

此時最基本的專案檔案就建立好，`main.js` 和 `index.html` 的完整程式碼在範例程式碼包的以下位置找到：

</>	ch1/00/index.html
</>	ch1/00/main.js

讀者也可以參考下圖 1-4，以 Visual Studio Code 為例打開這兩個檔案的畫面：

---

[4] ES Module 是 ECMAScript 規定的一系列寫法，使用這種方式引入的 Javascript 可以使用 `import`、`export` 等關鍵字，使得 Javascript 可以更容易地模組化；同時這樣引入也有 `defer` 的效果，也就是 Javascript 會在 HTML 解析完畢之後才執行。

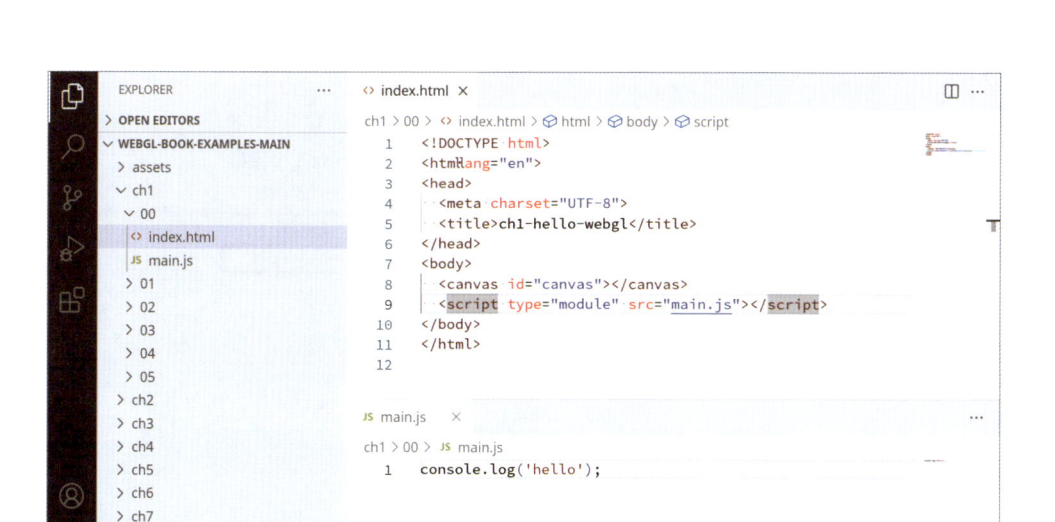

▲ 圖 1-4 起始點工作畫面

接著啟動網站開發伺服器來測試並觀察結果，如果使用 Visual Studio Code 搭配 Live Server 的話，延伸模組會自動開啟瀏覽器，若瀏覽器沒有直接開啟 HTML 檔案而是列出專案頂層的資料夾以及檔案，可以點擊 `ch1` 以及 `00` 進入資料夾開啟 `index.html`；Live Server 運作中的狀態可以在 Visual Studio Code 視窗的右下角看到，再次點擊即可關閉開發伺服器。

瀏覽器打開 `index.html` 之後，畫面一片白是正常的，按下 F12 或是右鍵檢視元件開啟開發者工具，接著應該可以在 Console 看到 `console.log()` 印出的 hello 文字，表示一切正常如圖 1-5：

1　Hello WebGL

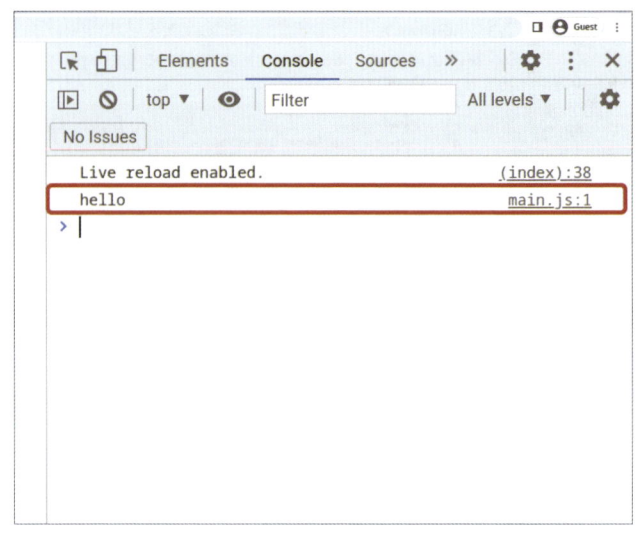

▲ 圖 1-5　起始點之 console.log('hello') 表示運作正常

　　為了確保之後對原始碼的更動能夠反應在瀏覽器上，建議切換至 Network tab 關閉快取，以 Chrome 為例如下圖 1-6。

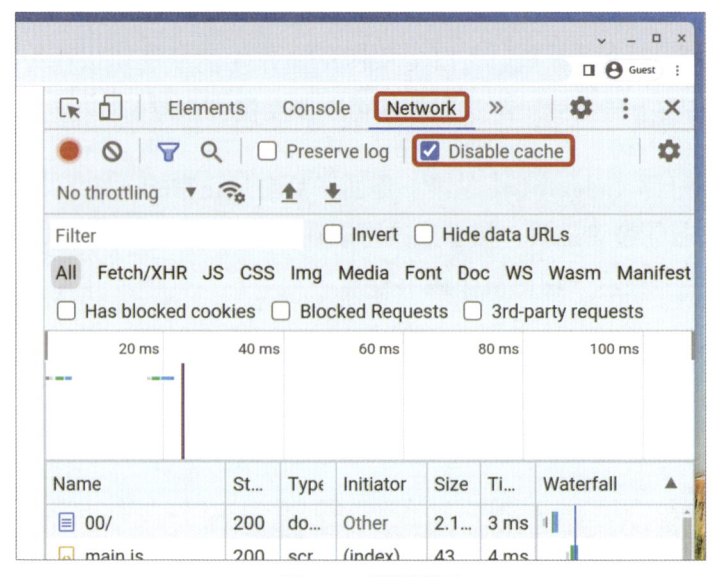

▲ 圖 1-6　關閉快取

1-8

⌘ 取得 WebGL2 Instance

好了，基本的 HTML 以及 Javascript 準備就緒，接著我們要取得一個 WebGL 實體，在一個 `<canvas />` 元素中有各種繪製 API 可供挑選，稱之為『context』，取得該 DOM 元素，並透過 `.getContext()` 傳入字串 `'webgl2'` 來指定並取得 WebGL2 實體，像是這樣（如果要取得 WebGL1，則傳入 `'webgl'`）：

```javascript
const canvas = document.getElementById('canvas');
const gl = canvas.getContext('webgl2');
window.gl = gl;
```

第一行透過 id 取得元素，使用 `.getContext()` 取得 gl 之後也設定到 `window.gl` 上方便在開發工具 Console 中玩轉，存檔後回到瀏覽器，讀者可以像下圖 1-7 一樣使用 Chrome 的提示瀏覽一下 gl 的屬性以及方法（methods）。

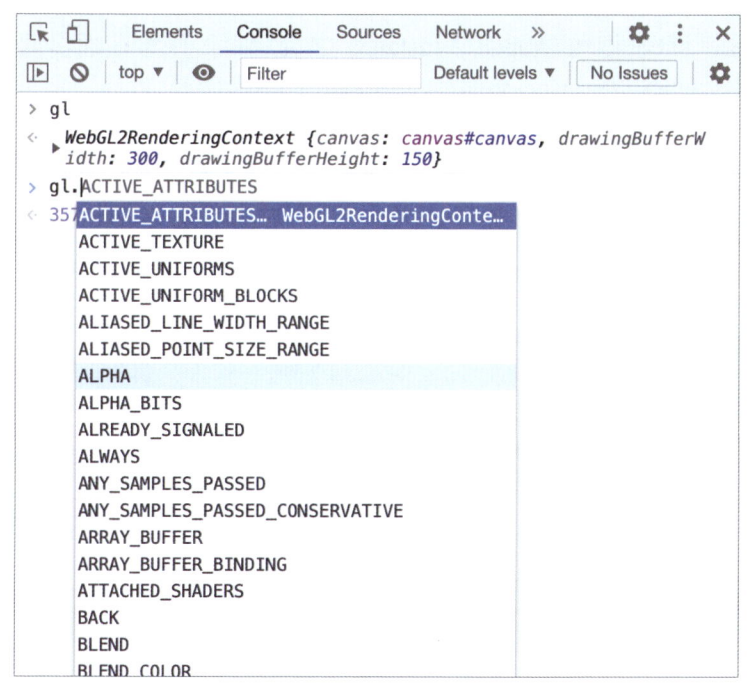

▲ 圖 1-7　使用 Chrome 的提示，瀏覽 gl 的屬性及方法

1 Hello WebGL

⌘ 好歹畫點東西吧

老實說，我們距離繪製一些有意義的東西還有點遙遠，不過倒是可以先找個顏色「清除」畫面，讓讀者們初嘗使用 WebGL API 做點操作，這邊說的「清除」其實有點像是重設整個畫面的 pixel 成為指定的顏色，產生類似油漆工具填滿整個畫面的效果。

首先要透過 `gl.clearColor(red, green, blue, alpha)` 設定清除（Clear）用的顏色，這邊 `red`、`green`、`blue`、`alpha` 是介於 0 到 1 之間的浮點數，設定好之後，呼叫 `gl.clear(gl.COLOR_BUFFER_BIT)` 進行清除（Clear），而 `gl.COLOR_BUFFER_BIT` 是用來指定清除顏色的部份，以筆者的主題色 `#6bde99`（`rgb(108, 225, 153)`）為例：

```
gl.clearColor(108/255, 225/255, 153/255, 1);
gl.clear(gl.COLOR_BUFFER_BIT);
```

把這兩行放在取得 WebGL2 Instance 之後執行，便會在網頁讀取之後把 `<canvas />` 清除為 #6bde99 的顏色，如下圖 1-8 所示：

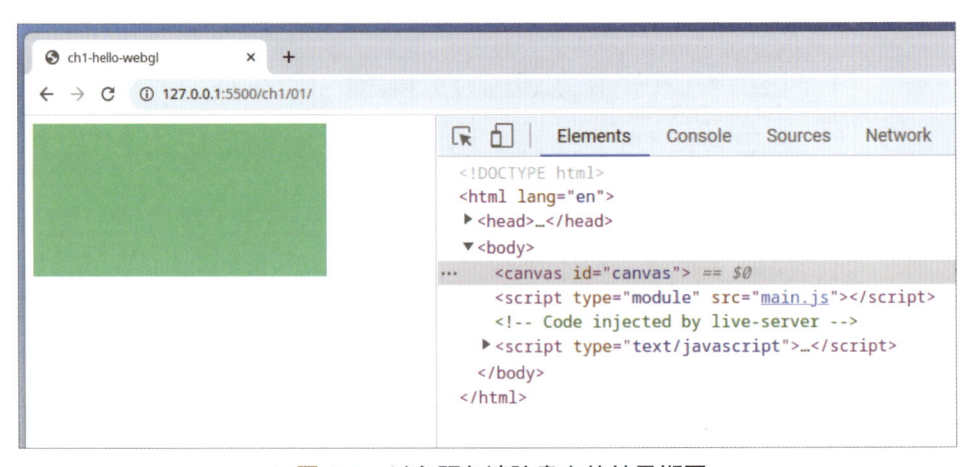

▲ 圖 1-8　以主題色清除畫布的結果擷圖

1-10

雖然只是拿油漆工具填滿整張畫布，但是至少讓畫面有點東西了，畫面上的綠色區塊就是 `<canvas />`，因為我們沒有設定長寬，在 Chrome 上預設的大小是寬 300px、高 150px。

不知道讀著們是否有一個疑問，為何不是直接呼叫 `gl.clear()` 同時傳入顏色，而是先呼叫 `gl.clearColor()` 先設定顏色呢？在 WebGL 中操作 GPU，事實上是從 CPU 下指令到 GPU 去，在 GPU 可能 `gl.clear()` 做繪製是一個指令，`gl.clearColor()` 設定顏色也是一個指令，就像是 GPU 有一個操作面板，CPU 作為操作員要先設定好清除用的顏色，接著按下清除按鈕；除此之外，在 GPU 這邊許多東西都是介於 0 到 1 之間的浮點數、`gl.COLOR_BUFFER_BIT` 更是體現跟底層甚至硬體溝通用的 bit flag。

從網站工程師的角度來看，WebGL 像是來自另外一個世界的底層 API，想當然爾若不是要學習 GPU 繪圖的原理，而是進行實際應用程式的開發時，開發者們不是使用寫好的工具、框架，就是自己開發出一套自己的包裝來加快開發速度。

到此進度的完整程式碼可以在這邊找到：

</>	ch1/01/index.html
</>	ch1/01/main.js

也可以參考上線版本，透過瀏覽器打開此頁面：
https://webgl-book.pastleo.me/ch1/01/index.html

1 Hello WebGL

1-2 畫一個三角形（上）

準備好開發環境以及最基本的 HTML 以及 `<canvas>`，接下來將從一個三角形開始，介紹 WebGL 操作 GPU 進行繪製的基本工作流程。

為什麼是從三角形開始呢？在讓電腦繪製一個三維場景時，我們實際在做的事情把這三維場景中物體的『表面』畫在畫面上，而構成一個面最少需要三個點，三個點構成一個三角形，而所有更複雜的形狀或是表面都可以用複數個三角形做出來，因此使用 3D 繪製相關的工具時基本的單位往往是三角形，我們就來使用 WebGL 畫一個純色三角形吧！

⌘ WebGL 的繪製流程

在使用 WebGL 時，我們寫的 Javascript 主程式在 CPU 上跑，透過 WebGL API 對 GPU『一個口令，一個動作』；不像是 HTML/CSS 那樣，給系統一個結構，然後系統會根據這個結構直接生成畫面。而且我們還要先告訴好 GPU『怎麼畫』、『畫什麼』，講好之後再叫 GPU 進行『畫』這個動作。

Step 1 『怎麼畫』

我們會把一種特定格式的程式（program）傳送到 GPU 上，在『畫』的動作時執行，這段程式稱為 shader，而且分成 vertex（頂點）及 fragment（片段）兩種 shader，vertex shader 負責計算每個形狀（三角形）的每個頂點在畫布上的位置、fragment shader 負責計算填滿形狀時每個 pixel 使用的顏色，兩者組成所謂『怎麼畫』的 program 讓 GPU 去執行。

Step 2 『畫什麼』

除了 shader 之外，還要傳送資料給程式（主要是 vertex shader）使用，在 shader 中這些資料叫做 attribute，並且透過 buffer 來傳送到 GPU 上，通常最基本會包含各個頂點的位置資訊。

1-2 畫一個三角形（上）

Step 3 『畫』這個動作

怎麼畫、畫什麼都設定完成之後，重頭戲來了，下了這個繪製指令（Draw call）後，GPU 首先執行 vertex shader，每執行一次產生一個頂點，且每次執行只會從 buffer 中拿出對應的片段作為 attribute，接著把每三個頂點組成三角形，並且點陣化[5]（rasterization）以對應螢幕的 pixel，最後為每個 pixel 分別執行 fragment shader。

以接下來要畫的三角形為例，筆者畫了簡易的示意圖 1-9 表示『畫』這個動作在 GPU 內發生的事情：

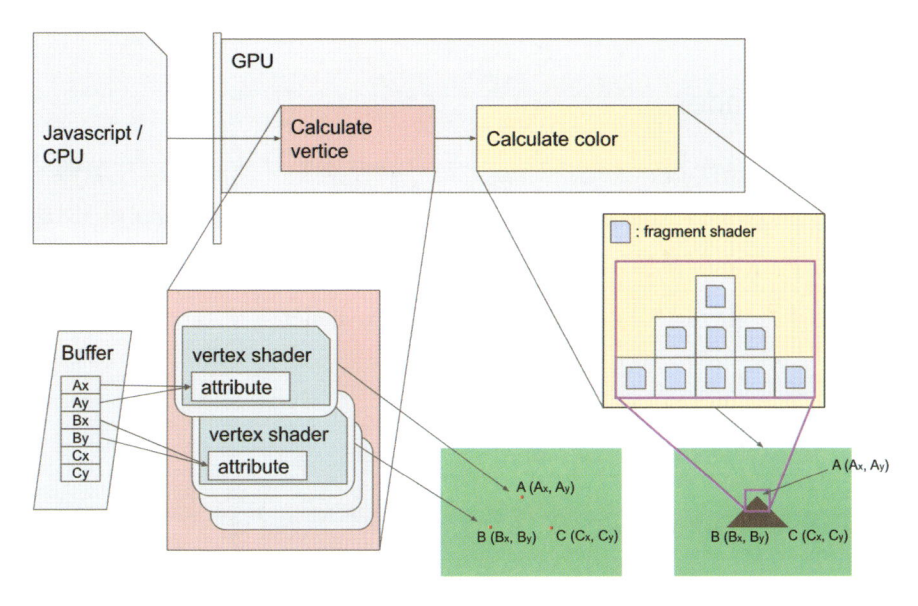

▲ 圖 1-9 『畫』這個動作在 GPU 內發生的事情

---

5 點陣化：根據維基百科 https://zh.wikipedia.org/wiki/柵格化，是是將向量圖形格式表示的圖像轉換成點陣圖以用於顯示器或者印表機輸出的過程。

1 Hello WebGL

　　圖中之 buffer 即為『畫什麼』的資料來源，紅色區塊為 vertex shader，每次執行從 buffer 拿出一至數個資料作為一個 attribute 向量的值，圖中拿出 2 筆作為一個 2 維向量（`[Ax, Ay]`），算出頂點在螢幕（左綠色區塊）上的位置；黃色區域為 fragment shader，螢幕上（右綠色區塊）著色之三角形之頂部放大表示其每個 pixel 之顏色是由 fragment shader 個別執行得到的結果。

　　為什麼是這樣的流程其實筆者也不得而知，或許就是維基百科 – openGL 之設計[6]這邊所說的：『它是為大部分或者全部使用硬體加速而設計的』，稍微想像一下，每個頂點位置以及每個 pixel 著色的計算工作可以高度平行化，而在顯示卡硬體上可以針對這個特性使這些工作平行地在大量的 ALU[7] / FPU[8] 上同時計算以達到加速效果。

⌘ 建立 shader

　　當筆者第一次看到這個的時候，第一個反應是『原來可以在瀏覽器裡面寫 C 呀』，這個語言稱為 OpenGL Shading Language，簡稱 GLSL[9]，雖然看起來很像 C 語言，但是不能直接當成 C 來寫，他有自己的資料格式[10]，我們直接來看本書第一個、無法再更簡單的 vertex shader：

```glsl
#version 300 es
in vec2 a_position;

void main() {
  gl_Position = vec4(a_position, 0, 1);
}
```

　　這個 shader 其實沒做什麼事，只是把輸入到 buffer 的位置資料放到

6　https://zh.wikipedia.org/wiki/OpenGL#設計
7　ALU: https://zh.wikipedia.org/wiki/算術邏輯單元
8　FPU: https://zh.wikipedia.org/wiki/浮點運算器
9　GLSL: https://zh.wikipedia.org/wiki/GLSL
10　GLSL 的資料格式：https://www.khronos.org/opengl/wiki/Data_Type_(GLSL)

`gl_Position`，這邊解釋一下各個語法所代表的意義：

- `#version 300 es` 表示使用 GLSL 3.00 ES 版本。
- 每次 shader 執行時跑 `void main()`。
- `in vec2 a_position` 的 in 表示此 shader 的輸入，我們稱之為 attribute，執行時會從 buffer 拿出對應的部份作為變數 `a_position` 的值。
  - 型別 vec2 表示二維向量，存放兩個浮點數。
  - 接下來要繪製的三角形在 2D 上，只需要 x、y 即可，因此使用 `vec2`。
- `gl_Position` 是 GLSL 規定用來輸出在畫布上位置的變數，其型別是 vec4。
  - 這個變數的第一到第三個元素分別是 x、y、z，而 x、y 的部份理所當然地表示螢幕或畫布上的位置，要注意的是這些值必須介於 -1 至 +1 才會落在畫布中，這個範圍稱為 clip space。
  - `vec4()` 建構一個 `vec4`，理論上應該寫成 `vec4(a_position[0], a_position[1], 0, 1)`，因為 `a_position` 是 `vec2`，這邊有語法糖[11]自動展開，所以可以寫成 `vec4(a_position, 0, 1)`。
  - 第四個元素我們先傳 1，到後面的章節再討論。

⌘ 關於 GLSL 從 vector 取值的語法

假設今天有個 `vec4` 的變數叫做 `bar`，不僅可以使用 `bar[i]` 這樣的寫法取得第 i 個元素（當然，從 0 開始），還可以用 `bar.x` / `bar.y` / `bar.z` / `bar.w` 取得第一、第二、第三、第四個元素，甚至有種叫做 swizzling 的寫法：`bar.xzz` 等同於 `vec3(var[0], var[2], var[2])`。

---

11 若電腦語言中的一個語法沒有為系統或軟體增加功能，但是更方便程式設計師使用，那麼我們會稱這種語法為『語法糖』。

1 Hello WebGL

⌘ Fragment Shader

因為要繪製的是『純色』三角形，fragment shader 只需要輸出一個固定的顏色：

```glsl
#version 300 es
precision highp float;

out vec4 outColor;

void main() {
  outColor = vec4(0.4745, 0.3333, 0.2823, 1);
}
```

同樣地解釋一下各個語法所代表的意義：

- `#version 300 es` 表示使用 GLSL 3.00 ES 版本。
- `precision highp float;` 表示使用高精準度的浮點數。
- 對於這個三角形圍出來區域內之每個 pixel，各執行一次 `void main()`。
- `out vec4 outColor;` 宣告所輸出之顏色變數為 `outColor`。
  - 因為是變數名稱，`outColor` 這個名字其實是可以自訂的，重點是 `out` 來表示這個變數為此 fragment shader 之輸出。
  - `outColor` 之各個元素分別是介於 0 到 1 之間的 red、green、blue、alpha
    - 此範例使用 `vec4(0.4745, 0.3333, 0.2823, 1);`，為棕色，hex 色碼為 `#795548`。

⌘ 編譯、連結 shader 成為 program

上方撰寫的兩段 GLSL 程式原始碼接下來要以 Javascript 字串來存放原始

1-16

碼，這邊可以使用樣板字面值[12]字串來寫：

```
// 把上方 GLSL 寫的 vertex shader 以 `` 字串包在 Javascript 中：
const vertexShaderSource = `#version 300 es
in vec2 a_position;

void main() {
  gl_Position = vec4(a_position, 0, 1);
}
`;

// 把上方 GLSL 寫的 fragment shader 以 `` 字串包在 Javascript 中：
const fragmentShaderSource = `#version 300 es
precision highp float;

out vec4 outColor;

void main() {
  outColor = vec4(0.4745, 0.3333, 0.2823, 1);
}
`;
```

這邊要注意 #version 300 es 必須在原始碼的一開始，因此樣板字面值的 backtick (`) 立刻接著 #version 300 es，連換行、空白都不能有；把原始碼字串定義好，接著傳給 WebGL API 進行編譯並且連結：

```
const vertexShader =
  createShader(gl, gl.VERTEX_SHADER, vertexShaderSource);

const fragmentShader =
  createShader(gl, gl.FRAGMENT_SHADER, fragmentShaderSource);

const program =
  createProgram(gl, vertexShader, fragmentShader);
```

12　樣板字面值：『使用 backtick (`) 包起來的字串』，因為可以跨越多行，適合存放 GLSL 原始碼，其 MDN 文件請參考：https://developer.mozilla.org/zh-TW/docs/Web/JavaScript/Reference/Template_literals

1-17

1 Hello WebGL

一個描述『怎麼畫』的 program 由一個 vertex shader 及一個 fragment shader 組成，這邊分別把 vertex、fragment shader 原始碼傳入 `createShader()` 進行編譯建立 shader，接著把兩個 shader 傳入 `createProgram()` 進行連結獲得 program。

可惜的是 `createShader()`、`createProgram()` 並不是內建的 function，其內部是一連串繁瑣的 WebGL API 指令，筆者直接包成 function 方便理解：

```javascript
function createShader(gl, type, source) {
  const shader = gl.createShader(type);

  // 設定 shader 原始碼，也就是上面用 `` 包起來的 GLSL 程式：
  gl.shaderSource(shader, source);

  gl.compileShader(shader); // 編譯 shader

  // 檢查編譯狀況
  const ok = gl.getShaderParameter(shader, gl.COMPILE_STATUS);
  if (ok) return shader; // 如果成功才回傳 shader

  // 如果編譯過程有出問題，將問題顯示在 Console 上
  console.log(gl.getShaderInfoLog(shader));
  gl.deleteShader(shader);
}
```

可以看到 `createShader` 內包裝了一連串建立 shader 的流程；`createProgram` 也是同樣繁瑣，建立空 program 要分別對 vertex、fragment shader 呼叫 `gl.attachShader()` 再使用 `gl.linkProgram()` 進行連結：

```javascript
function createProgram(gl, vertexShader, fragmentShader) {
  const program = gl.createProgram();

  // 設定 vertex shader 到新建立的 program：
  gl.attachShader(program, vertexShader);

  // 設定 fragment shader 到新建立的 program：
```

```
gl.attachShader(program, fragmentShader);

gl.linkProgram(program); // 連結 shader

// 檢查連結狀況
const ok = gl.getProgramParameter(program, gl.LINK_STATUS);
if (ok) return program; // 如果成功才回傳 program

// 如果連結過程有出問題,將問題顯示在 Console 上
console.log(gl.getProgramInfoLog(program));
gl.deleteProgram(program);
}
```

這樣一來 program 就建立完成!簡單的觀察一下究竟建立的 `vertexShader`、`fragmentShader`、`program` 是什麼樣子:

```
console.log({
  vertexShader,
  fragmentShader,
  program,
});
```

在瀏覽器上進行測試,因為還沒進行『畫』的動作,畫布沒有變化還是之前用綠色清空的樣子,而在 Console 上應可看到類似圖 1-10 的結果,編譯好的 shaders 以及連結好的 program 是『瀏覽器 WebGL 內部物件』,沒有透漏什麼更多資訊:

▲ 圖 1-10　『怎麼畫』準備工作之執行結果

1 Hello WebGL

到此進度的完整程式碼可以在這邊找到：

</>	ch1/02/index.html
</>	ch1/02/main.js

也可以參考線上版本，透過瀏覽器打開此頁面：
https://webgl-book.pastleo.me/ch1/02/index.html

⌘ 補充：若 WebGL 的版本是 WebGL1（GLSL 1.00 ES）

在 shader 中開頭寫的 `#version 300 es` 使用之 GLSL 3.00 ES，必須是 WebGL2 才支援，如果在 WebGL2 模式下沒寫這行，那麼就會使用 WebGL1 所使用的 GLSL 1.00 ES 解析原始碼。

GLSL 1.00 ES（WebGL1）中，vertex shader 的 attribute 關鍵字是 attribute，而非 in；同時 fragment shader 輸出顏色用的變數是固定的 `gl_FragColor`，不用用 out 定義一個輸出變數，以下這段為等價的 GLSL 1.00 ES 版 shader：

```
const vertexShaderSource = `
attribute vec2 a_position;

void main() {
  gl_Position = vec4(a_position, 0, 1);
}
`;

const fragmentShaderSource = `
precision highp float;

void main() {
  gl_FragColor = vec4(0.4745, 0.3333, 0.2823, 1);
}
`;
```

1-20

1-3 畫一個三角形（下）

建立 shader 並連結成 program，『怎麼畫』的部份就準備完成，接下來要告訴 GPU『畫什麼』，精確來說，就是提供 vertex shader 中 `a_position` 所需的資料。

⌘ 取得 Attribute 位置

這個有點指標的感覺，我們可以透過呼叫 `gl.getAttribLocation()` 取得一個 attribute 在 program 中的位置，同時也把取得的值印出來看看：

```
const positionAttributeLocation = gl.getAttribLocation(
  program, 'a_position',
);
console.log({ positionAttributeLocation });
// => {positionAttributeLocation: 0}
```

此處取得的『位置』就是一個單純的數字，待會這個數字會用來跟 buffer 綁定。

> **當數字為 -1 時**
>
> 找不到的時候這個數字會是 -1，如果 GLSL 裡面寫了一些沒有被使用到的 attribute 變數，那在 GPU 編譯的過程中會消失，所以就算 GLSL 原始碼有宣告 attribute，也有可能因為被判定沒有使用到變數導致 gl.getAttribLocation() 拿到 -1。

⌘ 建立並使用 Buffer

```
const positionBuffer = gl.createBuffer();
gl.bindBuffer(gl.ARRAY_BUFFER, positionBuffer);
console.log({ positionBuffer });
// => {positionBuffer: WebGLBuffer}
```

使用 `gl.createBuffer()` 便可建立 buffer，`console.log()` 顯示之 `WebGLBuffer` 表示 `positionBuffer` 為瀏覽器包裝 WebGL 之內建物件，然後使用 gl.bindBuffer() 『設定目前使用中的 array buffer』；在 WebGL API 中，有許多內部的『對準的目標』（binding point），而看到 bind 的字眼時，他們的功能往往是去設定這些『對準的目標』，設定完成後，接下來呼叫的其他 WebGL API 就會對著設定好的目標做事。

除此之外，`gl.bindBuffer()` 第一個參數傳入了 `gl.ARRAY_BUFFER`，在 mdn[13] 上這個參數叫做 target，表示 buffer 不只有一種，而 `gl.ARRAY_BUFFER` 這種 buffer 才能與 vertex shader 的 attribute 連結，描述這層連結關係的功能叫做 vertex attribute array。

⌘ Vertex Attribute Array

首先在 attribute `a_position` 位置上啟用這個功能：

```
gl.enableVertexAttribArray(positionAttributeLocation);
```

啟用之後，設定 attribute 拿資料的方法：

```
gl.vertexAttribPointer(
  positionAttributeLocation, // index
  2, // size
  gl.FLOAT, // type
  false, // normalize
  0, // stride
  0, // offset
);
```

---

13 https://developer.mozilla.org/en-US/docs/Web/API/WebGLRenderingContext/bindBuffer

雖然看似沒有提到任何 buffer 的東西，但是這行執行下去的時候會讓 attribute 設定成與目前『對準的 ARRAY_BUFFER 目標』關聯，也就是剛剛 `gl.bindBuffer(gl.ARRAY_BUFFER, positionBuffer)` 的 `positionBuffer`，同時來看一下各個參數所代表的意義：

- 第一個參數 `index`：看到傳入 `positionAttributeLocation` 應該可以猜到，就是要設定的 attribute 位置。

- 第二個參數 `size`：筆者認為這是這個 API 最重要的參數，設定了每次 vertex shader 執行時該 attribute 要從 buffer 中拿出多少個數值，依序填入 vecX 的各個元素，這邊使用 2 剛好填滿 shader 中的 `in vec2 a_position`。

  - 事實上，就算 attribute 是 `vec4` 而 size 只餵 2 進去也是可以的，剩下的空間 WebGL 會有預設值填上，預設值的部份與之後 3D 相關，之後再來討論。

- 第三個參數 type 與第四個參數 `normalized`：設定原始資料與 attribute 的轉換，type 指的是原始資料的型別，此範例直接一點傳入 `gl.FLOAT`，而 `normalize` 在整數型別時可以把資料除以該型別的最大值使 attribute 變成介於 -1 ~ +1 之間的浮點數，此範例不使用此功能傳 false 進去即可。

  - 假設今天原始資料是 0~255 整數表示的 RGB，那麼就可以將 `type` 設定成 `gl.UNSIGNED_BYTE` 搭配 `normalize` 為 `true` 使用，在 shader 中 attribute 就會直接是符合 `gl_FragColor` 的顏色資料。

- 第五個參數 `stride` 與第六個參數 `offset`：控制讀取 buffer 時的位置，`stride` 表示這次與下次 vertex shader 執行時 attribute 讀取的起始位置的距離，設定為 0 表示每份資料是緊密排列的，`offset` 則是第一份資料距離開始位置的距離，這兩個參數的單位皆為 byte。

圖 1-11 表示這個範例呼叫 `gl.vertexAttribPointer()` 後 buffer 與 attribute 的運作關係：

1 Hello WebGL

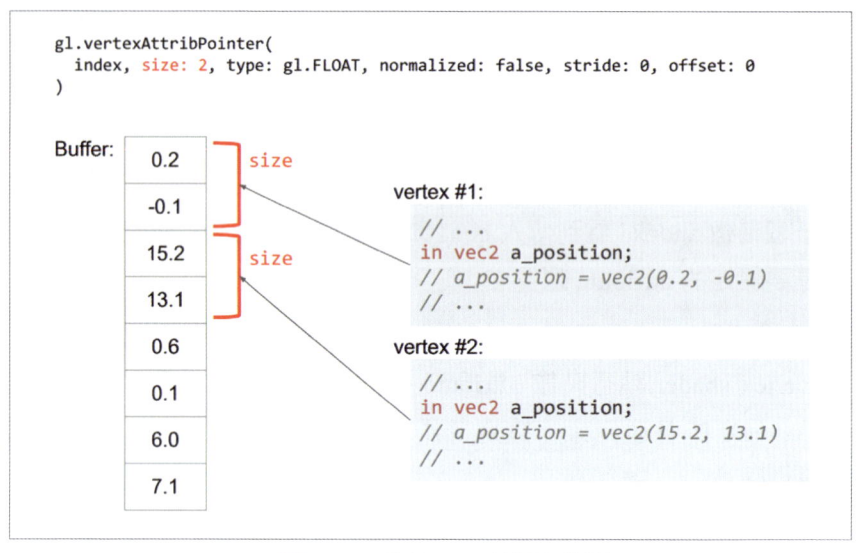

```
gl.vertexAttribPointer(
  index, size: 2, type: gl.FLOAT, normalized: false, stride: 0, offset: 0
)
```

Buffer:
| 0.2 | } size
| -0.1 |
| 15.2 | } size
| 13.1 |
| 0.6 |
| 0.1 |
| 6.0 |
| 7.1 |

vertex #1:
```
// ...
in vec2 a_position;
// a_position = vec2(0.2, -0.1)
// ...
```

vertex #2:
```
// ...
in vec2 a_position;
// a_position = vec2(15.2, 13.1)
// ...
```

▲ 圖 1-11　參數 size 的運作關係

本範例雖然沒有使用到 stride 與 offset，還是舉個使用 stride 與 offset 的狀況，下圖 1-12 中可以看到第一個頂點讀取的位置因為 offset 移動了兩格，下個讀取的位置會因為 stride 間隔四格：

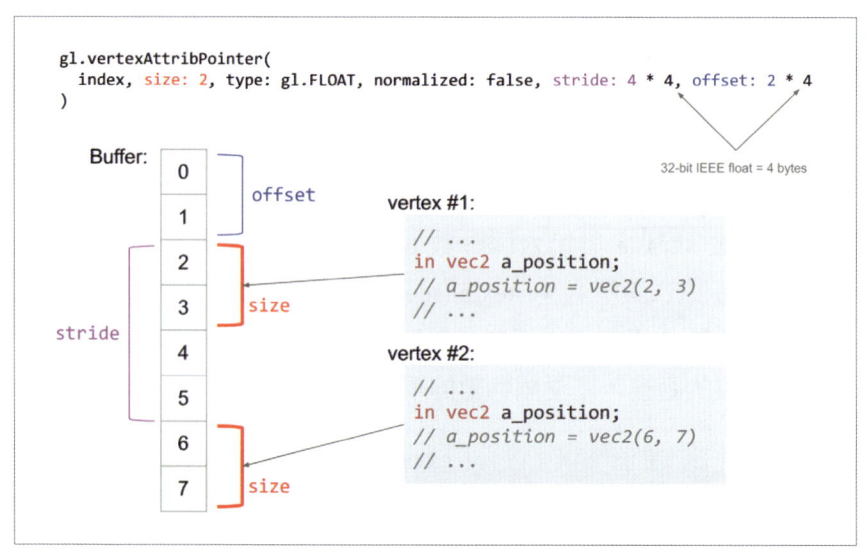

```
gl.vertexAttribPointer(
  index, size: 2, type: gl.FLOAT, normalized: false, stride: 4 * 4, offset: 2 * 4
)
```
32-bit IEEE float = 4 bytes

Buffer:
| 0 | } offset
| 1 |
| 2 | } size
| 3 |
| 4 |
| 5 |
| 6 | } size
| 7 |

stride

vertex #1:
```
// ...
in vec2 a_position;
// a_position = vec2(2, 3)
// ...
```

vertex #2:
```
// ...
in vec2 a_position;
// a_position = vec2(6, 7)
// ...
```

▲ 圖 1-12　參數加入 stride 與 offset 的運作關係

傳入資料到 buffer，設定三角形的位置

在上面已經使用 `gl.bindBuffer()` 設定好『對準的 ARRAY_BUFFER 目標』，接下來呼叫 `gl.bufferData()` 對 buffer 輸入資料。

```
gl.bufferData(
  gl.ARRAY_BUFFER,
  new Float32Array([
    0, 0.2,
    0.2, -0.1,
    -0.2, -0.1,
  ]),
  gl.STATIC_DRAW,
);
```

第二個參數即為 buffer 的資料，也就是三角形頂點的位置，注意要傳入與 `gl.vertexAttribPointer() type` 符合的 TypedArray[14]，關於這些數值：

- 三個點形成一個三角形，每個點有兩個浮點數表示 x、y 數值，共 6 個浮點數。

- 每個點的座標必須在 (-1, -1) 至 (1, 1) 才會落在畫布中，在 x 軸方向左方邊界為 -1、右方邊界為 +1；y 軸方向上方邊界為 +1，下方邊界為 -1；正中間座標為 (0, 0)，這個 (-1, -1) 至 (1, 1) 的範圍稱為 clip space。

假設三角形頂點分別依序為 A、B、C，示意如圖 1-13

---

14 TypedArray 表示一段連續記憶體（ArrayBuffer），例如 `Float32Array` 即為一種 TypedArray，以一連串的 4 byte（32 bit）浮點數『看待』這一段連續記憶體，另外還有 `Uint8Array`, `Int32Array`，其 mdn 文件請見：https://developer.mozilla.org/zh-TW/docs/Web/JavaScript/Reference/Global_Objects/TypedArray

1 Hello WebGL

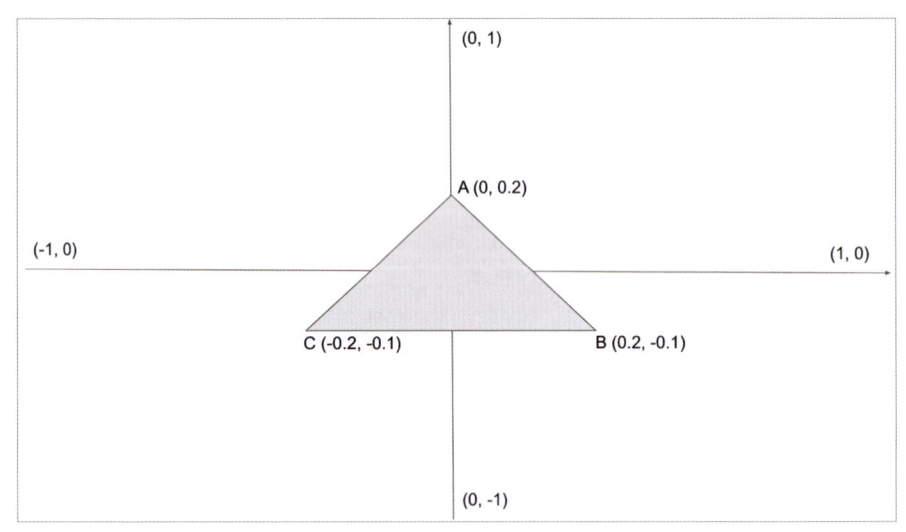

▲ 圖 1-13　數值範圍、Clip space 與頂點位置之示意圖

好的，這麼一來『畫什麼』的資料也輸入到 buffer 準備完成了。

⌘ 終於，『畫』這個動作

```
gl.useProgram(program);
gl.drawArrays(
  gl.TRIANGLES, // mode
  0, // first
  3, // count
);
```

`gl.useProgram()` 設定使用建立好的 `program`，這是在『如何畫』篇章使用 `createProgram(gl, vertexShader, fragmentShader)` 建立好的變數，接著 `gl.drawArrays()` 就是『畫』這個動作，其參數功能：

- 第一個參數 `mode`：透過這個參數可以請 WebGL 畫點、線，而面的部份就是三角形 `gl.TRIANGLES` 了。
- 第二個參數 `first`: 類似上面的 `offset`，精確來說是『略過多少個頂點』
- 第三個參數 `count`: 有多少個頂點，我們畫一個三角形，共三個頂點。

1-3 畫一個三角形（下）

好事多磨，我們的第一個三角形終於出現在畫布上，如圖 1-14

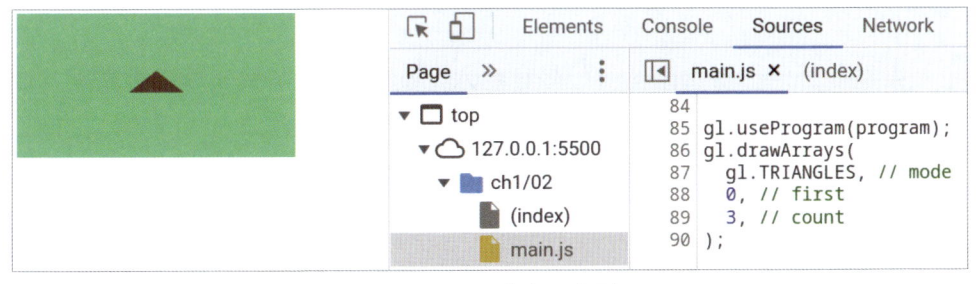

▲ 圖 1-14　畫出三角形

到最後一刻才呼叫 `gl.useProgram()`，整個資料設定的過程還是透過 attribute 的『位置』設定的，這表示資料在一定程度上可以與 shader 脫鉤，打個比方，同一個 3D 物件可以根據情況使用不同的 shader 來渲染達成不同的效果，但是在記憶體中這個 3D 物件只需要一份就好。

三角形的顏色是寫死在 fragment shader 內，讀者們可以試著調顏色、頂點位置玩玩看，本篇的完整程式碼可以在這邊找到：

</>	ch1/03/index.html
</>	ch1/03/main.js

也可以參考上線版本，透過瀏覽器打開此頁面：

https://webgl-book.pastleo.me/ch1/03/index.html

花了這麼多力氣，就只是一個純色的三角形，而且定位還得先用畫布的 -1～1 來算，但別氣餒，接下來繼續介紹更多 shader 接收資料、參數的方式來使繪製更加靈活！

1 Hello WebGL

1-4　Uniform – shader 之參數

雖然把三角形畫出來了，但是在傳入 `a_position` 時要先算出頂點在 clip space 中 -1 ~ +1 的值，如果今天三角形頂點位置可以透過以 pixel 為單位的 x、y 座標來定位，類似於 CSS 的 `position: absolute;` 並使用 `left: Xpx; top: Ypx;` 定位，那麼就三角形頂點位置的輸入就可以變得比較直覺；同時又想要讓畫布填滿整個網頁，表示畫布大小將會由網頁畫面大小決定，開始脫離 `<canvas />`（在 Chrome 上）預設的 300x150 寬高。

我們將以這兩個目標進行修改，進而學習到下一個 WebGL 的基本功能 — uniform。

⌘ 什麼是 uniform

Uniform 類似 attribute，可以把資料傳到 shader 內，但是在一次『畫』動作中每個頂點、每格著色計算時的 uniform 值都一樣，所以才叫做 uniform 吧，同時因為 uniform 是直接設定在 program 上的，因此不會有各個頂點讀取 buffer 中哪個位置的問題，使用上比 attribute 簡單許多。

使三角形的定位使用『網頁中的 pixel 位置』，要先算出頂點在 clip space 中 -1 ~ +1 的值，要做到這件事情當然可以寫一個簡單的 function 在 `gl.bufferData()` 之前對座標做一些處理，但是這邊是一個很適合使用 uniform 解決的問題：當傳入的頂點座標是畫布上 x、y 軸的 pixel 位置，而輸出給 `gl_Position` 的值必須介於 -1 ~ +1，shader 需要知道的資訊就是畫布寬高，畫布的寬高不論在哪個頂點值都相同，故適合以 uniform 來處理。

⌘ Uniform 的使用

在 vertex / fragment shader 中宣告 uniform 與 attribute 很像，只是關鍵字從 `in` 換成 `uniform`，從我們要傳入的畫布寬高為例，在 vertex shader 中這樣宣告：

```glsl
#version 300 es
in vec2 a_position;

uniform vec2 u_resolution;

void main() {
  // ...
}
```

型別為二維向量 `vec2`，變數名稱為 `u_resolution`，接著跟 attribute 一樣，先取得變數位置：

```js
const resolutionUniformLocation =
  gl.getUniformLocation(program, 'u_resolution');
```

有了 uniform 在 GPU 中的位置，我們就可以對著『使用中』的 program 設定 uniform：

```js
gl.uniform2f(resolutionUniformLocation, canvas.width, canvas.height);
```

注意，是『使用中』的 program，所以要在呼叫 `gl.useProgram()` 設定好使用中的 program 之後再設定 uniform 的數值。

> 這邊的 `canvas` 變數是一開始 `document.getElementById('canvas')` 取得的元素，其身上就有 `.width`、`.height` 可用

有發現到 `gl.uniform2f` 的 2f 嗎？我們可以打開 Chrome 的 Console，輸入 `gl.uniform` 可以看到如圖 1-15 一般，有這麼多可呼叫的 API：

1 Hello WebGL

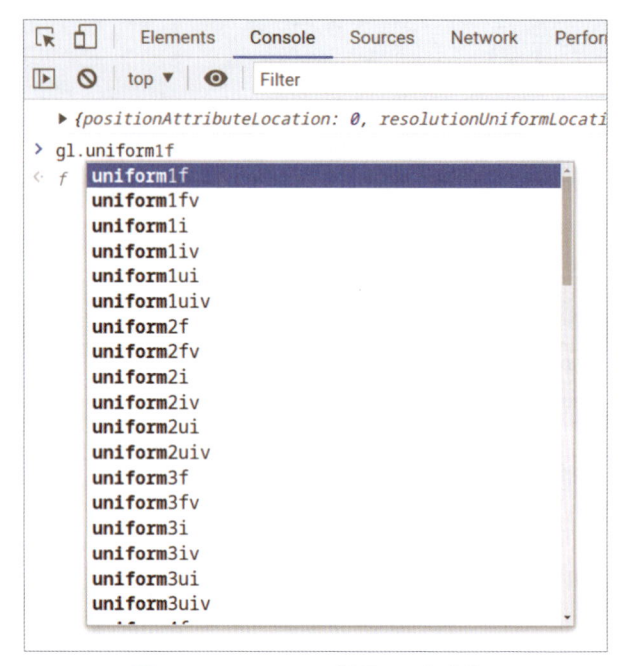

▲ 圖 1-15　Console 提示可呼叫的 API

這些 `gl.uniformXXX` 是針對不同型別所使用的，像是筆者上面使用的 `gl.uniform2f` 的 2f 表示 2 個元素的浮點數，也就是二維向量 `vec2`，這邊可以看到型別一路從 1f 單個 float 到 `Matrix4f` 整個 4x4 矩陣都有。除此之外，對於每種資料型別分別還有一個結尾多了 v 的版本（以 `gl.uniform2f` 為例：`gl.uniform2fv`），其實功能沒什麼不同，只是 function 接收參數的方式改變，從 `gl.uniform2f(index, x, y)` 變成 `gl.uniform2fv(index, [x ,y])`。

把畫布的寬高傳送進入之後，當然也得來修改 vertex shader 好好地使用 `u_resolution` 做轉換：

```
void main() {
  gl_Position = vec4(
    // 把作為畫面中 pixel 位置的 a_position 換算成 clip space
    a_position / u_resolution * vec2(2, -2) + vec2(-1, 1),
    0, 1
  );
);
```

1-30

1-4　Uniform – shader 之參數

```
  // = vec4(
  //   a_position.x / u_resolution.x * 2.0 - 1.0,
  //   a_position.y / u_resolution.y * -2.0 + 1.0,
  //   0,
  //   1
  // );
}
```

　　針對這段轉換程式，如果畫面寬高是 300x150，一個頂點位置 `a_position` 為 (150, 90) 應在畫面中間偏下的位置，除以 `u_resolution` 得到 (0.5, 0.6)，數值範圍介於 0 ~ 1，分別對 x 座標乘以 2 減 1、對 y 座標乘以 -2 加 1 得到 (0.0, -0.2) 介於 -1 ~ +1 之間的值，也就是給 `gl_Position` 在 clip space 中的位置。這邊我想讀者會有兩個疑問：

- `vec2` 可以跟 vec2 直接做加減乘除運算？對，相當於每個元素分別做運算，以加法為例像是這樣：`vec2(x1, y1) + vec2(x2, y2) = vec2(x1+x2, y1+y2)`。筆者看到這樣的寫法第一個瞬間也是『這樣會動？』像 Javascript `[1,2] * [3,4]` 只會得到 NaN，畢竟一般常見的程式語言的用途比較通用，不像 GLSL 很常有這樣的運算，因而特化出 vec 之間加減乘除的寫法。

- 對 x 座標乘以 2 減 1，而對 y 座標要乘以 -2 加 1？因為在畫布（精確來說是 clip space）中，上方為 y = 1、下方為 y = -1，因此 y 軸正向指著上方的，這個方向和我們在電腦中圖片、網頁的 y 軸方向是相反的，因此要乘以 -1 反向，並且也因此最後兩軸一個要減 1 一個要加 1。

1-31

1　Hello WebGL

⌘ 傳入的頂點位置 `a_position` 可以改用 pixel 座標了

筆者用上面的公式拿原本的值做反向運算可以得知在 300x150 的 pixel 座標：

```
gl.bufferData(
  gl.ARRAY_BUFFER,
  new Float32Array([
    150, 60,
    180, 82.5,
    120, 82.5,
  ]),
  gl.STATIC_DRAW,
);
```

因為最後算出來的位置與呈現與一開始的三角形是等價的，因此到目前為止不會有變化。

⌘ 使畫布填滿整個網頁

如果讀者使用過 CSS，而且 `<canvas />` 元素呈現上就是一張圖，類似於 `<img />`，那麼應該可以想到簡單的這幾行 CSS，筆者直接寫在 HTML 上：

```
<style>
  html, body {
    margin: 0;
    height: 100%;
  }
  #canvas {
    width: 100%;
    height: 100%;
  }
</style>
```

但是重整之後看到的只是放大的樣子，就像是把圖片放大的感覺，如圖 1-16。

1-32

1-4　Uniform – shader 之參數

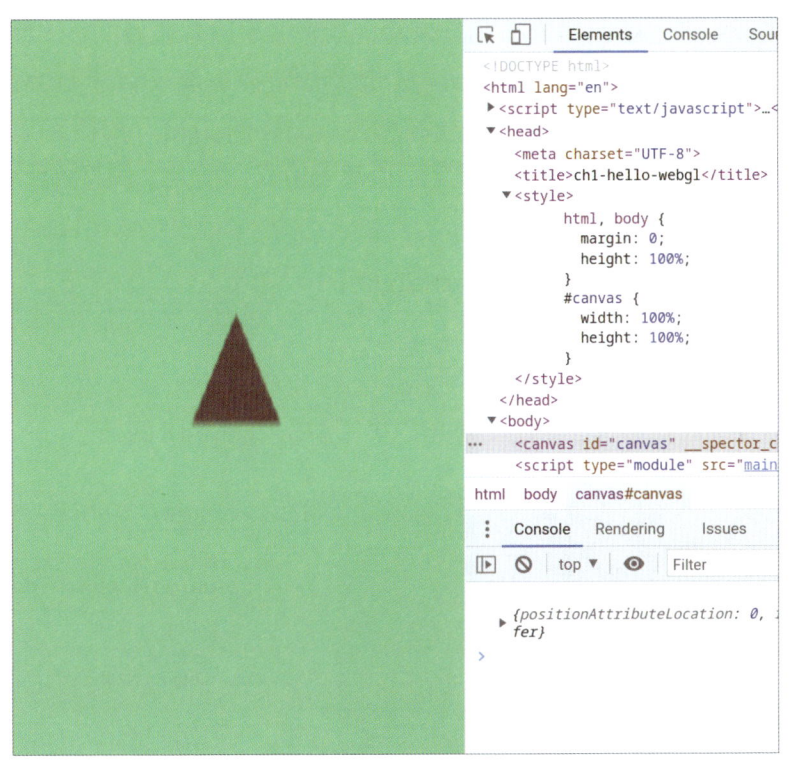

▲ 圖 1-16　使用 CSS 將 <canvas /> 撐滿整個頁面

　　事實上，在 `<canvas />` 元素上有自己的寬高資訊，類似於圖片的原始大小，可以在 Console 上輸入 `gl.canvas.width` 從 WebGL instance 找回 canvas 元素並取得『原始大小』的寬度，如圖 1-17。

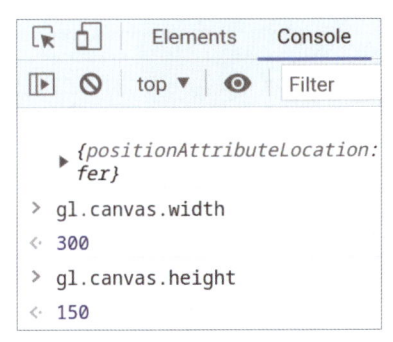

▲ 圖 1-17　Console 上找回 canvas『原始大小』的寬高

1-33

1　Hello WebGL

　　從 Console 上顯示的畫布寬高來看，顯然還是原本預設的值，我們得手動去設定這些數值來符合 `<canvas />` 元素的實際寬高，幸好 DOM API 有另外一組提供實際的寬高 `.clientWidth`、`.clientHeight`，我們可以直接把 `.clientWidth` 與 `.clientHeight` 設定回這個 canvas『圖片』的原始大小：

```
canvas.width = canvas.clientWidth;
canvas.height = canvas.clientHeight;

// before gl.clearColor(...);
```

　　想說問題應該解決了，但是重整之後卻看到如圖 1-18 的樣子：

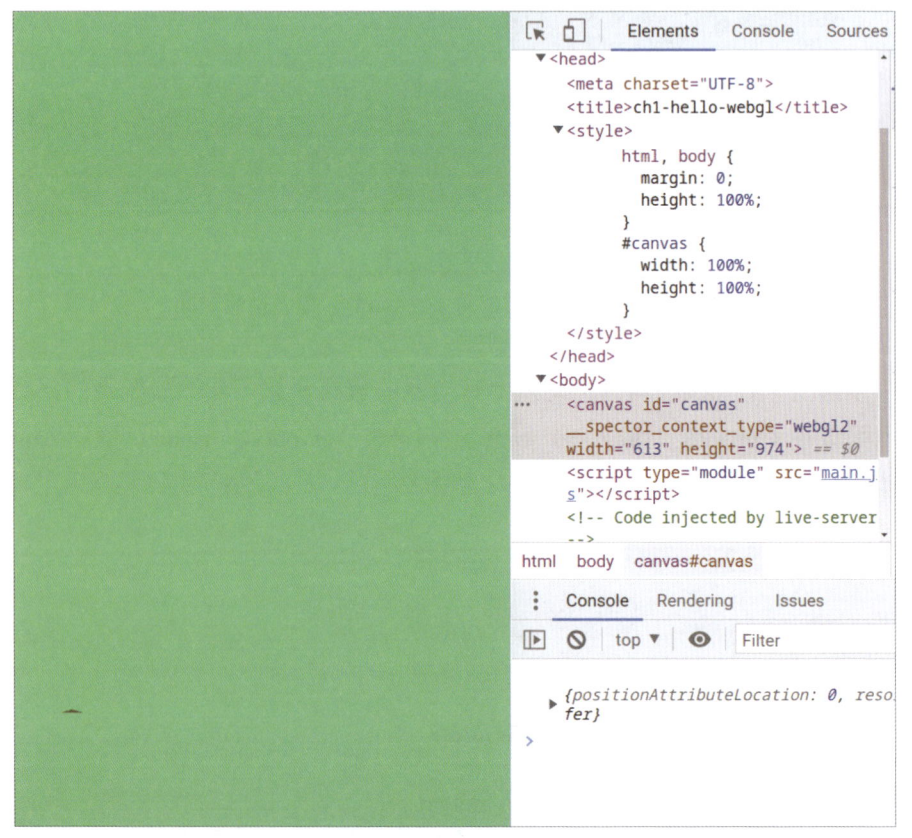

▲ 圖 1-17　把『實際寬高』同步回『原始寬高』

1-34

模糊的現象消失了，看起來實際大小的更動有效，但是那個三角形的位置顯然不太對，這是因為 WebGL 還有一個內部的『繪製區域』設定，這件事情也是需要手動設定的，設定的 API 為 `gl.viewport()`：

```javascript
// after canvas.height = canvas.clientHeight;
gl.viewport(
  0, // x
  0, // y
  canvas.width, // width
  canvas.height, // height
);

// before gl.clearColor(...);
```

`gl.viewport()` 的參數分別為 `x`、`y`、`width`、`height`，這個 `x`、`y` 是指左下角在畫布中的位置，這邊我們要填滿整張畫布，給 0 即可，並把 `width`、`height` 給滿。大家可能會想說，為什麼 WebGL 有內部的『繪製區域』的設定？不知道各位有沒有玩過馬力歐賽車的多人同樂模式，這種在同一個螢幕『分割畫面』的狀況，就可以用 `gl.viewport()` 設定繪製區域為一個玩家繪製畫面，繪製完再呼叫 `gl.viewport()` 繪製另外一位玩家的畫面，WebGL 沒有幫開發者預設使用情境，因此需要自行呼叫 `gl.viewport()` 來修正。

> **假設有一個左右分割畫面的雙人遊戲**
>
> 首先為 1 號玩家繪製左半邊畫面，此時設定繪製區：
>
> `gl.viewport(0, 0, canvas.width / 2, canvas.height);`
>
> 接著為 2 號玩家繪製右半邊畫面，設定繪製區：
>
> `gl.viewport(canvas.width / 2, 0, canvas.width, canvas.height);`
>
> 也就是繪製範圍 x 軸的部分 1 號玩家從最左（0）到正中間（canvas.width 除以 2），而 2 號玩家從正中間（canvas.width 除以 2）到最右邊（canvas.width）

1 Hello WebGL

經過這麼多次針對寬高的設定之後，終於如圖 1-18 達到我們要的效果：

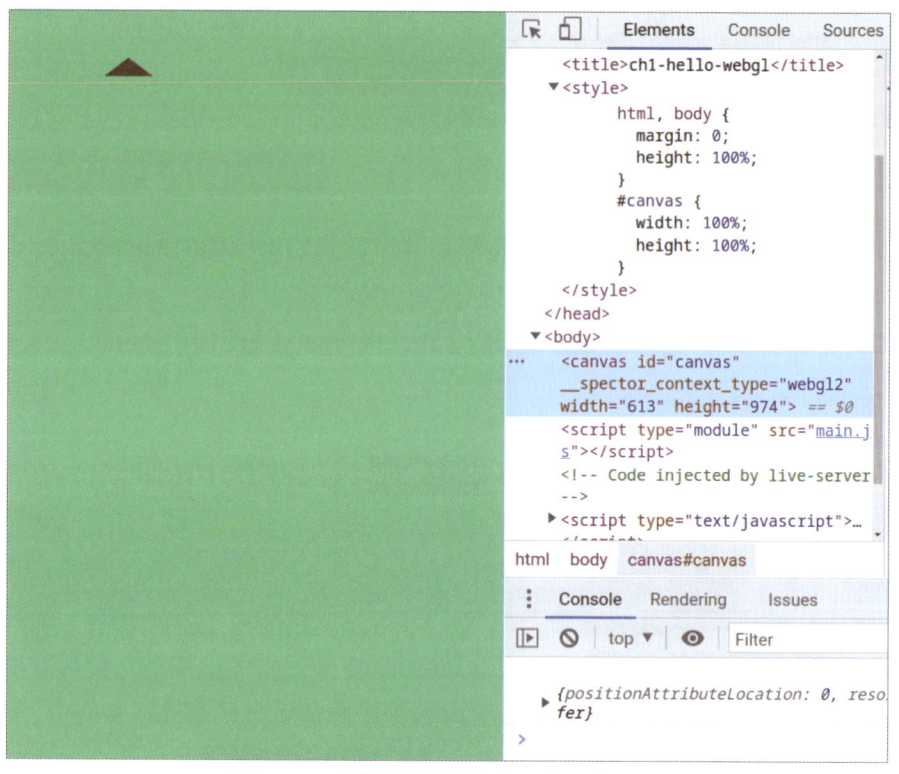

▲ 圖 1-18 撐滿網頁的同時正確渲染三角形

若在網頁載入渲染完成後調整視窗大小，一樣會發生拉伸的狀況，這是因為 WebGL 不會自行重新繪製；若要解決這個問題，必須再設定一次畫布的寬高、WebGL 繪製區域並重新繪製。

好的，我們完整地讓 buffer 可以直接寫『網頁』上的絕對位置來決定三角形的頂點位置，並讓 vertex shader 將 `a_position` 透過 unifrom `u_resolution` 計算出對應的 clip space 與最終呈現的位置；到此進度的完整程式碼可以在這邊找到：

1-36

</>	ch1/04/index.html
</>	ch1/04/main.js

也可以參考上線版本,透過瀏覽器打開此頁面:

https://webgl-book.pastleo.me/ch1/04/index.html

1-5 畫多個三角形

到目前為止畫三角形時,在 buffer 中傳入了 3 個頂點,每個頂點分別有兩個值 (x, y) 表示座標乘下來共 6 個值,並且在 `gl.drawArrays()` 的最後一個參數 `count` 參數傳入 3 表示畫三個頂點;若要畫更多三角形,我們只需要對 buffer 傳入更多『組』三角形的每個頂點座標,並且修改 `gl.drawArrays()` 的 `count` 即可,筆者這邊以畫三個三角形為範例,首先修改傳入 buffer 的部份:

```
gl.bufferData(
  gl.ARRAY_BUFFER,
  new Float32Array([
    150, 60,
    180, 82.5,
    120, 82.5,

    100, 60,
    80, 40,
    120, 40,

    200, 60,
    180, 40,
    220, 40,
  ]),
  gl.STATIC_DRAW,
);
```

1 Hello WebGL

第一個三角形的 3 個頂點位置與先前相同，除此之外再加入 2 個三角形（6 個頂點），共有 9 個頂點；接著修改『畫』指令的頂點數量到 9：

```
gl.drawArrays(
  gl.TRIANGLES, // mode
  0, // first
  9, // count
);
```

存檔重整之後可以看到 3 個棕色三角形，如圖 1-19：

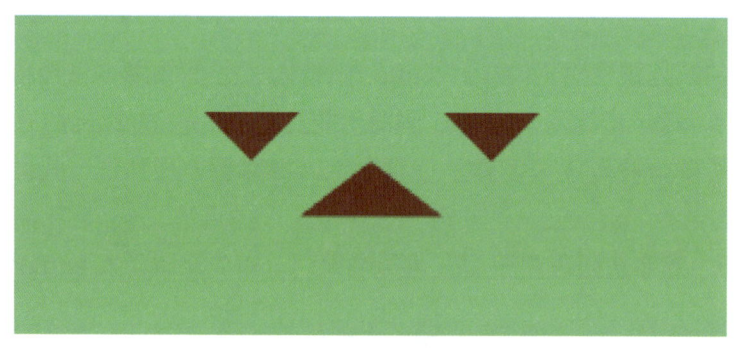

▲ 圖 1-19　畫出 3 個棕色三角形

這些只是筆者隨便想到的圖案，讀者們可以自行發揮想像力調整頂點座標。

如果 buffer 資料長度不足，或是 count 不是三的倍數，那麼會怎麼樣呢？如果把最後一組頂點刪除（220, 40 那組），count 保持為 9，不僅不會有什麼陣列超出的錯誤，感覺上 vertex attribute 還給不足的部份填上預設值 0，如圖 1-20 所示：

1-38

▲ 圖 1-20　把最後一組頂點刪除

或者把 count 改成 8，也不會有錯誤，只是最後一組三角形沒有完整，兩個點湊不出一個面，因此最後一個三角形就消失了，如圖 1-21：

▲ 圖 1-21　把 count 改成 8

1-6　Varying – fragment shader 之資料

現在我們要來回顧一下 fragment shader，我們實做的 fragment shader 只顧著把一個固定的顏色指定到輸出顏色的變數（outColor），所以現在不論畫幾個三角形，顏色都是當初寫死在 fragment shader 中的顏色：

```
outColor = vec4(0.4745, 0.3333, 0.2823, 1);
```

1 Hello WebGL

　　如果接下來想要讓三角形可以有不同的顏色，那麼該怎麼做呢？因為 fragment shader 會對每個 pixel 執行一次來算出該 pixel 所要被畫上的顏色，所以要想辦法讓指定到 `outColor` 上的顏色能夠根據某種因子而有所不同，而這些因子又希望可以由某種輸入的資料或是參數來指定，可能我們會想說能不能在 fragment shader 內使用 attribute 這種『每次 shader 執行從 buffer 內取用不同組資料』的功能作為輸入？可惜不行，attribute 是給 vertex shader 使用的，而且 vertex shader 是以頂點為單位執行，取用 buffer 的方式顯然會與 fragment shader 對不起來，因此需要另外一種傳輸工具 — varying。

> 不知道讀到這邊有沒有讀者想到另一種方法：使用 uniform，一次繪製一個三角形，『畫』完一個三角形設定 unifrom 再『畫』下一個三角形，這樣下來確實可以獲得多個不同顏色三角形的效果，但是如果今天要繪製成千上萬個三角形，每次的『畫』（draw call）都是一次 CPU 與 GPU 的來回，我們很可能會失去 GPU 平行運算的加速效果。

⌘ Varying

　　varying 這功能說白了就是讓 vertex shader 輸出資料給 fragment shader 使用，但是兩者執行的回合不同，究竟一個 vertex shader 輸出的 varying 資料是給哪個 fragment shader 拿到？

　　答案是：WebGL 會把頂點與頂點之間輸出的 varying 做內插（interpolation）！[15]

　　內插是什麼意思？接下來拿一個低解析度三角形的狀況來舉例，如圖 1-22，vertex shader 執行三次得到三個頂點，灰色的方格每格（pixel）執行一次 fragment shader 計算顏色。

15　https://zh.wikipedia.org/wiki/插值

1-6　Varying – fragment shader 之資料

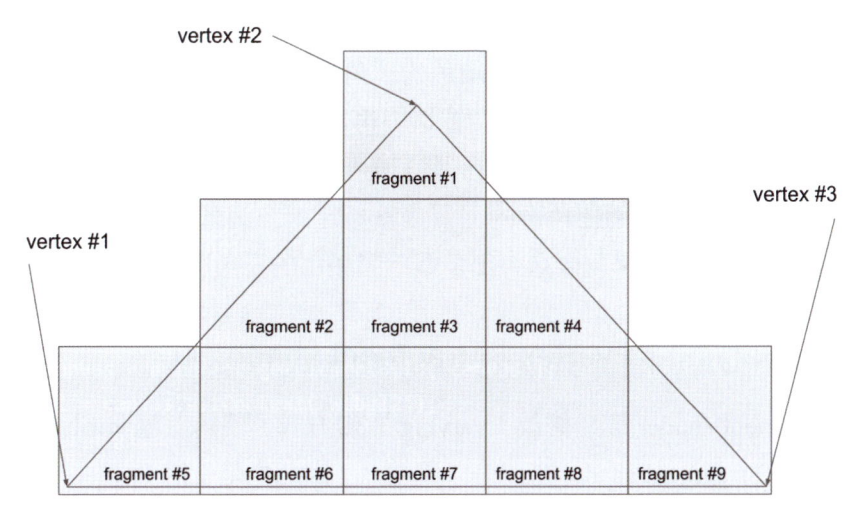

▲ 圖 1-22　低解析度三角形中 vertex 與 fragment shader 的計算關係

vertex #1 輸出 `v_number = 0.2`、vertex #2 輸出 `v_number = 1.1`，那麼介於 vertex #1, #2 之間的 fragment #2 將拿到 `v_number` 為兩個點輸出的中間值，也就是 `(0.2 + 1.1) / 2 = 0.65`，並且越接近某個頂點的 pixel 就會得到越接近該頂點輸出的 varying，若 vertex #3 輸出 `v_number = -0.4`，整個三角形的 fragment shader 個別拿到的 `v_number` 如圖 1-23 所示：

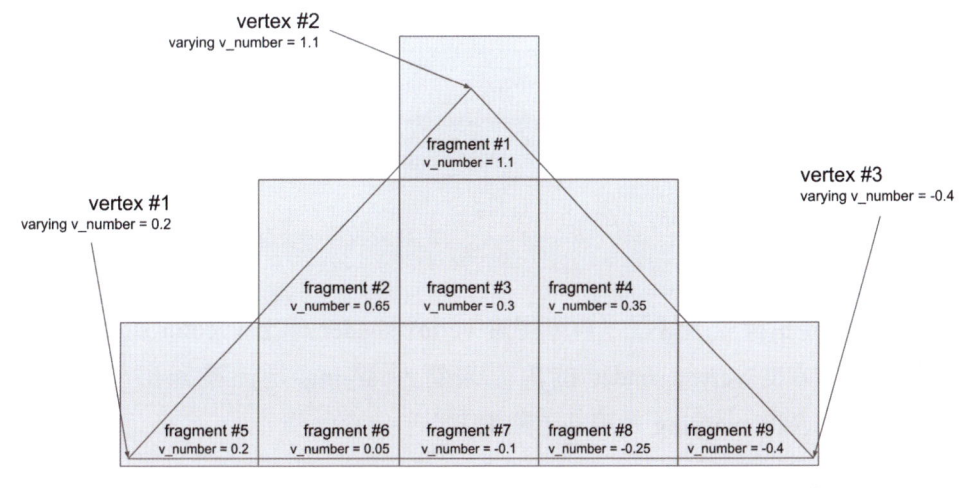

▲ 圖 1-22　頂點與頂點間數值的平行化給 fragment shader 做使用

1-41

1 Hello WebGL

　　這個特性不僅解決問題，也讓筆者覺得相當有意思，有種當初玩 flash 移動補間動畫的感覺，為了渲染不同顏色的三角形，我們讓每個頂點，也就是 vertex shader 每次執行，除了座標位置之外再拿到一個顏色，並使用 varying 功能傳送顏色給 fragment shader，第一個三角形的三個頂點指定一個顏色獲得一個純色三角形，第二個三角形的三個頂點指定第二個顏色獲得第二個顏色的純色三角形，這樣一來就能選染出多個不同顏色的三角形了。

⌘ 實做 varying 到 vertex 與 fragment shader 中

　　從 fragment shader 開始修改，varying 宣告方式、使用上跟 attribute 差不多，而且對於 fragment shader 來說是一種輸入，因此也是使用 in 這個關鍵字：

```
in vec3 v_color;
```

　　定義這個這個 varying 變數稱為 v_color，因為顏色有 rgb 3 個 channel 型別為 vec3，alpha 的部分目前先直接指定為 1，如果想要繪製半透明會需要不少額外的設定，日後有用到時再來深入討論；而原本寫死的顏色也就改用 v_color 加上 alpha 輸出到 outColor，完整的 fragment shader 如下：

```
#version 300 es
precision highp float;

in vec3 v_color;

out vec4 outColor;

void main() {
  outColor = vec4(v_color, 1);
}
```

　　fragment 這樣就改好了，再來得讓 vertex shader 輸出 v_color 給 fragment shader 使用，從 vertex shader 的角度來看，varying 是一種輸出，因此在 vertex shader 宣告 varying 是用 out 關鍵字：

```
out vec3 v_color;
```

同時也得讓 vertex shader 使用 attribute 接收每個頂點的顏色資訊，這邊稱為 a_color，vertex shader 只需要直接把 a_color 複製到 v_color 即可，修改後的完整 vertex shader 如下：

```
#version 300 es
in vec2 a_position;
in vec3 a_color;

uniform vec2 u_resolution;

out vec3 v_color;

void main() {
  gl_Position = vec4(
    a_position / u_resolution * vec2(2, -2) + vec2(-1, 1),
    0, 1
  );
  v_color = a_color;
}
```

shader 修改完成，接著對於 attribute a_color 進行『畫什麼』的變數位置取得、建立、設定綁定 buffer：

```
const colorAttributeLocation =
  gl.getAttribLocation(program, 'a_color');

// a_position: buffer, vertexAttribPointer, bufferData
// ...

const colorBuffer = gl.createBuffer();
gl.bindBuffer(gl.ARRAY_BUFFER, colorBuffer);

gl.enableVertexAttribArray(colorAttributeLocation);
gl.vertexAttribPointer(
  colorAttributeLocation,
  3, // size
  gl.UNSIGNED_BYTE, // type
  true, // normalize
```

1-43

1 Hello WebGL

```
  0, // stride
  0, // offset
);
```

這段程式碼原理在『畫一個三角形（下）』章節都有提過，比較需要注意幾點：

- `gl.vertexAttribPointer()` 以及接下來 `gl.bufferData()` 這些指令執行的當下要注意 bind 的 `ARRAY_BUFFER` 是哪個，要不然會對著錯誤的目標做事，當然最好的就是把對於一個 attribute 的操作清楚分好，日後也比較好看出該區域在操作的對象。

- `gl.vertexAttribPointer()` 的 `size` 為 3，也就是顏色的 rgb 3 個 channel，`a_color` 為 `vec3` 接收 3 個元素的向量，因此傳送 buffer 時對於每個頂點 `gl.bufferData()` 要給 3 個值。

- `gl.vertexAttribPointer()` 使用 `gl.UNSIGNED_BYTE` 配合 `normalize` 設定為 `true` 來使用，在『畫一個三角形（下）』章節有提到：`normalize` 配合整數型別時可以把資料除以該型別的最大值使資料轉換成介於（包含）0 到 1 的浮點數，下面在 gl.bufferData() 傳送顏色資料時候使用型別 `Uint8Array`，`Uint8` 的值域為 0 至 255，剛好使我們可以在資料內容寫熟悉的 rgb 值。

最後傳送 `a_color` 的資料到其 buffer 中，為每個頂點指定一個顏色：

```
gl.bufferData(
  gl.ARRAY_BUFFER,
  new Uint8Array([
    121, 85, 72,
    121, 85, 72,
    121, 85, 72,

    0, 76, 213,
    0, 76, 213,
    0, 76, 213,
```

```
    0, 0, 0,
    255, 255, 255,
    255, 255, 255,
  ]),
  gl.STATIC_DRAW,
);
```

資料中的第一個三角形位置在渲染結果中央，三個頂點相同顏色為原本的棕色，第二個三角形在左方，同樣為純色，顏色使用藍色 `rgb(0, 76, 213)`，色碼為 `#004cd5`，而第三個就有趣了，筆者特意其第一個頂點為黑色，剩下兩個頂點為白色，因為內插的緣故，會得到漸層的效果，整體渲染結果如圖 1-23。

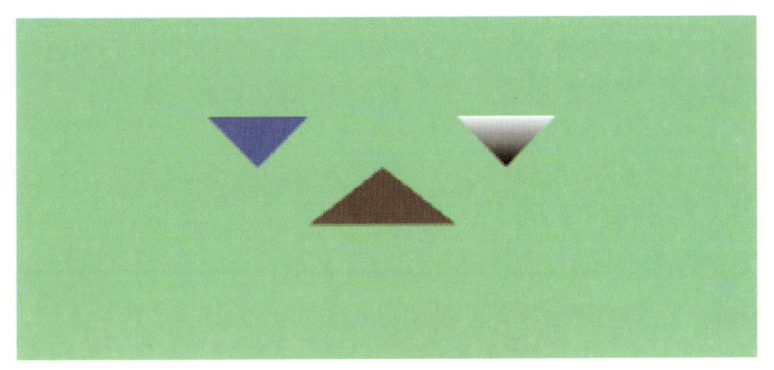

▲ 圖 1-23　整體渲染結果圖

最後再總結一下，整體資料流：

1. 每個頂點有一組 (x, y) 座標值 `a_position` 以及顏色 rgb 資料 `a_color`。

2. 在 vertex shader 除了計算 clip space 座標外，複製 attribute `a_color` 到 varying `v_color`。

3. 在各個頂點之間 `v_color` 會做內插，約接近一個頂點之 pixel 其 `v_color` 就越接近該頂點當初設定的 `a_color`。

4. fragment shader 對於每個 pixel 執行，使用 `v_color` 並直接輸出該顏色。

1 Hello WebGL

到此進度的完整程式碼可以在這邊找到：

</>	ch1/05/index.html
</>	ch1/05/main.js

也可以參考上線版本，透過瀏覽器打開此頁面：

https://webgl-book.pastleo.me/ch1/05/index.html

如果讀者好奇去修改傳入 `gl.bufferData()` 的資料玩玩的話，應該很快就會發現要自己去對 `a_position` 的第幾組資料跟 `a_color` 的第幾組資料是屬於同一個頂點的，他們在程式碼上有點距離，沒有那種 `{position: [1, 2], color: '#abcdef'}` 清楚的感覺，真的要做些應用程式，很快就得自己對這部份做點抽象開始包裝，要不然程式碼一轉眼就會讓人難以摸著頭緒。

WebGL1 中 varying 寫法

在 GLSL 1.00 ES（WebGL1）中，varying 的關鍵字不是 `in`、`out`，而是直接寫 `varying`，下面這段為等價的 GLSL 1.00 ES 版 shader：

```
const vertexShaderSource = `
attribute vec2 a_position;
attribute vec3 a_color;

uniform vec2 u_resolution;

varying vec3 v_color;

void main() {
  gl_Position = vec4(
    a_position / u_resolution * vec2(2, -2) + vec2(-1, 1),
    0, 1
  );
```

1-46

```
  v_color = a_color;
}
`;

const fragmentShaderSource = `
precision highp float;

varying vec3 v_color;

void main() {
  gl_FragColor = vec4(v_color, 1);
}
`;
```

筆者認為 WebGL API 最核心的基石（building block）其實就是 CH1 的內容，接下來除了 texture、framebuffer、skybox 之外，幾乎可以說是用這些基石，搭配線性代數，建構出 3D、光影等效果。下一章就來介紹 texture 並讓 shader 真的開始做一些運算。

1 Hello WebGL

Texture & 2D

有在玩遊戲的讀者們在討論 3D 遊戲描述選染效果的時候，不知道有沒有聽到『3D 貼圖』這樣的字眼？遊戲 3D 物件表面常常不會是純色或是漸層單調的樣子，而是有一張圖片貼在這個物件表面的感覺，所以才叫做『3D 貼圖』吧，而且也可以用在圖案重複的『材質』顯示上，因此在英文叫做 texture。雖然現在還沒進入 3D 的部份，但是 3D 貼圖、texture 其實就是把圖片做縮放、旋轉等變形來顯示，此章節就來介紹 WebGL 取用、顯示圖片並開始讓形狀產生變形吧！

2 Texture & 2D

2-1　在 WebGL 取用、顯示圖片 – Textures

在 CH2 的一開始，我們將使用新的 index.html 以及 main.js 作為開始，起始點完整程式碼可以在這邊找到：

</>	ch2/00/index.html
</>	ch2/00/main.js

也可以參考上線版本，透過瀏覽器打開此頁面：

https://webgl-book.pastleo.me/ch2/00/index.html

筆者將 createShader、createProgram 移動到工具箱 lib/utils.js，裡面有 loadImage 用來下載並回傳 Image 元素（注意，這是一個 async function[1]），並且頂點位置與 CH1 最後相同使用 pixel 座標定位，看起來如圖 2-1 所示：

▲ 圖 2-1　ch2/00/index.html 起始點跑起來的樣子

1　https://developer.mozilla.org/zh-TW/docs/Web/JavaScript/Reference/Statements/async_function

2-1 在 WebGL 取用、顯示圖片 – Textures

這一個灰色的正方形是由兩個三角形組成的，讀者可以在 `main.js` 中的 `gl.bufferData()` 中看到每個頂點後面有一個註解字母，其對應被渲染在畫布上的位置請見示意圖 2-2：

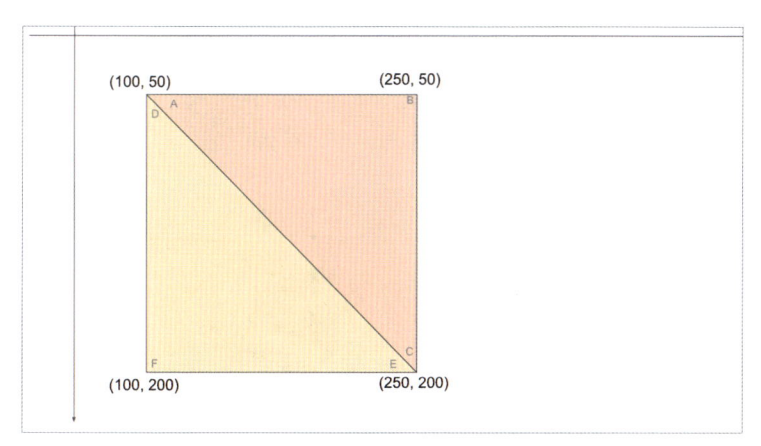

▲ 圖 2-2 頂點被渲染在畫布上的位置

接下來以此為起點，讓灰色方形區域顯示圖片吧！

⌘ 建立 WebGL texture

建立之前，把來源圖片下載好，直接呼叫 `loadImage` 並傳入圖片網址，因為圖片讀取往往需要一些時間，在 Web 上通常使用非同步[2]的方式讀取讓使用者界面保持流暢，這邊也不例外，使用 `await`（也是因此得寫在 `async function main()` 內）確保圖片已經下載好：

```
const image = await loadImage('/assets/pastleo.jpg');
```

這張圖片是筆者的大頭貼，在『範例程式碼包』中的這個位置：

</>	assets/pastleo.jpg

---

2   https://developer.mozilla.org/zh-TW/docs/Learn/JavaScript/Asynchronous/Introducing

2　Texture & 2D

此 image 是 Web API 的 Image[3] 物件，其看起來如圖 2-3 所示：

▲ 圖 2-3　筆者大頭貼圖片

接著建立、bind（對準）並設定 texture：

```javascript
const texture = gl.createTexture();
gl.bindTexture(gl.TEXTURE_2D, texture);
gl.texImage2D(
  gl.TEXTURE_2D,
  0, // level
  gl.RGB, // internalFormat
  gl.RGB, // format
  gl.UNSIGNED_BYTE, // type
  image, // data
);
```

可以發現 `gl.createTexture()` / `gl.bindTexture()` 這個組合與 `gl.createBuffer()` / `gl.bindBuffer()` 這個組合神似，建立並且把 `gl.TEXTURE_2D` 目標對準 texture；接下來透過 `gl.texImage2D()`[4] 設定 texture 所使用的圖片：

3　https://developer.mozilla.org/en-US/docs/Web/API/HTMLImageElement/Image
4　mdn 文件請見：
　https://developer.mozilla.org/en-US/docs/Web/API/WebGLRenderingContext/texImage2D

- **level**：對於一個 `texture` 其實有許多縮放等級，與接下來的 `gl.generateMipmap()` 有一定的關係，通常是填 0 表示輸入的是原始尺寸，在下面會深入解釋。

- **internalFormat**、**format**、**type**：這三個參數告訴 WebGL 如何解讀、使用輸入的圖片資料，可用的組合可以參考 WebGL 規格上的表格[5]；顯然來源圖片有 RGB 三個 channel，`format` 選 RGB，而 `type` 沒有也特別的需求選擇 `UNSIGNED_BYTE` 即可。

> 在 WebGL1 時，`format` 必須與 `internalFormat` 相同，而且 `internalFormat`、`format`、`type` 能用的組合也少非常多。

要傳入的圖片以 `data` 參數傳入，這邊支援直接傳入 Web API 的 `Image` 物件，因此直接放前面用 `loadImage` 讀取好的 `image`，圖片就以 RGB 的格式輸入到 GPU 內 `texture` 的 `level: 0` 位置上。

對於 `level` 的部份，讀者們可以想像一下，在 3D 的世界中鏡頭可能距離貼圖很遙遠，顯示時為了效率沒辦法當下做完整圖片的縮放，因此需要事先把各個尺寸的縮圖做好放在記憶體裡，這樣的東西叫做 mipmap[6]，而 `level` 表示縮放的等級，0 表示沒有縮小的版本；所以開發者得自己把各個縮放尺寸做好分別輸入嗎？幸好 WebGL 有內建方法一行對著目前的 texture 產生所有尺寸：

```
gl.generateMipmap(gl.TEXTURE_2D);
```

5　https://www.khronos.org/registry/webgl/specs/latest/2.0/#TEXTURE_TYPES_FORMATS_FROM_DOM_ELEMENTS_TABLE

6　https://zh.wikipedia.org/wiki/Mipmap

2 Texture & 2D

⌘ 如何在 shader 中使用 texture

回想 fragment shader 的運作方式：在每個 pixel 運算其顏色。那麼如果要顯示 `texture`，就會變成『在每個 pixel 運算時從 texture 圖片上的某個位置取出其顏色來輸出』。在 GLSL 中可以透過 uniform 傳輸一種叫做 `sampler2D` 的資料型別：

```glsl
uniform sampler2D u_texture;
```

把這個 uniform 變數叫做 `u_texture`。接著 GLSL 的內建 function `texture()`[7] 可以進行上面所說的『從 texture 圖片上的某個位置取出其顏色』：

```glsl
outColor = texture(u_texture, v_texcoord);
```

我們直接把 `texture()` 取回『指定』『位置』的顏色（型別為 `vec4`）放置到輸出之顏色上，而這個 `v_texcoord` 即為『指定』的『位置』，既然是 `sampler2D`，`v_texcoord` 型別必須是 `vec2` 表示在 `u_texture` 圖片上的 (x, y) 座標，並且 (0.0, 0.0) 為圖片左上角，(1.0, 1.0) 為圖片右下角。腦筋快的應該已經想到，`v_texcoord` 是一個 varying，因為每個頂點之間所要取用的 `u_texture` 圖片座標是連續的。最後完整的 fragment shader 長這樣：

```glsl
#version 300 es
precision highp float;

in vec2 v_texcoord;

uniform sampler2D u_texture;

out vec4 outColor;

void main() {
  outColor = texture(u_texture, v_texcoord);
}
```

7　`texture()` 文件：https://www.khronos.org/registry/OpenGL-Refpages/gl4/html/texture.xhtml

2-1　在 WebGL 取用、顯示圖片 – Textures

在 WebGL1 中，shader 中呼叫的取樣 function 叫做 `texture2D()` 而不是 `texture()`；神奇的是，在 WebGL2 中還是有 `texture2D()` function，但是回傳值變成 `float` 而非 `vec4`，功能顯然不同。

⌘ Varying `v_texcoord`

既然 `fragment shader` 需要 `varying`，那麼就得在 vertex shader 提供對應的 varying，vertex shader 又需要從 attribute 取得 texture 各個頂點需要取用的座標，對 vertex shader 加上這幾行：

```
#version 300 es
in vec2 a_position;
in vec2 a_texcoord;

uniform vec2 u_resolution;
out vec2 v_texcoord;

void main() {
  gl_Position = vec4(
    a_position / u_resolution * vec2(2, -2) + vec2(-1, 1),
    0, 1
  );
  v_texcoord = a_texcoord;
}
```

接下來又回到 CH1『畫什麼』的設定，取得 a_texcorrd attribute 位置：

```
const texcoordAttributeLocation = (
  gl.getAttribLocation(program, 'a_texcoord')
);
```

並且建立 buffer、設定好 vertex attribute array 使得 vertex shader 能正確取用：

2 Texture & 2D

```
// a_texcoord
const texcoordBuffer = gl.createBuffer();
gl.bindBuffer(gl.ARRAY_BUFFER, texcoordBuffer);

gl.enableVertexAttribArray(texcoordAttributeLocation);
gl.vertexAttribPointer(
  texcoordAttributeLocation,
  2, // size
  gl.FLOAT, // type
  false, // normalize
  0, // stride
  0, // offset
);
```

因為 texture (0.0, 0.0) 為圖片左上角，(1.0, 1.0) 為圖片右下角，對於每個頂點的填入的 texcoord 如示意圖 2-4：

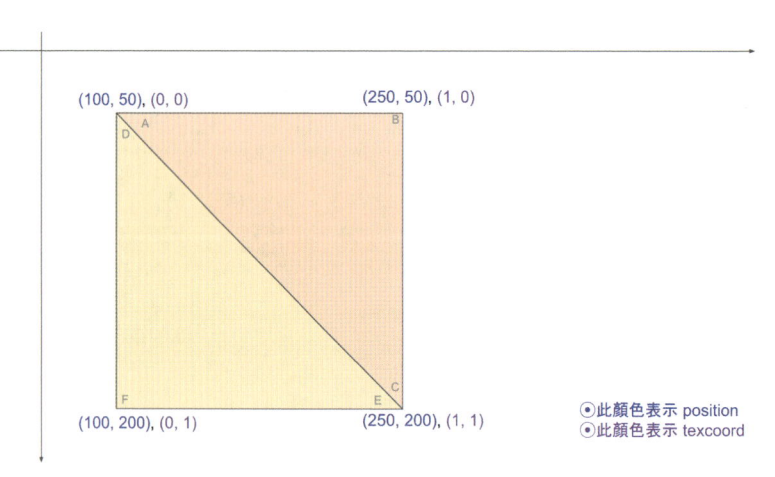

(100, 50), (0, 0)　　　(250, 50), (1, 0)

(100, 200), (0, 1)　　(250, 200), (1, 1)

⊙ 此顏色表示 position
⊙ 此顏色表示 texcoord

▲ 圖 2-4　頂點之 texcoord 示意圖

因此要給 **a_texcoord** 所傳入的 buffer 資料如下：

```
gl.bufferData(
  gl.ARRAY_BUFFER,
  new Float32Array([
    0, 0, // A
    1, 0, // B
```

2-8

2-1 在 WebGL 取用、顯示圖片 – Textures

```
    1, 1, // C

    0, 0, // D
    1, 1, // E
    0, 1, // F
  ]),
  gl.STATIC_DRAW,
);
```

好的，這麼一來 texcoord 的部份就準備妥當，剩下 `u_texture` 尚未設定 uniform 去使用建立好的 texture。

⌘ 提供 texture 給 shader 使用

理論上來說整張 texture 應該是個巨大的陣列資料，但是與 array buffer 不同，texture 必須是可以隨機存取的（random access），意思是說 fragment shader 不論在哪個 pixel 都可以取用 texture 任意位置的資料，因為這個原因，texture 使用 uniform 透過一個特別的 `sampler2D` 類似包裝起來的物件在 shader 中使用，而且可能因為早期 GPU 硬體的限制，在使用的時候會先把 texture 放置在一個『通道』，並且設定通道編號到 uniform 上，從 WebGL API 就可以看得出來通道的數量其實不多。

首先一樣取得 uniform 位置：

```
const textureUniformLocation = gl.getUniformLocation(program, 'u_texture');
```

然後設定把 texture 啟用在一個『通道』，並把這個通道的編號傳入 u_texture uniform：

```
// after gl.useProgram()...

const textureUnit = 0;
gl.bindTexture(gl.TEXTURE_2D, texture);
gl.activeTexture(gl.TEXTURE0 + textureUnit);
gl.uniform1i(textureUniformLocation, textureUnit);
```

透過這幾行程式碼放置 texture 到通道並設定於 uniform，詳細解釋一下：

- `textureUnit` 為通道的編號，設定為 0 使用第一個通道。
- `gl.bindTexture` 把目標指向建立好的 `texture`，如果有其他 texture 導致目標更換時，這邊要把目標設定正確，雖然本篇只有一個 `u_texture` 就是了。
- `gl.activeTexture()` 啟用通道並把目標 texture 設定到通道上，這邊還有神奇的 gl.TEXTURE0 + textureUnit 寫法：
    - 讀者可以嘗試在 Console 輸入 `gl.TEXTURE1 - gl.TEXTURE0`（相減結果為 1），或是 `gl.TEXTURE5 - gl.TEXTURE2`（相減結果為 3），就可以知道為什麼可以用 + 共用 textureUnit 指定通道了。
- 在 CH1 介紹 uniform 提到對於每種資料型別都有一個傳入 function，`gl.uniform1i` 傳的是 1 個整數，把通道的編號傳入，在 fragment shader 中 `sampler2D` 就能去使用對應的 texture。

一切順利的話，就可以看到圖片出現在 canvas 裡頭，fragment shader 成功地『在每個 pixel 運算時從 texture 圖片上的某個位置取出其顏色來輸出』，如圖 2-5：

▲ 圖 2-5　成功輸出的結果

讀者如果有興趣，可以修改 texcoord 的數字感受一下 `texture()`，像是把 C 點 texcoord 改成 (0.8, 0.8) 就變成如圖 2-6 的樣子：

▲ 圖 2-6　修改 texcoord 的數字

到此進度的完整程式碼可以在這邊找到：

</>	ch2/01/index.html
</>	ch2/01/main.js

也可以參考上線版本，透過瀏覽器打開此頁面：
https://webgl-book.pastleo.me/ch2/01/index.html

2 Texture & 2D

2-2　Texture 使用上的細節

繼上篇繪製出圖片後，不知道是不是有讀者覺得成像的品質好像沒有原本的 JPG 圖檔來的好？筆者大頭貼的圖檔的寬高為 1024x1024（如圖 2-7，於範例程式碼包 `assets/pastleo.jpg`），而渲染結果只要 150x150，因此圖片以『縮小』的方式呈現，根據 shader 內『從 texture 圖片上的某個位置取出其顏色』的方式，其實已經可以知道會有些 pixel 因為沒有被採用而失去一些細節。

▲ 圖 2-7　筆者大頭貼原始圖片

圖 2-7 這邊各位讀者看到的原圖，在現今大多顯示、輸出時會將原圖上的所有 pixel 盡量納入考量，儘管一堆細節被壓縮到一個 pixel 上，也是呈現一個平均的顏色，至少圖片看起來比較圓滑；而 WebGL 也有打算改善這個狀況，在先前建立 texture 的最後，我們有呼叫這一行來產生縮圖：

```
gl.generateMipmap(gl.TEXTURE_2D);
```

但是這個時候系統並不知道要產生的縮圖大小為何，只能產生出多個不同 `level` 的縮圖，假設原圖的一邊大小是 1024，那麼可能會產生出大小為

2-12

512、256、128、64、32⋯的縮圖，在實際 shader 執行時根據縮放狀況決定使用哪張縮圖來取樣，如果系統選用的縮圖與需要的大小有一段距離，那麼就會導致渲染出來的成像品質不這麼理想。

這樣的設計也是為了有更好的效能，shader 執行時通常是遊戲或是動畫的狀況，需要『即時』渲染，應該是事前就把縮圖準備好，渲染當下直接去對應的記憶體位置拿資料就好，這也是為什麼 WebGL 讓開發者可以手動提供 texture 的各個 `level` 縮圖。

也因為 WebGL 預設在縮小渲染 texture 的狀況就會去使用縮圖，如果沒有呼叫 gl.generateMipmap(gl.TEXTURE_2D); 產生縮圖或是手動指定縮圖，那麼渲染時就會因為縮圖的記憶體區塊沒有資料而導致成像一片黑。

> 筆者當時看到 mipmap 相關資料時有個疑惑，明明在 fragment shader 內只是使用 `texture()` 給予取樣的位置，是怎麼知道縮放比例的？查到這篇 stackoverflow[8] 回答，看起來因為 fragment shader 是平行運算的，所以各個鄰近 pixel 運算會同時呼叫 `texture()`，這樣 GPU 就可以知道縮放的比例。

⌘ Texture 縮放取樣方式

事實上，對於在 shader 執行時使用的縮圖以及取樣方式是可以設定的，而且若成像比原圖來的小，WebGL 預設就會選兩張縮圖分別取一個 pixel 來平均，除此之外還有許多的設定，透過下面這個 API 來進行。

```
gl.texParameteri(gl.TEXTURE_2D, pname, param);
```

`pname` 指設定值名稱（key），`param` 指設定值選項（value），筆者將幾個比較常用的設定值與對應的選項條列下來[9]：

8　https://stackoverflow.com/a/7391241
9　`gl.texParameteri()` 的完整參數表請見 mdn 文件頁面：https://developer.mozilla.org/en-US/docs/Web/API/WebGLRenderingContext/texParameter

2 Texture & 2D

- **pname** 為 `gl.TEXTURE_MIN_FILTER` 時,設定『顯示大小比原圖小』之取樣方式:
  - `gl.NEAREST`: 從原圖選擇 1 個 pixel。
  - `gl.LINEAR`: 從原圖選擇 4 個 pixel 平均。
  - `gl.NEAREST_MIPMAP_NEAREST`: 從 mipmap 中選最接近的縮圖,再選擇 1 個 pixel。
  - `gl.LINEAR_MIPMAP_NEAREST`: 從 mipmap 中選最接近的縮圖,再選擇 4 個 pixel 平均。
  - `gl.NEAREST_MIPMAP_LINEAR`: 從 mipmap 中選最接近的 2 張縮圖,分別選擇 1 個 pixel 平均,此為預設取樣方式。
  - `gl.LINEAR_MIPMAP_LINEAR`: 從 mipmap 中選最接近的 2 張縮圖,分別選擇 4 個 pixel 平均。
- **pname** 為 `gl.TEXTURE_MAG_FILTER` 時,設定『顯示大小比原圖大』之取樣方式:
  - `gl.NEAREST`: 從原圖選擇 1 個 pixel。
  - `gl.LINEAR`: 從原圖選擇 4 個 pixel 平均,此為預設取樣方式。

假設我們要讓 `gl.TEXTURE_MIN_FILTER` 選用 `gl.NEAREST`,因為只需要原圖,這樣的狀況下甚至不需要呼叫 `gl.generateMipmap()` 產生縮圖,寫成這樣:

```
// gl.generateMipmap(gl.TEXTURE_2D);
gl.texParameteri(
  gl.TEXTURE_2D,
  gl.TEXTURE_MIN_FILTER,
  gl.NEAREST,
);
```

這時候成像看起來如圖 2-8:

2-2　Texture 使用上的細節

▲ 圖 2-8　使用 `gl.NEAREST` 作為縮放取樣方式的成果

讀者可以觀察比較一下 `gl.TEXTURE_MIN_FILTER` 其他模式的視覺差異：

gl.LINEAR 從原圖選擇 4 個 pixel 內插 （不需要產生縮圖） 	gl.NEAREST_MIPMAP_NEAREST 從 mipmap 中選最接近的縮圖 再選擇 1 個 pixel 
gl.LINEAR_MIPMAP_LINEAR 從 mipmap 中選最接近的縮圖 再選擇 4 個 pixel 做內插 	gl.NEAREST_MIPMAP_LINEAR 從 mipmap 中選最接近的 2 張縮圖分別 選擇 1 個 pixel 做內插（預設取樣方式） 

> gl.LINEAR_MIPMAP_LINEAR
>
> 從 mipmap 中選最接近的 2 張縮圖
>
> 各自做完內插之後再內插一次，稱為三線性內插

　　使用越多 pixel 做平均的顯示出來的成像就越平滑，但是也比較消耗效能，最後這個 `gl.LINEAR_MIPMAP_LINEAR` 從 mipmap 中選最接近的 2 張縮圖，分別選擇 4 個 pixel 平均，意思就是一次 `texture()` 得讀取 texture 中 2x4 = 8 個 pixel 出來平均，結果也最平滑。

⌘ 繪製賽車格紋：重複 pattern 的 texture。

　　除了設定縮放取樣方式之外，還有另外一組可以設定『對 texture 取樣時，座標超出範圍的行為』，也就是 shader 內傳入 texture() 大於 1 或是小於 0 時候該如何處理，這組設定值在 `gl.texParameteri(gl.TEXTURE_2D, pname, param)` 的 `pname` 為 `gl.TEXTURE_WRAP_S`、`gl.TEXTURE_WRAP_T`，`_S`、`_T` 分別對 x、y 軸方向進行設定（在 texture 內的座標通常以 s、t 來表示 x、y 方向），可以填入的值：

- `gl.REPEAT`: 重複 pattern，此為預設運作模式
- `gl.CLAMP_TO_EDGE`: 延伸邊緣顏色
- `gl.MIRRORED_REPEAT`: 重複並鏡像 pattern

　　為了試玩 `gl.TEXTURE_MAG_FILTER` 以及 `gl.TEXTURE_WRAP` 重複 pattern 的 texture，接下來繪製如圖 2-9 的賽車格紋：

▲ 圖 2-9　賽車格紋

可以看得出來整張圖就是重複這邊紅色框起來的區域，如圖 2-10：

▲ 圖 2-10

　　也就是說，這張圖只需要 2x2 的大小，左上、右下為白色，右上、左下為黑色。其實輸入 texture 的 `gl.texImage2D()` 支援各式各樣的輸入來源，其中一個是 `ArrayBufferView`[10]，也就是可以透過型別陣列傳各個 pixel 的顏色資料進去：

10　ArrayBufferView 包含了 TypedArray，這邊傳入的 texture 資料使用 Uint8Array，為 TypedArray 的一種，關於 ArrayBufferView 的詳細文件請見：
　　https://developer.mozilla.org/en-US/docs/Web/API/ArrayBufferView

2-17

2 Texture & 2D

```javascript
// const image = await loadImage('/assets/pastleo.jpg');

const texture = gl.createTexture();
gl.bindTexture(gl.TEXTURE_2D, texture);

// gl.texImage2D(
//   ...
//   image, // data
// );

const whiteColor = [255, 255, 255, 255];
const blackColor = [0, 0, 0, 255];
gl.texImage2D(
  gl.TEXTURE_2D,
  0, // level
  gl.RGBA, // internalFormat
  2, // width
  2, // height
  0, // border
  gl.RGBA, // format
  gl.UNSIGNED_BYTE, // type
  new Uint8Array([
    ...whiteColor, ...blackColor,
    ...blackColor, ...whiteColor,
  ])
);

gl.generateMipmap(gl.TEXTURE_2D);
```

　　type 使用 `gl.UNSIGNED_BYTE`，也就是每個 pixel 的每個顏色 channel 為一個 `Uint8Array` 的元素（1 byte = 8 bit，值域為 0-255），白色 RGBA 即為 `[255, 255, 255, 255]`、黑色 RGBA 即為 `[0, 0, 0, 255]`；另外直接傳陣列進去時，需要額外給定 `width, height, border`，這張圖為 2x2 且沒有 border，給予的參數如上所示。

> 為什麼要使用 `gl.RGBA`？因為有個 `gl.UNPACK_ALIGNMENT` 的設定值，這個值預設為 4，表示每行的儲存單位為 4 bytes，如果這樣 2x2 的小 texture 要使用 RGB 就得透過 `gl.pixelStorei()` 改成 1，我們使用 `gl.RGBA` 符合預設值就好。

把原本使用外部圖片的 `loadImage()` 以及 `texImage2D()` 註解起來，存檔重整可以看到如圖 2-11 模糊的格紋：

▲ 圖 2-11　預設取樣模式渲染出模糊的格紋

現在顯示大小顯然比原圖來的大，所以是『放大』取樣模式 `gl.TEXTURE_MAG_FILTER` 預設的 `gl.LINEAR` 導致的內插效果，但是現在的狀況不想要內插效果，因此設定改用 `gl.NEAREST`，成像如圖 2-12：

```
gl.texParameteri(
  gl.TEXTURE_2D,
  gl.TEXTURE_MAG_FILTER,
  gl.NEAREST,
);
```

▲ 圖 2-12　由 gl.LINEAR 改為 gl.NEAREST 的效果

以賽車格紋來說顯然不會用到 mipmap，可以把 `gl.generateMipmap()` 註解起來並設定 `gl.TEXTURE_MIN_FILTER` 為 `gl.NEAREST` 以停用 mipmap，整組 texture 取樣設定如下：

```
// gl.generateMipmap(gl.TEXTURE_2D);
gl.texParameteri(
  gl.TEXTURE_2D,
  gl.TEXTURE_MAG_FILTER,
  gl.NEAREST,
);
gl.texParameteri(
  gl.TEXTURE_2D,
  gl.TEXTURE_MIN_FILTER,
  gl.NEAREST,
);
```

接下來讓圖片重複，直接修改 texcoord `gl.bufferData()` 時傳入的值：

```
// a_texcoord
const texcoordBuffer = gl.createBuffer();
gl.bindBuffer(gl.ARRAY_BUFFER, texcoordBuffer);

// setup vertex attribute array...

gl.bufferData(
  gl.ARRAY_BUFFER,
  new Float32Array([
    0, 0, // A
    8, 0, // B
    8, 8, // C

    0, 0, // D
    8, 8, // E
    0, 8, // F
  ]),
  gl.STATIC_DRAW,
);
```

筆者把 1 改成 8，所以 x、y 軸皆將重複 8 次。

雖然 `gl.TEXTURE_WRAP` 預設就是 `gl.REPEAT`，不過我們還是明確的寫下我們使用重複圖案的渲染方式：

```
gl.texParameteri(
  gl.TEXTURE_2D,
  gl.TEXTURE_WRAP_S,
  gl.REPEAT,
);
gl.texParameteri(
  gl.TEXTURE_2D,
  gl.TEXTURE_WRAP_T,
  gl.REPEAT,
);
```

賽車格紋圖案就被渲染出來囉，如圖 2-13：

▲ 圖 2-13　渲染完成的賽車格紋

到此進度的完整程式碼可以在這邊找到：

</>	ch2/02/index.html
</>	ch2/02/main.js

也可以參考上線版本，透過瀏覽器打開此頁面：

https://webgl-book.pastleo.me/ch2/02/index.html

不知道有沒有讀者好奇 `gl.CLAMP_TO_EDGE` 以及 `gl.MIRRORED_REPEAT` 的結果？前者成像如圖 2-14、後者如圖 2-15 所示：

```
gl.texParameteri(
  gl.TEXTURE_2D,
  gl.TEXTURE_WRAP_S,
  gl.CLAMP_TO_EDGE, // gl.MIRRORED_REPEAT
);
gl.texParameteri(
  gl.TEXTURE_2D,
  gl.TEXTURE_WRAP_T,
  gl.CLAMP_TO_EDGE, // gl.MIRRORED_REPEAT
);
```

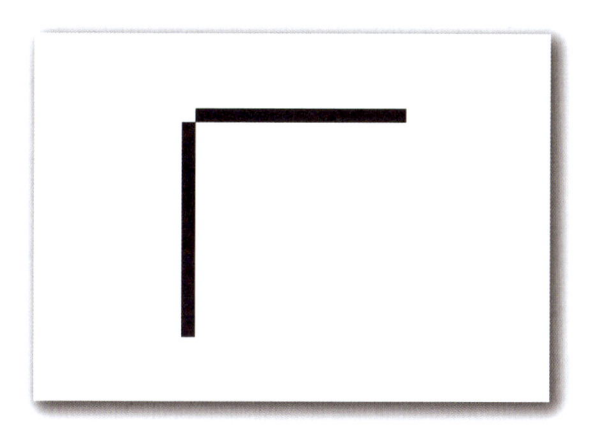

▲ 圖 2-14　使用 `gl.CLAMP_TO_EDGE` 結果

▲ 圖 2-15　使用 `gl.MIRRORED_REPEAT` 結果

⌘ WebGL1 的 NPOT 問題

雖然本書以 WebGL2 為平台，這邊還是要特別提及一下，因為這是筆者認為 WebGL1 最大的痛處；在所有 WebGL2 與 WebGL1 的差異中，除了 GLSL 版本導致的 shader 語法差異之外，常用但 WebGL1 缺乏的功能大多都有 WebGL extenstion 可以補足，所謂 WebGL extenstion 不是指要從 Chrome web store 或是 Firefox Add-ons 下載的瀏覽器擴充套件，比較像是 WebGL spec 上沒有指定要支援，但是各家瀏覽器可以自行加入的功能，所以得看各家瀏

2 Texture & 2D

覽器的臉色來決定特定功能能不能用，通常常用的功能各家瀏覽器在 WebGL1 都有透過 WebGL extenstion 來支援，所以通常比較像是『換個可能比較繞的寫法，但是可以達到一樣的效果』。

不過 WebGL1 的 NPOT 問題就不一樣了，當你打算在 WebGL1 使用一張長寬非 2 的次方倍的圖當成 texture 時，往往會看到如圖 2-16 的錯誤：

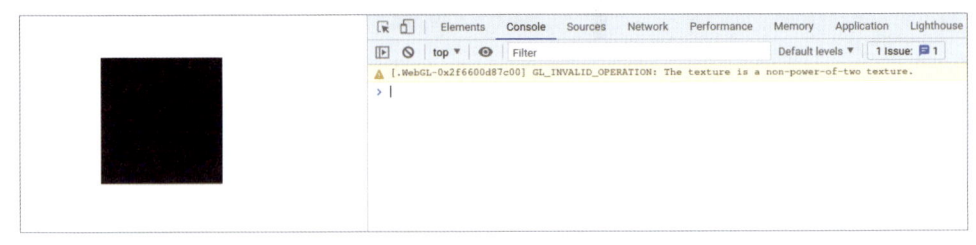

▲ 圖 2-16　在 WebGL1 環境中，圖長寬非為 2 次方倍的常見錯誤

例如一張 1024x768 的圖，寬是 2 的次方（2^{10}），但是高不是，因此產生了錯誤：

```
GL_INVALID_OPERATION: The texture is a non-power-of-two texture.
```

non-power-of-two，簡稱 NPOT，WebGL1 的 mipmap 以及 texture 許多取樣模式僅支援寬高皆為 2 次方的圖，包含了會使用到任何縮圖的 texture 取樣模式以及製作縮圖的功能；如果真的想要在 WebGL1 使用 NPOT 的圖作為 texture，需要避免執行 `gl.generateMipmap(gl.TEXTURE_2D)`、同時使用以下取樣模式：

```
// don't run this on NPOT texture:
// gl.generateMipmap(gl.TEXTURE_2D);
gl.texParameteri(
  gl.TEXTURE_2D,
  gl.TEXTURE_MIN_FILTER,
  gl.LINEAR, // or gl.NEAREST
);
gl.texParameteri(
```

2-24

```
  gl.TEXTURE_2D,
  gl.TEXTURE_WRAP_S,
  gl.CLAMP_TO_EDGE,
);
gl.texParameteri(
  gl.TEXTURE_2D,
  gl.TEXTURE_WRAP_T,
  gl.CLAMP_TO_EDGE,
);
```

`gl.TEXTURE_MIN_FILTER` 為 `gl.NEAREST` 或 `gl.LINEAR` 的狀況下,只有使用原圖,因此沒有使用到 mipmap 相關功能,這樣一來就能支援 NPOT 圖片;這樣的限制可能是效能考量,或是說 WebGL1 基於的 openGL 版本已經古老,當時的 GPU 硬體可能只能在 2 次方寬高的圖片上做運算。

使用純 WebGL1 所遇到的 NPOT 問題真的是讓開發上增加不少手續,開發者通常得將圖片調整好再提供給 WebGL1 使用,比起其他語法上的差異,此為功能限制上從現今觀點來看感覺很落後的缺陷;幸好這樣的限制在 WebGL2 已經取消,所有的 mipmap 以及 texture 取樣模式都已經完整支援 NPOT 尺寸的圖片。

介紹 texture 功能至此,之後甚至可以讓 GPU 渲染到 texture 上做更進階的應用,屆時再接續進行說明。

2-3 互動 & 動畫

講到現在我們渲染出來的畫面都是靜態的,準確的說是網頁載入後進行一次的繪製,如果使用者打算調整視窗大小還會導致成像的拉伸;然而用到 WebGL 繪製的遊戲或是特效網頁通常都是『會動』的,有許多會根據使用者操作反應在畫面上,或者是有根據時間產生變化的動畫,而要動起來就是得重新渲染刷新畫面,本篇將基於先前用 texture 渲染的畫面,加入簡易的 WebGL 互動、動畫功能。

2 Texture & 2D

⌘ 調整程式碼架構

在加入互動、動畫之前，我們回到渲染一張圖片的狀態，並調整一下程式碼架構，先前的實做都是從上到下一次執行完畢，畢竟也就只渲染這麼一次，但是接下來會開始有重畫的動作，所以要分成只有一開始要執行一次的初始化程式、更新狀態以及執行『畫』這個動作的程式。

這個只執行一次的初始化程式可以叫它 `setup()`，從建立 WebGL context、編譯連結 GLSL shaders、取得 GLSL 變數位置、下載圖片並建立 texture，最後到設立 buffer 及 vertex attribute 並輸入資料，這些都是一開始初始化要做的工作，因此把這些工作從原本的 `main()` 抽出來；同時也把初始化時建立的 Javascript 物件像是 `gl`、`program`、`xxxAttributeLocation`、`xxxUniformLocation`、`texture`、`xxxBuffer` 整理起來作為 setup() 的回傳值。

```javascript
async function setup() {
  const canvas = document.getElementById('canvas');
  const gl = canvas.getContext('webgl2');

  // createShader(); const program = createProgram()...

  const attributes = {
    position: gl.getAttribLocation(program, 'a_position'),
    texcoord: gl.getAttribLocation(program, 'a_texcoord'),
  };
  const uniforms = {
    resolution: gl.getUniformLocation(program, 'u_resolution'),
    texture: gl.getUniformLocation(program, 'u_texture'),
  };

  const image = await loadImage('/assets/cat-1.jpg');
  const texture = gl.createTexture();
  gl.bindTexture(gl.TEXTURE_2D, texture);

  // gl.texImage2D( ... image)
  gl.generateMipmap(gl.TEXTURE_2D);
```

```
const buffers = {};

// a_position
buffers.position = gl.createBuffer();
gl.bindBuffer(gl.ARRAY_BUFFER, buffers.position);

gl.enableVertexAttribArray(attributes.position);
// gl.vertexAttribPointer(attributes.position, ...
// gl.bufferData( ... )

// a_texcoord
buffers.texcoord = gl.createBuffer();
gl.bindBuffer(gl.ARRAY_BUFFER, buffers.texcoord);

gl.enableVertexAttribArray(attributes.texcoord);
// gl.vertexAttribPointer(attributes.texcoord, ...
// gl.bufferData( ... )

return {
  gl,
  program, attributes, uniforms,
  buffers, texture,
};
}
```

上方程式碼有用註解表示較冗長的程式，這樣整體比較容易看出這段程式的架構，同時筆者也改用了一張貓圖，如圖 2-17：

| </> | assets/cat-1.jpg |

2 Texture & 2D

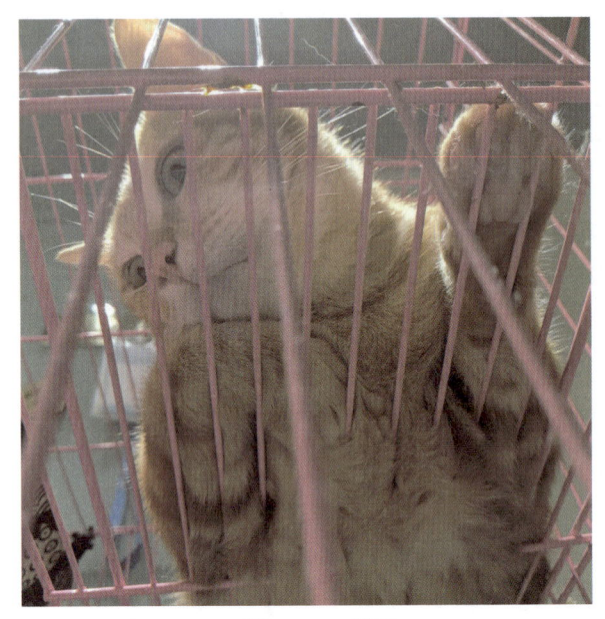

▲ 圖 2-17　貓圖

　　也請記得把輸入到 `buffers.texcoord` 的資料從超出邊界的 8 改回 1，否則待會成像會變成重複的貓圖，如圖 2-18。

▲ 圖 2-18　texcoord 超出邊界產生重複的貓圖

一開始要執行一次的初始化程式 `setup()` 準備完成，另外一部份就是每次執行『畫』這個動作要做的事情，雖然『畫』這個動作就是 `gl.drawArrays()` 這行，但是總是要改變些設定，要不然每次畫出來的東西都是一樣的，而 uniform 資料量小，所以常常作為每次繪製不同結果的參數設定，這些工作抽出來叫做 `render()`，但是因為會需要 `setup()` 回傳的 WebGL 物件，筆者把 `setup()` 回傳的整包東西叫做 app，這邊作為參數接收：

```javascript
function render(app) {
  const {
    gl,
    program, uniforms,
    texture,
  } = app;

  gl.canvas.width = gl.canvas.clientWidth;
  gl.canvas.height = gl.canvas.clientHeight;
  gl.viewport(0, 0, canvas.width, canvas.height);

  gl.useProgram(program);

  gl.uniform2f(uniforms.resolution, gl.canvas.width, gl.canvas.height);

  // texture uniform
  const textureUnit = 0;
  gl.bindTexture(gl.TEXTURE_2D, texture);
  gl.activeTexture(gl.TEXTURE0 + textureUnit);
  gl.uniform1i(uniforms.texture, textureUnit);

  gl.drawArrays(gl.TRIANGLES, 0, 6);
}
```

可以注意到這邊除了設定 uniform 以及最後的 `gl.drawArrays()` 之外，還包含了調整 `canvas` 大小、繪製區域的程式，這樣就可以在重畫的時候解決調整視窗大小時造成的拉伸問題。最後 `main()` 就是負責把 `setup()` 以及 `render()` 串起來：

Texture & 2D

```
async function main() {
  const app = await setup();
  window.app = app;
  window.gl = app.gl;

  render(app);
}

main();
```

調整後用瀏覽器測試，渲染結果如圖 2-19：

▲ 圖 2-19　架構調整完成後，渲染之結果

到此進度的完整程式碼可以在這邊找到：

</>	ch2/03/index.html
</>	ch2/03/main.js

也可以參考上線版本，透過瀏覽器打開此頁面：

https://webgl-book.pastleo.me/ch2/03/index.html

2-30

⌘ 互動：選擇 texture 圖片

在上篇文章中，我們嘗試了幾種不同的 texture，但是都要修改程式碼來更換，接下來來改成可以透過一組 radio input 來控制要顯示的 texture 圖片，除了剛剛的貓圖之外，筆者也準備了另一張貓圖、一張企鵝圖來切換，分別如圖 2-20、圖 2-21：

</>	assets/cat-2.jpg
</>	assets/penguins.jpg

▲ 圖 2-20

2 Texture & 2D

▲ 圖 2-21

原本在 `setup()` 中只建立一個 texture，筆者透過 `Promise.all()`[11] 以及 `async/await` 下載並建立 3 張圖片以及其 texture：

```javascript
const textures = await Promise.all([
  '/assets/cat-1.jpg',
  '/assets/cat-2.jpg',
  '/assets/penguins.jpg',
].map(async url => {
  const image = await loadImage(url);
  const texture = gl.createTexture();
  gl.bindTexture(gl.TEXTURE_2D, texture);
  gl.texImage2D(
    gl.TEXTURE_2D,
    0, // level
    gl.RGB, // internalFormat
    gl.RGB, // format
    gl.UNSIGNED_BYTE, // type
    image, // data
```

---

11 https://developer.mozilla.org/zh-TW/docs/Web/JavaScript/Reference/Global_Objects/Promise/all

```
  );

  gl.generateMipmap(gl.TEXTURE_2D);

  return texture;
}));
```

這樣一來 `textures` 就是一個包含 3 個 texture 的陣列，分別存入了不同的照片。然後加入一個叫做 `state` 的 Javascript object 來儲存目前的狀態，放上 `texture: 0` 表示一開始使用第一個 texture，整個 `setup()` 回傳的 app 也就改成這樣：

```
async function setup() {
  // ...

  return {
    gl,
    program, attributes, uniforms,
    buffers, textures,
    state: {
      texture: 0,
    }
  };
}
```

在 `render()` 這邊做出對應的修改，在 `gl.bindTexture()` 的地方根據 `state.texture` 選取要顯示的 texture：

```
function render(app) {
  const {
    gl,
    program, uniforms,
    textures,
    state,
  } = app;

  // ...
```

2　Texture & 2D

```
const textureUnit = 0;
gl.bindTexture(gl.TEXTURE_2D, textures[state.texture]);
gl.activeTexture(gl.TEXTURE0 + textureUnit);
gl.uniform1i(uniforms.texture, textureUnit);

gl.drawArrays(gl.TRIANGLES, 0, 6);
}
```

目前還沒實做控制，只會顯示出第一張圖，與原本渲染出相同的結果，如圖 2-22：

▲ 圖 2-22　顯示出第一張圖

為了控制顯示的圖片，我們可以借助 HTML 眾多互動元件的幫助，一個簡單的 form 以及一組 radio input，在加上一點 CSS 把整個 form#controls 固定於網頁右上角：

```
<head>
  <!-- ... -->
  <style>
    /* html, body, #canvas ... */
    #controls {
      position: fixed;
      top: 0;
```

2-34

```
      right: 0;
      margin: 1rem;
    }
  </style>
</head>
<body>
  <canvas id="canvas"></canvas>
  <form id="controls">
    <div>
      <input type="radio" id="cat1" name="texture" value="0" checked>
      <label for="cat1">Cat 1</label>
      <input type="radio" id="cat2" name="texture" value="1">
      <label for="cat2">Cat 2</label>
      <input type="radio" id="penguin" name="texture" value="2">
      <label for="penguin">Penguin</label>
    </div>
  </form>
  <script type="module" src="main.js"></script>
</body>
```

使用 HTML，意思就是可以使用 DOM API，在 main() 裡頭進行事件監聽 input 事件：

```
async function main() {
  // const app = ...

  const controlsForm = document.getElementById('controls');
  controlsForm.addEventListener('input', () => {
    const formData = new FormData(controlsForm);
    app.state.texture = parseInt(formData.get('texture'));

    render(app);
  });

  render(app);
}
```

2 Texture & 2D

筆者使用 `new FormData(form)` 直接蒐集整個 form 的資料，之後要加入其他控制項會更方便，使用 `app.state.texture = …` 改變要顯示的圖片之後，呼叫 `render(app)` 重新進行『畫』這個動作，存檔重整之後就會在選擇不同的 radio input 時重新渲染所選的圖片了，如圖 2-23：

▲ 圖 2-23 互動成果之示意圖

到此進度的完整程式碼可以在這邊找到：

</>	ch2/04/index.html
</>	ch2/04/main.js

也可以參考上線版本，透過瀏覽器打開此頁面：
https://webgl-book.pastleo.me/ch2/04/index.html

⌘ 動畫：隨著時間移動的圖片

可以接受事件重新渲染之後，下一步來讓圖片隨著時間移動，像是這個小時候 DVD 播放器的待機畫面：碰到邊緣會反彈的 DVD logo[12]。

12　Youtube 影片連結：https://youtu.be/5mGuCdlCcNM

為了讓圖片位置隨著時間移動,需要在每次渲染時透過某種因子去平移每個頂點 gl_Position 的輸出位置,因為每個頂點的平移量都相同,所以就等同平移整張圖,接著讓這個因子隨著時間改變,並搭配不斷重新渲染的畫面,就能達到『圖片隨著時間移動』的動畫;而這個因子最合適的 WebGL 功能即為 uniform,我們將用 Javascript 隨著螢幕刷新率更新此 uniform、重畫畫面,渲染時透過 uniform 偏移所有頂點位置以平移圖片,最後達到所要的動畫效果。

首先讓圖片位置可以透過一個 uniform 控制『平移』,我們叫它 u_offset,修改控制頂點位置的 vertex shader:

```
uniform vec2 u_resolution;
uniform vec2 u_offset;
out vec2 v_texcoord;

void main() {
  vec2 position = a_position + u_offset;
  gl_Position = vec4(
    position / u_resolution * vec2(2, -2) + vec2(-1, 1),
    0, 1
  );
  v_texcoord = a_texcoord;
}
```

加入 uniform vec2 u_offset 表示圖片的平移量後,建立 vec2 position 變數運算輸入的頂點位置 a_position 加上圖片平移量 u_offset,既然加上了一個 uniform,記得先取得其變數在 shader 中的位置:

```
async function setup() {
  // ...

  const uniforms = {
    resolution: gl.getUniformLocation(program, 'u_resolution'),
    texture: gl.getUniformLocation(program, 'u_texture'),
    offset: gl.getUniformLocation(program, 'u_offset'),
```

2 Texture & 2D

```
};

// ...
}
```

為了讓之後 u_offset 直接表示『畫面上的位置』,調整輸入的頂點座標讓一開始圖片位置在 (0, 0) 最左上角的位置:

```
// a_position ...
gl.bufferData(
  gl.ARRAY_BUFFER,
  new Float32Array([
    0, 0, // A
    150, 0, // B
    150, 150, // C

    0, 0, // D
    150, 150, // E
    0, 150, // F
  ]),
  gl.STATIC_DRAW,
);
```

接下來準備一些狀態,例如移動方向、速度以及現在的位置,就可以透過這些狀態算出下一個 frame 的 u_offset,我們在 setup() 回傳的初始 state 中加上這些狀態,筆者在這邊先隨機產生一個角度 directionDeg,再運用三角函式算出角度對應的方向向量,同時寫上預設速度:

```
async function setup() {
  // ...

  const directionDeg = Math.random() * 2 * Math.PI;

  return {
    gl,
    program, attributes, uniforms,
```

```
    buffers, textures,
    state: {
      texture: 0,
      offset: [0, 0],
      direction: [Math.cos(directionDeg), Math.sin(directionDeg)],
      speed: 0.08,
    },
    time: 0,
  };
}
```

可以發現這邊還有多輸出一個 `time: 0`，與待會『隨著時間』移動相關。在 `render()` 內，輸入剛才的 `u_offset uniform`：

```
function render(app) {
  // ...

  gl.uniform2fv(uniforms.offset, state.offset);

  // ...

  gl.drawArrays(gl.TRIANGLES, 0, 6);
}
```

WebGL 繪製的修改算是都準備好了，要做的事情就是請 WebGL 用一定的頻率重新渲染以產生動畫效果，這個最好的頻率就是與螢幕更新頻率同步，讓每次更新都是有意義可以反應在螢幕上，Web API 有個 function 叫做 `requestAnimationFrame`[13]，傳入一個 callback function，在下次螢幕更新（稱為 frame）時執行，若在此 callback 內再呼叫一次 `requestAnimationFrame` 在下一次螢幕更新時再執行一次，就形成隨著時間更新、重畫的迴圈，因此加上 `loop()` function：

13　https://developer.mozilla.org/zh-TW/docs/Web/API/window/requestAnimationFrame

Texture & 2D

```
1   function loop(app, now = 0) {
2     const { state, gl } = app;
3     const timeDiff = now - app.time;
4     app.time = now;
5
6     state.offset = state.offset.map(
7       (v, i) => v + state.direction[i] * timeDiff * state.speed
8     );
9
10    if (state.offset[0] + 150 > gl.canvas.width) {
11      state.direction[0] *= -1;
12      state.offset[0] = gl.canvas.width - 150;
13    } else if (state.offset[0] < 0) {
14      state.direction[0] *= -1;
15      state.offset[0] = 0;
16    }
17
18    if (state.offset[1] + 150 > gl.canvas.height) {
19      state.direction[1] *= -1;
20      state.offset[1] = gl.canvas.height - 150;
21    } else if (state.offset[1] < 0) {
22      state.direction[1] *= -1;
23      state.offset[1] = 0;
24    }
25
26    render(app);
27    requestAnimationFrame(now => loop(app, now));
28  }
```

講解一下 loop()：

- 第 6 - 8 行用來更新 `offset`，也就是下一個 frame 圖片的左上角位置。
- 第 10 - 24 用來做碰撞測試，碰到邊緣時把 `direction` 反向進行反彈。
- 第 26 行呼叫 `render()` 重畫畫面。
- 第 27 行呼叫 `requestAnimationFrame` 並傳入一個匿名函式，可以注意到這個匿名函式接收一個參數叫做 now，表示此匿名函式執行的時間，在匿名函式內執行 `loop()` 進行下次更新、渲染形成迴圈。

- 上方第 3 - 4 行透過接收到的 now 計算這次畫面更新與上次更新之間的時間差,並運用在第 7 行平移量的長度,為什麼要這樣做呢?因為每個裝置的螢幕更新頻率不一定都是 60Hz,現在有許多手機或是螢幕支援 120Hz 甚至更快的螢幕更新速度,又或者裝置的性能不足,只有 40Hz 之類的,使用 `requestAnimationFrame` 更新的我們如果一律每回合移動一單位,那麼在不同的裝置上動畫的速度會不一樣。

如果想要更了解更多關於 `requestAnimationFrame` 所謂『下一次螢幕更新』的時間點、其與 Javascript event loop 的關係,筆者先前看到一個解釋很好的演講:

In The Loop - JSConf.Asia by Jake Archibald。

網址為 https://youtu.be/cCOL7MC4Pl0?t=529,甚至在最後還有解釋 macro task 什麼時候、如何執行,推薦前端工程師把這場演講完整看一次。

`loop()` 準備好,但需要一個進入點呼叫一次來啟動,修改 `main()` 從原本呼叫一次 `render(app)` 改為 `loop(app)` 啟動動畫迴圈,同時因為已經會在每次螢幕更新時重新渲染,那麼就不用在接收事件時重新渲染了:

```js
async function main() {
  const app = await setup();
  window.app = app;
  window.gl = app.gl;

  const controlsForm = document.getElementById('controls');
  controlsForm.addEventListener('input', () => {
    const formData = new FormData(controlsForm);
    app.state.texture = parseInt(formData.get('texture'));
    app.state.speed = parseFloat(formData.get('speed'));

    render(app);
  });

  render(app);
  loop(app);
}
```

2 Texture & 2D

存檔回去瀏覽器測試，這時候圖片就動起來囉！

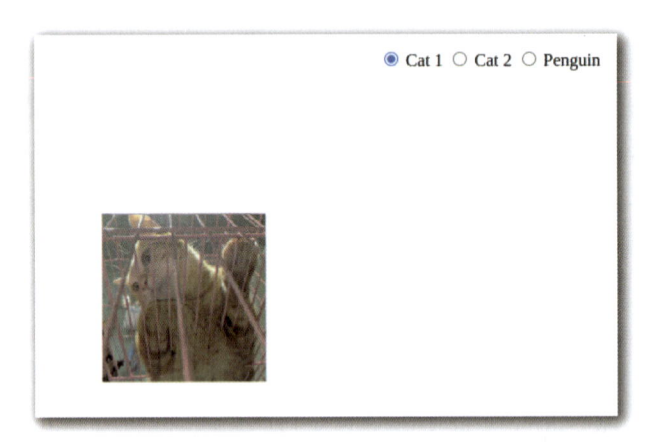

▲ 圖 2-24　成果示意圖

最後，還記得我們的狀態中有個『速度』嗎？我們也可以在 HTML Form 中加入拉桿串連這個狀態來進行速度控制，實做方式與前面 texture 圖片切換類似，先修改 HTML 加入 `type="range"` 的 `<input />`[14]：

```html
<body>
  <canvas id="canvas"></canvas>
  <form id="controls">
    <div>
      <!-- texture radio group -->
    </div>
    <div>
      <label for="speed">Speed</label>
      <input
        type="range" id="speed" name="speed"
        min="0" max="0.5" step="0.01" value="0.08"
      >
    </div>
  </form>
  <script type="module" src="main.js"></script>
```

14　https://developer.mozilla.org/en-US/docs/Web/HTML/Element/input/range

```
</body>
```

運用已經建立好了 form 輸入事件監聽 callback，更新速度到狀態中：

```
async function main() {
  // setup() ...

  const controlsForm = document.getElementById('controls');
  controlsForm.addEventListener('input', () => {
    const formData = new FormData(controlsForm);
    app.state.texture = parseInt(formData.get('texture'));
    app.state.speed = parseFloat(formData.get('speed'));
  });

  loop(app);
}
```

看起來如圖 2-25，可惜書本無法呈現移動動畫，到此進度的完整程式碼可以在這邊找到：

</>	ch2/05/index.html
</>	ch2/05/main.js

也可以參考上線版本，透過瀏覽器打開此頁面：

https://webgl-book.pastleo.me/ch2/05/index.html

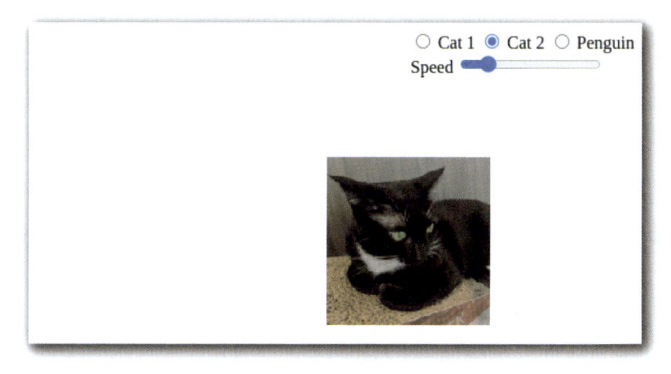

▲ 圖 2-25 成果示意圖

2 Texture & 2D

這邊使用了 offset 進行圖片的平移，並使用 `requestAnimationFrame` 在每次螢幕更新時進行重新渲染，組合出最基本的平移動畫，而我們實做的 `setup()` 以及 `loop()` 就是現今遊戲以及動畫引擎範例中常見的基礎架構，一個負責初始化、一個負責更新（因此大多稱此 function 為 `update()`），本書接下來也會以這樣的架構繼續實做，製作更多、更複雜的動畫以及互動。

2-4　2D Transform

為了控制物件的平移，前面使用一個 uniform `u_offset` 來控制，如果我們現在想要進行縮放、旋轉，那麼就得在傳入更多 uniform，而且這些動作會有誰先作用的差別，比方説，先往右旋轉 90 度、接著往右平移 30px，與先往右平移 30px、再往右旋轉 90 度，這兩著會獲得不一樣的結果，如果只是傳平移量、縮放量、旋轉度數，在 vertex shader 內得有一個寫死的的先後作用順序，那麼在應用層面上就會受到這個先後順序的限制，因此在 2D, 3D 渲染的世界中，尤其是 3D，常常利用線性代數方式控制物件的位置，只需要傳入一個矩陣，這個矩陣就能包含任意先後順序的平移、縮放、旋轉動作。

⌘ 矩陣運算

為什麼一個矩陣就能夠包含任意先後順序的平移、縮放、旋轉動作？筆者覺得筆者不論怎麼用文字怎麼解釋，都不會比這部 3Blue1Brown 的 Youtube 影片解釋得好：作為合成的矩陣乘法：https://youtu.be/XkY2DOUCWMU

如果覺得有需要，也可以從這系列影片[15]的第一部影片開始看。總之，所有的平移、縮放、旋轉動作（現在開始應該説轉換 – transform）都可以用一個矩陣表示，而把這些 transform 矩陣與向量相乘後，可以得到平移、縮放或旋轉後的結果，而且重點來了，我們也可以事先把多個矩陣相乘，這個相乘

---

[15] https://youtube.com/playlist?list=PLZHQObOWTQDPD3MizzM2xVFitgF8hE_ab

2-4　2D Transform

的結果與向量相乘與個別依序與向量相乘會得到相同的 transform；舉例來說，先旋轉 45 度、往右平移 30px、再旋轉 45 度，這三個 transform 依序作用在一個向量上，與把這三個 transform 代表的矩陣先相乘，再跟向量相乘，會得到一樣的結果，基於這個特性，我們只要傳送這個相乘出來的矩陣即可。

但是在數學界的慣例中（3Blue1Brown 的影片就使用這樣的慣例），一個二維的向量的表示方式如圖 2-26：

$$\vec{v} = \begin{bmatrix} x \\ y \end{bmatrix}$$

▲ 圖 2-26　一個二維的向量的表示方式

然而，在程式語言（如 Javascript）中也只能這樣寫：`[x, y]`，因此在數學式與程式語言寫法之間要把行列做對調，一個像這樣的 2x2 矩陣在程式語言中的寫法如下方圖 2-27 藍色文字所示：

$$M = \begin{bmatrix} 1 & 3 \\ 2 & 4 \end{bmatrix} = \text{[1, 2, 3, 4]} = \text{[1, 2, 3, 4]}$$

▲ 圖 2-27　數學式與程式語言的表示方法差異

對向量進行 transform 的『矩陣乘以向量』數學上慣用是由右到左算，假設有個向量為 `[5, 6]`，與上方的 M 進行矩陣乘法（或是說以 M 矩陣對向量進行 transform），運算起來如圖 2-28 所示：

$$M\vec{v} = \begin{bmatrix} 1 & 3 \\ 2 & 4 \end{bmatrix} \begin{bmatrix} 5 \\ 6 \end{bmatrix} = \begin{bmatrix} 1*5+3*6 \\ 2*5+4*6 \end{bmatrix} = \begin{bmatrix} 23 \\ 34 \end{bmatrix}$$

▲ 圖 2-28　數學上矩陣乘向量的運算

2 Texture & 2D

在 shader 或是 Javascript 內一樣是右到左，大致會變成這樣寫：

```
v = [5, 6]
m = [
  1, 2,
  3, 4,
]
multiply(m, v) => [23, 34]
```

如果在學校有學過矩陣運算或是線性代數，那麼可能會知道有個矩陣乘法的記憶方式，下圖中左方矩陣為相乘結果，其相乘結果中塗色的元素為右邊兩個矩陣塗色長條形區域左上到右下依序相乘後相加的結果，如圖 2-29、2-30：

$$\begin{bmatrix} P_{(1,1)} & P_{(1,2)} & P_{(1,3)} \\ P_{(2,1)} & P_{(2,2)} & P_{(2,3)} \\ P_{(3,1)} & P_{(3,2)} & P_{(3,3)} \end{bmatrix} = \begin{bmatrix} A_{(1,1)} & A_{(1,2)} & A_{(1,3)} \\ A_{(2,1)} & A_{(2,2)} & A_{(2,3)} \\ A_{(3,1)} & A_{(3,2)} & A_{(3,3)} \end{bmatrix} \begin{bmatrix} B_{(1,1)} & B_{(1,2)} & B_{(1,3)} \\ B_{(2,1)} & B_{(2,2)} & B_{(2,3)} \\ B_{(3,1)} & B_{(3,2)} & B_{(3,3)} \end{bmatrix}$$

$$P_{(1,1)} = A_{(1,1)}B_{(1,1)} + A_{(1,2)}B_{(2,1)} + A_{(1,3)}B_{(3,1)}$$

▲ 圖 2-29　數學上矩陣運算（線性代數）運算示意圖

$$\begin{bmatrix} P_{(1,1)} & P_{(1,2)} & P_{(1,3)} \\ P_{(2,1)} & P_{(2,2)} & P_{(2,3)} \\ P_{(3,1)} & P_{(3,2)} & P_{(3,3)} \end{bmatrix} = \begin{bmatrix} A_{(1,1)} & A_{(1,2)} & A_{(1,3)} \\ A_{(2,1)} & A_{(2,2)} & A_{(2,3)} \\ A_{(3,1)} & A_{(3,2)} & A_{(3,3)} \end{bmatrix} \begin{bmatrix} B_{(1,1)} & B_{(1,2)} & B_{(1,3)} \\ B_{(2,1)} & B_{(2,2)} & B_{(2,3)} \\ B_{(3,1)} & B_{(3,2)} & B_{(3,3)} \end{bmatrix}$$

$$P_{(2,3)} = A_{(2,1)}B_{(1,3)} + A_{(2,2)}B_{(2,3)} + A_{(2,3)}B_{(3,3)}$$

▲ 圖 2-30　數學上矩陣運算（線性代數）運算示意圖

在用這樣的記憶方式模擬電腦中的矩陣乘法運算時，在紙筆試算跟程式語言表示之間每次都要做行列轉換，如果覺得這樣很麻煩，那麼也可以改變一下矩陣計算的方式，塗色長條形的行列對調，就可以全部使用程式語言中的表示方式，如圖 2-31、圖 2-32：

2-4　2D Transform

$$[p_{[0]}, p_{[1]}, p_{[2]}, p_{[3]}, p_{[4]}, p_{[5]}, p_{[6]}, p_{[7]}, p_{[8]}] = \text{multiply}\left(\begin{bmatrix} a_{[0]}, a_{[1]}, a_{[2]}, \\ a_{[3]}, a_{[4]}, a_{[5]}, \\ a_{[6]}, a_{[7]}, a_{[8]} \end{bmatrix}, \begin{bmatrix} b_{[0]}, b_{[1]}, b_{[2]}, \\ b_{[3]}, b_{[4]}, b_{[5]}, \\ b_{[6]}, b_{[7]}, b_{[8]} \end{bmatrix}\right)$$

$$p_{[0]} = a_{[0]}b_{[0]} + a_{[3]}b_{[1]} + a_{[6]}b_{[2]}$$

$$[p_{[0]}, p_{[1]}, p_{[2]}, p_{[3]}, p_{[4]}, p_{[5]}, p_{[6]}, p_{[7]}, p_{[8]}] = \text{multiply}\left(\begin{bmatrix} a_{[0]}, a_{[1]}, a_{[2]}, \\ a_{[3]}, a_{[4]}, a_{[5]}, \\ a_{[6]}, a_{[7]}, a_{[8]} \end{bmatrix}, \begin{bmatrix} b_{[0]}, b_{[1]}, b_{[2]}, \\ b_{[3]}, b_{[4]}, b_{[5]}, \\ b_{[6]}, b_{[7]}, b_{[8]} \end{bmatrix}\right)$$

$$p_{[5]} = a_{[2]}b_{[3]} + a_{[5]}b_{[4]} + a_{[8]}b_{[5]}$$

▲ 圖 2-31　程式語言中矩陣乘法運算示意圖

$$[r_{[0]}, r_{[1]}, r_{[2]}] = \text{transform}\left(\begin{bmatrix} m_{[0]}, m_{[1]}, m_{[2]}, \\ m_{[3]}, m_{[4]}, m_{[5]}, \\ m_{[6]}, m_{[7]}, m_{[8]} \end{bmatrix}, [v_{[0]}, v_{[1]}, v_{[2]}]\right)$$

$$r_{[1]} = m_{[1]}v_{[0]} + m_{[4]}v_{[1]} + m_{[7]}v_{[2]}$$

▲ 圖 2-32　程式語言中矩陣與向量相乘進行轉換之運算示意圖

圖 2-32 中，為矩陣與向量相乘，也就是以矩陣對向量做轉換，程式碼中向量的表示是橫向的，運算方向由右向左，向量放在最右邊，因此也可以用這樣來提醒自己模擬電腦矩陣相乘、向量轉換時行列的方向與數學上慣例的差異。

⌘ 調整 vertex shader 使用矩陣運算

矩陣可以一次表示多個對於向量的轉換，為了了解對 WebGL 渲染的應用有如何的幫助，接下來將平移從原本直接用一個二維向量改用矩陣的方式傳遞，以利往後加上更多轉換。

2 Texture & 2D

　　首先建立 `u_matrix` 取代 `u_offset` 以及 `u_resolution`，然後讓 `u_matrix` 與 `a_position` 相乘，相乘的結果即為任意順序的平移、縮放、旋轉做完後的座標位置。

```
#version 300 es
in vec2 a_position;
in vec2 a_texcoord;

uniform vec2 u_resolution;
uniform vec2 u_offset;
uniform mat3 u_matrix;
out vec2 v_texcoord;

void main() {
  vec2 position = a_position + u_offset;
  gl_Position = vec4(
    position / u_resolution * vec2(2, -2) + vec2(-1, 1),
    0, 1
  );
  vec3 position = u_matrix * vec3(a_position.xy, 1);
  gl_Position = vec4(position.xy, 0, 1);
  v_texcoord = a_texcoord;
}
```

　　沒錯，針對螢幕寬高調整 clip space 的 `u_resolution` 也可以在矩陣中順便帶著。

　　在 GLSL 中 `*` 運算子可以接受 mat3 與 vec3 相乘，將 `u_matrix * vec3(a_position.xy, 1)` 就可以得到『使用矩陣轉換向量』後的結果，可以說是把矩陣運算直接內建在語法中了。不知道大家有沒有注意到 `u_matrix` 為 3x3 矩陣，我們還在 2D，為何要用 mat3 呢？

　　因為 2x2 矩陣在向量轉換上，是無法『平移 (translate)』的，數學上平移並不是線性轉換。我們必須使用 3x3 矩陣，在運算時增加一個維度，並填上 1，

2-48

2-4　2D Transform

使得向量為 [x, y, 1]，這樣的向量稱為勻相座標 (homogeneous coordinate)[16]，接著平移矩陣在多的維度中放上要平移的量，即可得到結果。想深入了解原因，可以觀察 2-33 數學上的例子：我們試著將向量 [4, 5]，平移 [2, 3]：

```
multiply(
  [
    1, 0, 0,
    0, 1, 0,
    2, 3, 1,
  ],
  [4, 5, 1]
)
// => [6, 8 ,1]
```

數學上寫起來像是這樣，如圖 2-33：

$$\begin{bmatrix} 1 & 0 & 2 \\ 0 & 1 & 3 \\ 0 & 0 & 1 \end{bmatrix} \begin{bmatrix} 4 \\ 5 \\ 1 \end{bmatrix} = \begin{bmatrix} 1*4+0*5+2*1 \\ 0*4+1*5+3*1 \\ 0*0+0*0+1*1 \end{bmatrix} = \begin{bmatrix} 6 \\ 8 \\ 1 \end{bmatrix}$$

▲ 圖 2-33　將 [4, 5] 平移 [2, 3] 成 [6, 8] 之平移矩陣的數學式

其實藉由心算，我們可以輕易地將 [4, 5] 加上平移量 [2, 3]，得到結果 [6, 8]；而在矩陣運算中，多的維度上的數字會與向量的 1 相乘加在原本的座標上，形成平移。我們將 [4, 5] 改為 [4, 5, 1]，並將 [2, 3] 改為專為平移所安排的 3x3 行列矩陣，計算兩個矩陣後，可以得到多一個維度但結果相同的 [6, 8, 1]。

當然，uniform 位置的取得得修改一下：

```
const uniforms = {
  matrix: gl.getUniformLocation(program, 'u_matrix'),
```

16　https://zh.wikipedia.org/wiki/齊次座標

```
  resolution: gl.getUniformLocation(program, 'u_resolution'),
  texture: gl.getUniformLocation(program, 'u_texture'),
  offset: gl.getUniformLocation(program, 'u_offset'),
};
```

⌘ 平移以及投影的矩陣

shader、uniform 部份準備好後，建立 `const matrix3 = { ⋯ }`，當成一個工具箱，產生各式各樣 transform 用的矩陣，同時也實做 Javascript 端的矩陣相乘運算，才能『事先』運算、合成好矩陣給 GPU 使用。除了矩陣相乘 `matrix3.multiply()` 之外，也把上面提到的平移矩陣 `matrix3.translate()` 實做好：

```
const matrix3 = {
  multiply: (a, b) => ([
    a[0]*b[0] + a[3]*b[1] + a[6]*b[2],
     a[1]*b[0] + a[4]*b[1] + a[7]*b[2],
      a[2]*b[0] + a[5]*b[1] + a[8]*b[2],
    a[0]*b[3] + a[3]*b[4] + a[6]*b[5],
     a[1]*b[3] + a[4]*b[4] + a[7]*b[5],
      a[2]*b[3] + a[5]*b[4] + a[8]*b[5],
    a[0]*b[6] + a[3]*b[7] + a[6]*b[8],
     a[1]*b[6] + a[4]*b[7] + a[7]*b[8],
      a[2]*b[6] + a[5]*b[7] + a[8]*b[8],
  ]),
  translate: (x, y) => ([
    1, 0, 0,
    0, 1, 0,
    x, y, 1,
  ]),
};
// vertexShaderSource, fragmentShaderSource, setup() ...
```

那麼就剩下針對螢幕寬高調整 clip space 的矩陣，這樣的矩陣稱為 `projection`，也就是把場景中一個寬高區域框起來，『投影』在 clip space – 畫布上，看著原本 shader 程式碼拆解一下：

```
position / u_resolution * vec2(2, -2) + vec2(-1, 1)
```

可以發現,我們要分別對 x 座標縮放 2 / u_resolution.x 倍、對 y 軸縮放 `-2 / u_resolution.y` 倍,做完縮放後平移 vec2(-1, 1),平移已經知道要怎麼做了,那麼縮放的矩陣要怎麼產生呢?觀察一下圖 2-34 這個算式:

$$\begin{bmatrix} 2 & 0 & 0 \\ 0 & 3 & 0 \\ 0 & 0 & 1 \end{bmatrix} \begin{bmatrix} 4 \\ 5 \\ 1 \end{bmatrix} = \begin{bmatrix} 2*4+0*5+0*1 \\ 0*4+3*5+0*1 \\ 0*0+0*0+1*1 \end{bmatrix} = \begin{bmatrix} 8 \\ 15 \\ 1 \end{bmatrix}$$

▲ 圖 2-34　[4, 5] 的 x 乘以 2、y 乘以 3 得到 [8, 15] 之縮放矩陣的數學式

在單位矩陣中對應維度的數字即為縮放倍率,在這邊把 x 座標乘以 2、y 座標乘以 3,其他欄位為零不會影響,因此縮放矩陣的產生 `matrix3.scale()` 這樣實做:

```
const matrix3 = {
  // multiply, translate

  scale: (sx, sy) => ([
    sx, 0,  0,
    0,  sy, 0,
    0,  0,  1,
  ]),
};
```

最後投影矩陣 `matrix3.projection()`,為平移與縮放相乘的矩陣,即為『縮放以及平移』兩個動作加在一起:

```
const matrix3 = {
  // multiply, translate, scale

  projection: (width, height) => (
    matrix3.multiply(
      matrix3.translate(-1, 1),
```

```
      matrix3.scale(2 / width, -2 / height),
    )
  ),
};
```

這樣一來 matrix3 就準備好，我們將用 `matrix3.translate()` 進行平移，並且使用 `matrix3.projection()` 『投影』到對應畫布尺寸上的 clip space 位置，因為 `matrix3.projection()` 是由 `matrix3.translate()` 以及對應螢幕轉換的縮放所組成，實做 `matrix3.scale()` 作為縮放使用，這些都是產生一個矩陣表示一個轉換（transform），透過矩陣相乘 `matrix3.multiply()`，來結合這些動作成為『一個』矩陣，腦筋快的讀者們是否已經能了解到矩陣對於 WebGL 物件位置的操作的威力了呢？

> 記得矩陣運算的動作效果發生方向與一般運算運算不同，向量放在最右邊，向左運算，因此 `matrix3.translate(-1, 1)` 雖然放在前面，但是其 transform 作用次序是在後面的。

畫龍點睛的時候來了，在 `render()` function 設定 uniform 的地方產生、運算矩陣：

```
function render(app) {
  const {
    gl,
    program, uniforms,
    textures,
    state,
  } = app;

  // gl.canvas.width, useProgram ...

  const viewMatrix = matrix3.projection(gl.canvas.width, gl.canvas.height);
  const worldMatrix = matrix3.translate(...state.offset);

  gl.uniformMatrix3fv(
```

```
    uniforms.matrix,
    false,
    matrix3.multiply(viewMatrix, worldMatrix),
  );
  gl.uniform2f(uniforms.resolution, gl.canvas.width, gl.canvas.height);
  gl.uniform2fv(uniforms.offset, state.offset);
}
```

筆者建立名叫 `viewMatrix` 的矩陣，包含 `matrix3.projection()` 負責投影到 clip space；以及另一個矩陣稱為 `worldMatrix`，表示該物件在場景中位置的 transform。最後把 `viewMatrix` 與 `worldMatrix` 相乘，得到包含所有 transform 的矩陣，並使用 `gl.uniformMatrix3fv()` 設定到 uniform 上，其第二個參數表示要不要做矩陣的轉置[17] `transpose`，我們沒有需要因此傳入 `false`。

存檔重整，使用 matrix 做 transform 的版本看起來跟先前 offset 的版本一模一樣，到此進度的完整程式碼可以在這邊找到：

</>	ch2/06/index.html
</>	ch2/06/main.js

也可以參考上線版本，透過瀏覽器打開此頁面：

https://webgl-book.pastleo.me/ch2/06/index.html

總結一下 transform，首先 `u_matrix` 為 `viewMatrix` 與 `worldMatrix` 相乘。`viewMatrix` 為 `matrix3.projection()`、`worldMatrix` 為 `matrix3.translate(offset)`，最後在 vertex shader 內 `u_matrix * a_position`，來展開一路從 `a_position` 到 `gl_Position` 的運算式（忽略維度的調整）：

```
gl_Position = u_matrix * a_position;
=> gl_Position = viewMatrix * worldMatrix * a_position;
```

---

17 https://zh.wikipedia.org/wiki/轉置矩陣

2 Texture & 2D

```
=> gl_Position =
  matrix3.projection() * matrix3.translate(offset) * a_position;
```

而這些 transform 在電腦中與數學上都是由右到左計算的，因此最後整個計算下來的效果就是：先對 `a_position` 平移 offset 量（translate），接著投影螢幕範圍到 clip space（projection）。

雖然看似沒有什麼新功能，且只有兩個 transform，但是建構出來的流程讓我們可以很容易地加入更多 transform，也更適合之後 3D 場景中複雜的物件位置到螢幕上位置的運算，待下篇再繼續加入更多 2D transform。

2-5 2D transform Continued

我們把平移、投影到 clip space 的 `u_offset`、`u_resolution` 替換成只用一個矩陣來做轉換，實做了矩陣的相乘（multiply）、平移（translate）以及縮放（scale），還剩下旋轉（rotation）以及一些常見的 transform、運算尚未實做，除此之外也應將 matrix3 改成在各個章節都可以共用的工具箱；我們將在本篇加上平移、縮放、旋轉之控制項，完整 transform 的概念以及工具，成為更往後進入 3D 的基石。

⌘ 旋轉 transform

相信很多人在學校數學課都有學過座標的旋轉[18]，如果原本一個向量為 (x, y)，旋轉 θ 角度後將變成 (x', y')，那麼公式為如圖 2-35：

$$x' = x\cos(\theta) - y\sin(\theta)$$
$$y' = x\sin(\theta) + y\cos(\theta)$$

▲ 圖 2-35 座標旋轉公式

---

18 https://zh.wikipedia.org/wiki/旋轉

寫成矩陣的話，矩陣如圖 2-36：

$$M_{rotation} = \begin{bmatrix} \cos(\theta) & -\sin(\theta) \\ \sin(\theta) & \cos(\theta) \end{bmatrix}$$

▲ 圖 2-36　旋轉矩陣

只不過我們需要的是 3x3 的矩陣，才能符合運算時的需要，多的維度跟單位矩陣一樣，同時記得行列轉換成電腦世界使用的慣例（假設要旋轉角度為 rad）：

```
[
  Math.cos(rad),  Math.sin(rad), 0,
  -Math.sin(rad), Math.cos(rad), 0,
  0,              0,             1,
]
```

最後實做在 `matrix3`：

```
const matrix3 = {
  // multiply, translate, scale

  rotate: rad => {
    const c = Math.cos(rad), s = Math.sin(rad);
    return [
      c, s, 0,
      -s, c, 0,
      0, 0, 1,
    ]
  },

  // projection
};
```

2 Texture & 2D

⌘ 加入平移、縮放、旋轉控制

像是速度控制那樣，在 HTML 中分別給 X 軸平移、Y 軸平移、縮放、旋轉一個拉桿：

```html
<form id="controls">
  <!-- texture, speed inputs -->
  <div>
    <label for="translate-x">TranslateX</label>
    <input
      type="range" id="translate-x" name="translate-x"
      min="-150" max="150" value="0"
    >
  </div>
  <div>
    <label for="translate-y">TranslateY</label>
    <input
      type="range" id="translate-y" name="translate-y"
      min="-150" max="150" value="0"
    >
  </div>
  <div>
    <label for="scale">Scale</label>
    <input
      type="range" id="scale" name="scale"
      min="0" max="10" value="1" step="0.1"
    >
  </div>
  <div>
    <label for="rotation">Rotation</label>
    <input
      type="range" id="rotation" name="rotation"
      min="0" max="360" value="0"
    >
  </div>
</form>
```

2-5　2D transform Continued

使得右上角控制 UI 看起來像是圖 2-37：

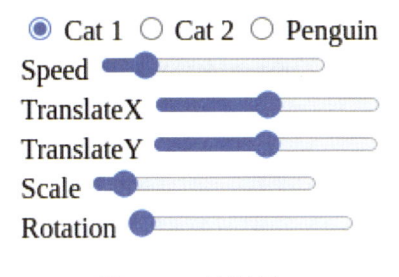

▲ 圖 2-37　控制項 UI

接下來在 setup() 中初始化的 `app.state` 中加入平移、縮放、旋轉：

```
async function setup() {
  return {
    // gl, program, buffers ...
    state: {
      texture: 0,
      offset: [0, 0],
      direction: [Math.cos(directionDeg), Math.sin(directionDeg)],
      translate: [0, 0],
      scale: 1,
      rotation: 0,
      speed: 0.08,
    },
    time: 0,
  };
}
```

修改矩陣計算之前，把狀態與使用者輸入事件串好：

```
async function main() {
  // ...
  controlsForm.addEventListener('input', () => {
    const formData = new FormData(controlsForm);
    app.state.texture = parseInt(formData.get('texture'));
    app.state.speed = parseFloat(formData.get('speed'));
```

2-57

Texture & 2D

```
  app.state.translate[0] = parseFloat(formData.get('translate-x'));
  app.state.translate[1] = parseFloat(formData.get('translate-y'));
  app.state.scale = parseFloat(formData.get('scale'));
  app.state.rotation =
    parseFloat(formData.get('rotation')) * Math.PI / 180;
});

loop(app);
}
```

⌘ 使用旋轉矩陣

在 `render()` 內，原本 `worldMatrix` 只有平移轉換 `translate(...state.offset)`，現在開始也要由多個矩陣相乘：

```
function render(app) {
  // ...

  const viewMatrix = matrix3.projection(gl.canvas.width, gl.canvas.height);
  const worldMatrix = matrix3.translate(...state.offset);
  const worldMatrix = matrix3.multiply(
    matrix3.translate(...state.offset),
    matrix3.rotate(state.rotation),
  );

  gl.uniformMatrix3fv(
    uniforms.matrix,
    false,
    matrix3.multiply(viewMatrix, worldMatrix),
  );
}
```

到這邊旋轉就串好囉，存檔後去瀏覽器試玩一下，貓圖可以透過 Rotation 拉桿旋轉，只是會以圖片左上角旋轉，如圖 2-38，圖中紅色圈選位置為旋轉圓心：

2-58

2-5　2D transform Continued

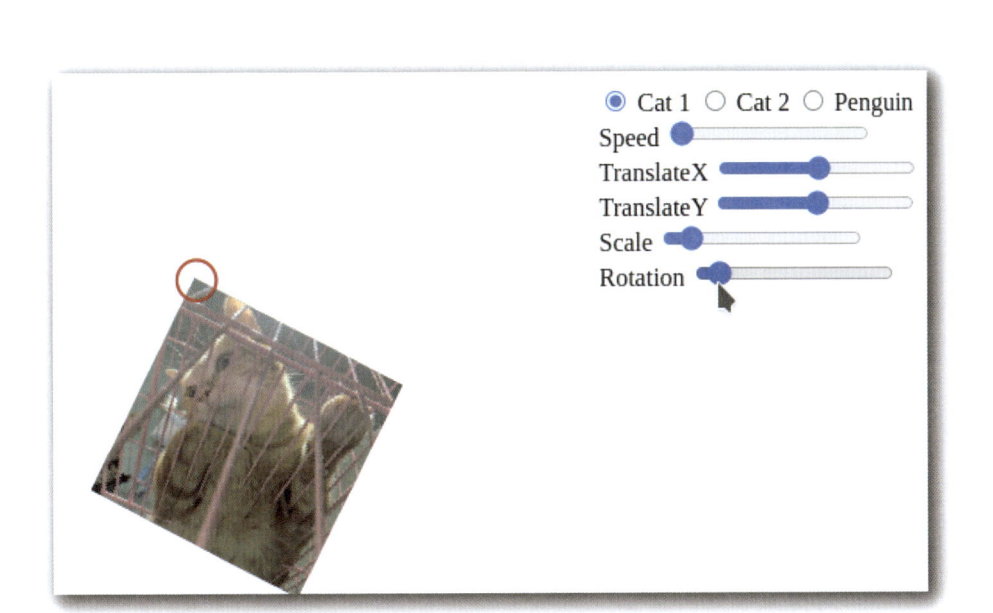

▲ 圖 2-38　旋轉功能使用範例，左上角紅色圈選處為旋轉中心

你會發現旋轉的基準點是圖片左上角，如果我們想要用圖片正中央來旋轉而不是左上角呢？在輸入頂點位置時，左上角的點為 (0, 0)，如圖 2-39：

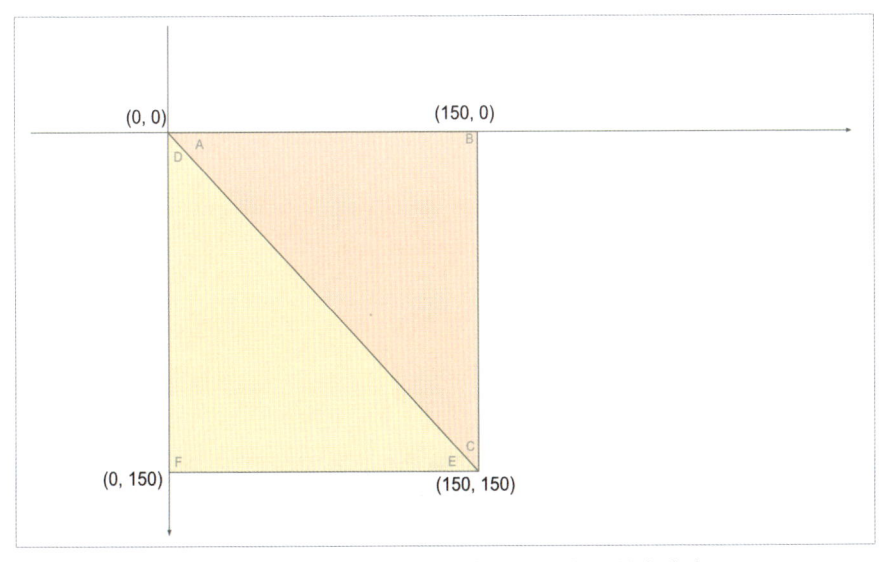

▲ 圖 2-39　原點、旋轉基準點在正方形的左上角

2-59

2 Texture & 2D

而 `matrix3.rotate()` 是基於原點做旋轉的，因此調整一下頂點位置，使得原點在正中間，如圖 2-40：

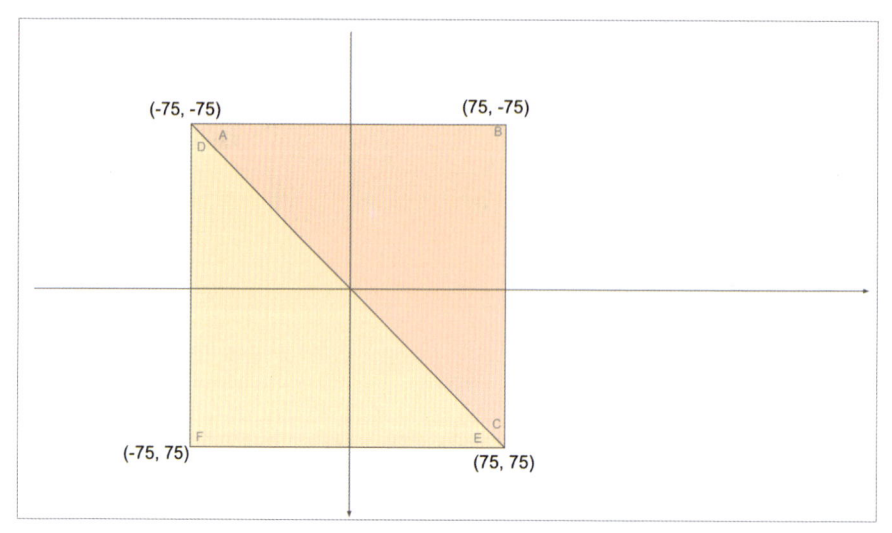

▲ 圖 2-40　原點、旋轉基準點改為正方形的正中間

還記得頂點位置要怎麼調整嗎？要修改的是『畫什麼』給 `a_position` buffer 輸入資料的地方：

```js
async function setup() {
  // ...
  // a_position ...

  gl.bufferData(
    gl.ARRAY_BUFFER,
    new Float32Array([
      -75, -75, // A
      75, -75, // B
      75, 75, // C

      -75, -75, // D
      75, 75, // E
      -75, 75, // F
    ]),
```

2-60

```
    gl.STATIC_DRAW,
  );

  // ...
}
```

不過就沒辦法做完美的邊緣碰撞測試了,筆者就用原點當成碰撞測試點:

```
function loop(app, now = 0) {
  // ...
  // state.offset = ...

  if (state.offset[0] > gl.canvas.width) {
    state.direction[0] *= -1;
    state.offset[0] = gl.canvas.width;
  } else if (state.offset[0] < 0) {
    state.direction[0] *= -1;
    state.offset[0] = 0;
  }

  if (state.offset[1] > gl.canvas.height) {
    state.direction[1] *= -1;
    state.offset[1] = gl.canvas.height;
  } else if (state.offset[1] < 0) {
    state.direction[1] *= -1;
    state.offset[1] = 0;
  }

  // ...
}
```

這麼一來圖片就乖乖的以中心點旋轉了,如圖 2-41,圖中紅色圈選位置為旋轉圓心:

2 Texture & 2D

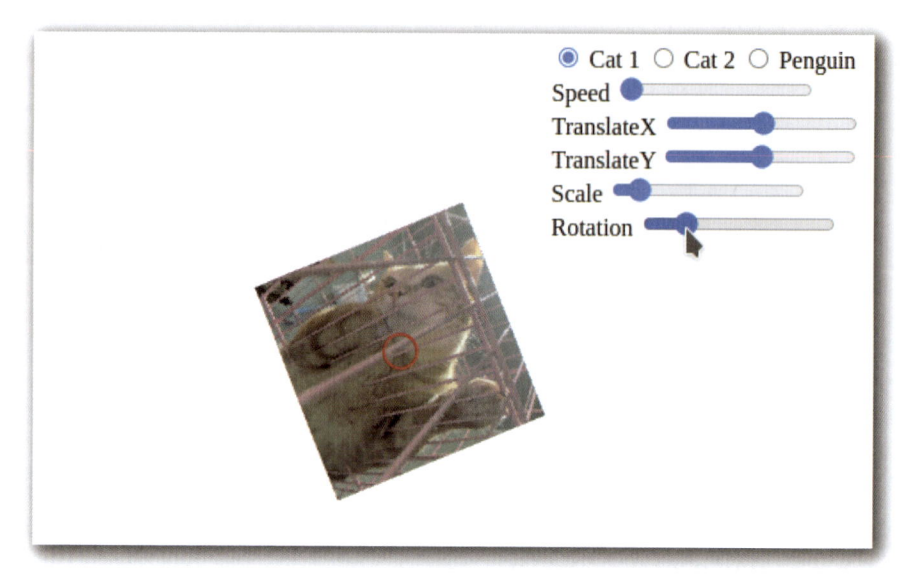

▲ 圖 2-41　使用正方形正中央為原點、旋轉中心之旋轉功能使用範例，
　　　　　中心紅色圈選處為旋轉中心

其實縮放也是從原點出發的，因此這個調整也可以修正待會加入縮放時變成從左上角縮放的問題。筆者學到矩陣 transform 時，似乎就可以感受到 WebGL 的世界為什麼很多東西都是以 -1～+1 作為範圍⋯這樣使得原點在正中間，可能在硬體或是 driver 層也更方便使用矩陣做 transform 運算吧。

⌘ 所有 Transform 我全都要

現在 `worldMatrix` 由 `matrix3.translate()` 與 `matrix3.rotate()` 相乘而成，要串上使用者控制的 `state.translate, state.scale`，假設 `worldMatrix` 要用下面的算式計算而成：

```
translate(...state.offset) *
  rotate(state.rotation) *
  scale(state.scale, state.scale) *
  translate(...state.translate)
```

2-62

以現成的 `matrix3.multiply()` 來看會變成這樣：

```javascript
function render(app) {
  // ...
  // viewMatrix = ...
  const worldMatrix = matrix3.multiply(
    matrix3.multiply(
      matrix3.multiply(
        matrix3.translate(...state.offset),
        matrix3.rotate(state.rotation),
      ),
      matrix3.scale(state.scale, state.scale),
    ),
    matrix3.translate(...state.translate),
  );

  gl.uniformMatrix3fv(
    uniforms.matrix,
    false,
    matrix3.multiply(viewMatrix, worldMatrix),
  );

  // ...
}
```

顯然可讀性已經大幅下降，經過好好換行至少還看的出來相乘的順序，沒換行更慘，之後也會有許多超過兩個矩陣依序相乘的狀況，因此修改 `matrix3.multiply()` 使之可以接收超過兩個矩陣，並遞迴依序做相乘：

```javascript
const matrix3 = {
  multiply: (a, b, ...rest) => {
    const multiplication = [
      a[0]*b[0] + a[3]*b[1] + a[6]*b[2],
        a[1]*b[0] + a[4]*b[1] + a[7]*b[2],
          a[2]*b[0] + a[5]*b[1] + a[8]*b[2],
      a[0]*b[3] + a[3]*b[4] + a[6]*b[5],
        a[1]*b[3] + a[4]*b[4] + a[7]*b[5],
          a[2]*b[3] + a[5]*b[4] + a[8]*b[5],
```

Texture & 2D

```
      a[0]*b[6] + a[3]*b[7] + a[6]*b[8],
      a[1]*b[6] + a[4]*b[7] + a[7]*b[8],
      a[2]*b[6] + a[5]*b[7] + a[8]*b[8],
    ];

    if (rest.length === 0) return multiplication;
    return matrix3.multiply(multiplication, ...rest);
  },

  // ...
};
```

`...rest` 的語法叫做 rest parameters[19],傳超過 2 個參數時再呼叫自己將這回合計算的結果繼續與剩下的矩陣做計算。

回到主程式,worldMatrix 就可以用清楚的語法寫了:

```
function render(app) {
  // ...
  // viewMatrix = ...
  const worldMatrix = matrix3.multiply(
    matrix3.translate(...state.offset),
    matrix3.rotate(state.rotation),
    matrix3.scale(state.scale, state.scale),
    matrix3.translate(...state.translate),
  );

  gl.uniformMatrix3fv(
    uniforms.matrix,
    false,
    matrix3.multiply(viewMatrix, worldMatrix),
  );

  // ...
}
```

---

19 https://developer.mozilla.org/zh-TW/docs/Web/JavaScript/Reference/Functions/rest_parameters

所有的控制就完成了，讀者也可以自行調整這些矩陣相乘的順序，玩玩看所謂『轉換順序』的差別，像是 `matrix3.scale()` 的縮放效果如圖 2-42：

▲ 圖 2-42　縮放功能使用範例

⌘ 『什麼都不做』轉換

在總結 2D transform 之前，給 `matrix3` 再補上一個 function：

```
const matrix3 = {
  // ...

  identity: () => ([
    1, 0, 0,
    0, 1, 0,
    0, 0, 1,
  ]),

  // ...
};
```

2-65

這是一個單位矩陣[20]，如果有時候要除錯想要暫時取消一些矩陣的轉換效果，但是不想修改程式結構：

```
const worldMatrix = matrix3.multiply(
  matrix3.translate(...state.offset),
  matrix3.rotate(state.rotation), // 想要暫時取消這行
);
```

其中一個暫時取消的方式是利用單位矩陣的特性：與其相乘不會改變任何東西，像是這樣：

```
const worldMatrix = matrix3.multiply(
  matrix3.translate(...state.offset),
  matrix3.identity(),
  // matrix3.rotate(state.rotation), // 想要暫時取消這行
);
```

為了驗證，回到主程式修改 `worldMatrix` 的計算：

```
const worldMatrix = matrix3.multiply(
  matrix3.identity(),
  matrix3.translate(...state.offset),
  matrix3.rotate(state.rotation),
  matrix3.scale(state.scale, state.scale),
  matrix3.translate(...state.translate),
);
```

不論 `matrix3.identity()` 放在哪個位置，都不會改變結果；上述用途只是其中一個舉例，之後可能會因為兩個物件共用同一個 shader，但是其中一個物件不需要特定轉換，那麼也會傳入單位矩陣來『什麼都不做』。

20　https://zh.wikipedia.org/wiki/單位矩陣

⌘ 獨立 `matrix3` 成為可共用的 ES module

這個 `matrix3` 工具箱在接下來的章節會被大量使用，因此我們來把它拆成一個獨立的檔案以利其他章節共用此工具箱；如果原本有在寫前端的讀者肯定不陌生，將實做之 `matrix3` 改放到 `lib/matrix.js`：

> `lib/matrix.js`

```javascript
export const matrix3 = {
  multiply: (a, b, ...rest) => // ...
  identity: () => // ...
  translate: (x, y) => // ...
  scale: (sx, sy) => // ...
  rotate: rad => // ...
  projection: (width, height) => // ...
};
```

記得加上 `export` 關鍵字表示 ES module 的輸出項目，完成之後回到主程式進行引入：

```javascript
import { matrix3 } from '../../lib/matrix.js';
```

回到瀏覽器測試，如果一切沒問題，那麼就證明 `matrix3` 在 `lib/matrix.js` 準備好囉！

到此進度的完整程式碼可以在這邊找到：

> `ch2/07/index.html`
>
> `ch2/07/main.js`

也可以參考上線版本，透過瀏覽器打開此頁面：

https://webgl-book.pastleo.me/ch2/07/index.html

2 Texture & 2D

3D & 物件

有了線性代數、矩陣來進行向量轉換（transform）的概念並準備好向量轉換的工具箱，只要將向量、矩陣增加一個維度，那麼我們就從 2D 前進到 3D 了！我們可以運用一樣的原理對於一個物件的各個頂點使用一個矩陣做轉換，一次控制平移、旋轉、縮放，就等同於控制這個物件在空間中的位置。

在這個章節我們將增加一個維度進入 3D，先使用 orthogonal 3D 投影，再使用 perspective 透視的方式成像，邁向 3D 物件渲染系統。

3 3D & 物件

3-1 Orthogonal 3D 投影

"Orthogonal" 查詢字典的話得到的意思是：直角的，正交的，垂直的，而 orthogonal 3D 投影的意思是：從 3D 場景中選取一個長方體區域作為 clip space 投影到畫布上，事實上與先前 2D 投影非常類似，只是多一個維度 z，在先前 2D 時就像是 z 永遠等於 0，在 orthogonal 3D 投影時叫做深度，本篇的目標將以 orthogonal 3D 投影的方式渲染一個 P 文字形狀的 3D 物件。

我們將使用新的 `index.html` 以及 `main.js` 作為開始，起始點完整程式碼可以在這邊找到：

</>	ch3/00/index.html
</>	ch3/00/main.js

絕大部分的程式碼在先前的章節都有相關的說明了，以下幾點比較值得注意的：

- vertex attribute 有：`a_position`、`a_color`，`a_color` 作為 fragment shader 輸出顏色使用，顯示的結果將會是一個一個由 `a_color` 指定的色塊（或是漸層，如果一個三角形內的顏色不同的話）。
- 從 `lib/matrix.js` 引入的工具箱為運算三維向量用的 `matrix4`，許多 `matrix3` 的 function 都可以在 `matrix4` 中找到對應的，同樣因為平移需要使用勻相座標（homogeneous coordinate），加上一個維度而成為四維矩陣。
  - 在三維場景中能夠以 x 軸、y 軸、z 軸旋轉，因此旋轉部份有 `xRotate`, `yRotate`, `zRotate` 三個 function。
- 此起始點繪製 0 個頂點，並且沒有做任何轉換，這些都將在本篇補上：
  - `a_position`、`a_color` 在起始點的 buffer 傳入空陣列。
  - `viewMatrix`、`worldMatrix` 起始點為 `matrix4.identity()`，也就是『什麼都不做』。

- ○ `gl.drawArrays(gl.TRIANGLES, 0, 0)`; 繪製 0 個頂點。
- 本篇不會用到動畫效果，因此 startLoop 被註解不會用到。

仔細看的話，會發現在 vertex shader 中的 `a_position` 宣告型別為 vec4，可以直接與 4x4 矩陣 `u_matrix` 相乘，但是在設定傳入的資料時候 `gl.vertexAttribPointer()` 指定的長度只有 3，那麼剩下的向量第四個元素（接下來稱為 w）的值怎麼辦？有意思的是，在提供的資料長度不足或是沒有提供時 x、y、z 預設值為 0，而 w 很有意思地預設值是 1，對於所有 `vec4` 的 vertex attribute 都是如此，這樣一來就可以符合平移時多餘維度為 1 的需求，很巧的是，如果今天這個 attribute 是顏色，那意義上就變成 alpha 預設值是 1。

⌘ P 文字形狀的 3D 物件

以下圖 3-1 為此 P 文字形狀 3D 物件的設計圖：

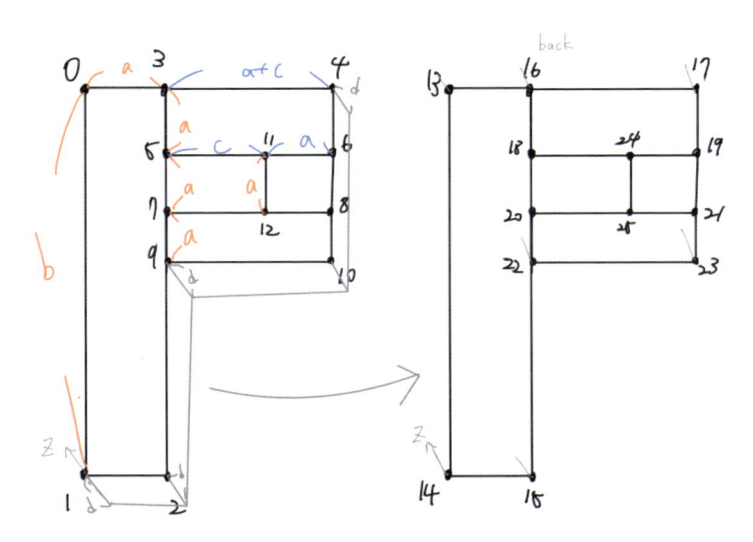

▲圖 3-1　左：物件正面；右：由上方透視的底面

圖 3-1 的左圖表示此物件正面的樣子，右圖為從上方透視的底面，這兩張表示了各個頂點的編號，同時以 a、b、c、d 表示特定邊長的長度，以便為各個頂點定位座標。

3 3D & 物件

圖 3-2 兩張圖表示各個長方形的編號，其右圖表示從上方透視的底面及編號：

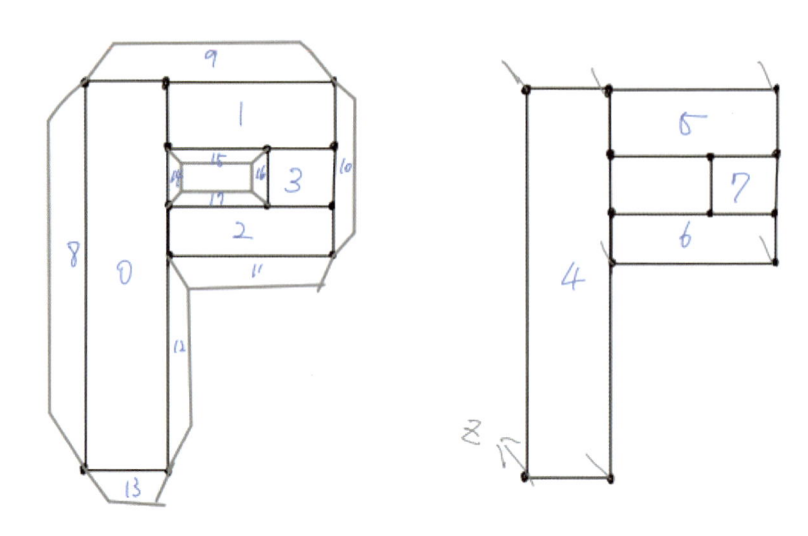

▲ 圖 3-2　左圖：各個長方形編號；右圖：由上方透視的底面及編號

這邊的任務是要為此 3D 物件產生對應的 `a_position`、`a_color` 陣列，可以想像要建立的資料不少，直接寫在 `setup()` 內很快就會讓 `setup()` 失去焦點，因此建立 function `createModelBufferArrays()`，選定 a、b、c、d 的數值後第一步驟就是產生各個頂點的座標：

```javascript
function createModelBufferArrays() {
  // a_positions
  const a = 40, b = 200, c = 60, d = 45;

  const points = [0, d].flatMap(z => ([
    [0, 0, z], // 0, 13
    [0, b, z],
    [a, b, z],
    [a, 0, z],
    [2*a+c, 0, z], // 4, 17
    [a, a, z],
    [2*a+c, a, z],
```

3-4

```
    [a, 2*a, z],
    [2*a+c, 2*a, z], // 8, 21
    [a, 3*a, z],
    [2*a+c, 3*a, z],
    [a+c, a, z],
    [a+c, 2*a, z], // 12, 25
  ]));
}
```

在圖 3-1 中頂點的編號對應此 `points` 陣列中的 index 值，因為正面、底面的座標位置只有 z 軸前後的差別，因此用 `flatMap`[1] 讓程式碼更少一點，筆者在程式碼中某些座標的右方有註解表示其對應的頂點編號。

不過 points 不能當成 `a_position` attribute buffer 用的陣列，`a_position` 陣列必須是一個個三角形的頂點，以 P 文字形狀的 3D 物件來說，所有的面都可以由長方形組成，兩個三角形可以形成一個長方形，因此寫一個 function 接受四個頂點座標，產生兩個三角形的 `a_position` 陣列：

```
function rectVertices(a, b, c, d) {
  return [
    ...a, ...b, ...c,
    ...a, ...c, ...d,
  ];
}
```

有了這個工具之後，`a_position` 就可以以長方形為單位寫成：

```
function createModelBufferArrays() {
  // a_positions ...

  const a_position = [
    ...rectVertices(points[0], points[1], points[2], points[3]), // 0
    ...rectVertices(points[3], points[5], points[6], points[4]),
```

---

[1] https://developer.mozilla.org/en-US/docs/Web/JavaScript/Reference/Global_Objects/Array/flatMap

```
    ...rectVertices(points[7], points[9], points[10], points[8]),
    ...rectVertices(points[11], points[12], points[8], points[6]),
    ...rectVertices(points[13], points[16], points[15], points[14]), // 4
    ...rectVertices(points[16], points[17], points[19], points[18]),
    ...rectVertices(points[20], points[21], points[23], points[22]),
    ...rectVertices(points[24], points[19], points[21], points[25]),
    ...rectVertices(points[0], points[13], points[14], points[1]), // 8
    ...rectVertices(points[0], points[4], points[17], points[13]),
    ...rectVertices(points[4], points[10], points[23], points[17]),
    ...rectVertices(points[9], points[22], points[23], points[10]),
    ...rectVertices(points[9], points[2], points[15], points[22]), // 12
    ...rectVertices(points[2], points[1], points[14], points[15]),
    ...rectVertices(points[5], points[7], points[20], points[18]),
    ...rectVertices(points[5], points[18], points[24], points[11]),
    ...rectVertices(points[11], points[24], points[25], points[12]), // 16
    ...rectVertices(points[7], points[12], points[25], points[20]),
  ];
}
```

同樣地，在 `a_position` 陣列中某些 `rectVertices()` 呼叫右方有註解表示其對應在上圖中的長方形。

完成 `a_position` 之後，`a_color` 也要為三角形每個頂點產生對應資料，筆者的設計是一個平面使用一個顏色，也就是一個長方形（兩個三角形，共 6 個頂點）的顏色至少都會一樣，同時希望面的顏色是隨機，因此寫了以下 function：

```
function rectColor(color) {
  return Array(6).fill(color).flat();
}

function randomColor() {
  return [Math.random(), Math.random(), Math.random()];
}
```

筆者私心想要正面顏色使用筆者的主題顏色（`#6bde99` / `rgb(108, 225, 153)`），背面四個長方形為同一個隨機顏色，除此之外的長方形顏色也隨機，

3-1　Orthogonal 3D 投影

因此 a_color 的產生如下：

```javascript
function createModelBufferArrays() {
  // a_positions ...

  // a_color
  const frontColor = [108/255, 225/255, 153/255];
  const backColor = randomColor();
  const a_color = [
    ...rectColor(frontColor), // 0
    ...rectColor(frontColor),
    ...rectColor(frontColor),
    ...rectColor(frontColor),
    ...rectColor(backColor), // 4
    ...rectColor(backColor),
    ...rectColor(backColor),
    ...rectColor(backColor),
    ...rectColor(randomColor()), // 8
    ...rectColor(randomColor()),
    ...rectColor(randomColor()),
    ...rectColor(randomColor()),
    ...rectColor(randomColor()), // 12
    ...rectColor(randomColor()),
    ...rectColor(randomColor()),
    ...rectColor(randomColor()),
    ...rectColor(randomColor()), // 16
    ...rectColor(randomColor()),
  ];
}
```

最後回傳整個『P 文字形狀的 3D 物件』相關的全部資料，除了 vertex attributes 之外，待會 `gl.drawArrays()` 需要知道要繪製的頂點數量，在這邊也以 `numElements` 回傳：

```javascript
function createModelBufferArrays() {
  // a_positions ...
  // a_color ...

  return {
```

3-7

```
      numElements: a_position.length / 3,
      vertexDataArrays: {
        a_position, a_color,
      },
    };
  }
```

⌘ Orthogonal 3D 繪製

辛苦寫好了 `createModelBufferArrays`，在 `setup()` 呼叫產生 P 文字形狀 3D 物件的頂點資料，把 attribute 資料傳送到對應的 buffer 內：

```
async function setup() {
  // ...

  const modelBufferArrays = createModelBufferArrays();

  // a_position ...
  gl.bufferData(
    gl.ARRAY_BUFFER,
    new Float32Array(modelBufferArrays.vertexDataArrays.a_position),
    gl.STATIC_DRAW,
  );

  // a_color ...
  gl.bufferData(
    gl.ARRAY_BUFFER,
    new Float32Array(modelBufferArrays.vertexDataArrays.a_color),
    gl.STATIC_DRAW,
  );

  return {
    gl,
    program, attributes, uniforms,
    buffers, modelBufferArrays,
    state: {
    },
    time: 0,
  };
}
```

3-1　Orthogonal 3D 投影

同時也把 `modelBufferArrays` 回傳，在 `render()` 使用，`viewMatrix` 的意含是『由場景位置轉換、投影到 clip space 到畫布上』，因此讓 `viewMatrix` 使用 `matrix4.projection()` 產生，orthogonal projection 就是在這行發生：

```javascript
function render(app) {
  const {
    gl,
    program, uniforms,
    modelBufferArrays,
    state,
  } = app;

  // ...

  const viewMatrix = matrix4.projection(
    gl.canvas.width, gl.canvas.height, 400,
  );
  const worldMatrix = matrix4.identity();

  gl.uniformMatrix4fv(
    uniforms.matrix,
    false,
    matrix4.multiply(viewMatrix, worldMatrix),
  );

  gl.drawArrays(gl.TRIANGLES, 0, modelBufferArrays.numElements);
}
```

存檔，來看看結果：

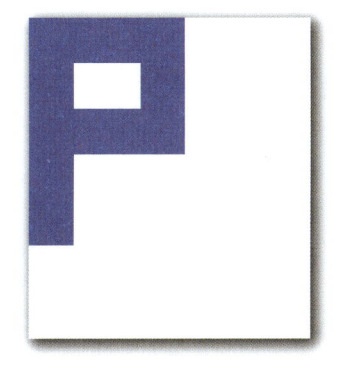

▲ 圖 3-3　P 形狀 3D 模型的投影成像

3-9

> 因為是其他顏色是隨機產生，讀者跟著跑到這邊的話看到的不一定是這個顏色。

有東西是有東西，但是正面顏色怎麼不是主題色？事實上，在 WebGL 繪製時，假設先繪製了一個三角形，然後再繪製一個三角形在同一個位置，那麼後面的三角形會覆蓋掉之前的顏色，也就是說當前看到的顏色是底面的顏色（上方程式碼中的 `backColor`），解決這個問題的其中一個方法是啟用『只繪製正面面向觀看者的三角形』功能 `gl.CULL_FACE`，當三角形的頂點順序符合右手開掌[2]的食指方向時，大拇指的方向即為三角形正面，如圖 3-4 所示的長方形（或是說兩個三角形）的正面朝觀看者：

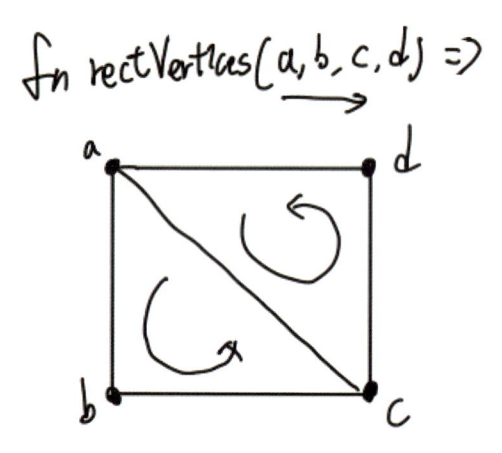

▲ 圖 3-4　右手定則正面示意圖

筆者已經在上方建立 `a_position` 時使得組成底面的三角形面向下，因此只要在繪製前加上這行設定『只繪製正面面向觀看者的三角形』：

```
function render(app) {
  // ...
```

2　https://zh.wikipedia.org/zh-tw/右手定則#右手開掌定則

```
gl.enable(gl.CULL_FACE);

// ...

gl.drawArrays(gl.TRIANGLES, 0, modelBufferArrays.numElements);
}
```

底面就會因為面向下，其正面沒有對著觀看者，不會被繪製，可以如圖 3-5 看到正面了：

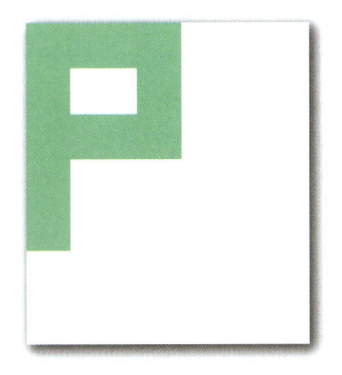

▲ 圖 3-5　底面朝下不被繪製，正面才不會被覆蓋掉

⌘ 轉一下，看起來比較 3D

因為使用 `matrix4.projection()` 做 orthogonal 投影，其實就是投影到 xy 平面上，這個 3D 模型投影下去看不出來是 3D 的，同時在 HTML 也已經準備好各種使用者控制項，接下來串上各個 transform，尤其是旋轉，使之看起來真的是 3D，首先在 `setup()` 回傳初始 transform 值：

```
async function setup() {
  // ...

  return {
    gl,
    program, attributes, uniforms,
    buffers, modelBufferArrays,
```

3-11

```
  state: {
    projectionZ: 400,
    translate: [150, 100, 0],
    rotate: [degToRad(30), degToRad(30), degToRad(0)],
    scale: [1, 1, 1],
  },
  time: 0,
};
```

可以注意到除了 `translate`、`rotate`、`scale` 之外，筆者也加上 `projectionZ`，之後讓使用者可以控制 z 軸 clip space；在旋轉矩陣產生的 function 以及其三角函數接收的參數單位為弧度[3]（radian），但是為了讓我們能比較直覺地用角度[4]（degree）描述、調整數值，在 `lib/utils.js` 有個 function `degToRad` 進行轉換：

```
export function degToRad(deg) {
  return deg * Math.PI / 180;
}
```

這個 `degToRad` 可以把原本以角度（Degree，一圈為 360 度）為單位的數值轉換成弧度（Radian，一圈為 2π），別忘了回來引入寫好的 `degToRad`：

```
import { createShader, createProgram, degToRad } from '../../lib/utils.js';
```

`worldMatrix` 的意含是『放置物件到場景中的轉換』，一開始使用 `matrix4.identity()` 直接放在場景、意指使用物件頂點自身 `position` 作為 world position 場景中位置；為了串上使用者控制的平移、旋轉以及縮放，在 `render()` 串上 `state` 與 `worldMatrix`、`viewMatrix` 矩陣的產生：

```
function render(app) {
  // ...
```

---

3 https://zh.wikipedia.org/zh-tw/弧度
4 https://zh.wikipedia.org/zh-tw/度_(角)

3-1　Orthogonal 3D 投影

```javascript
const viewMatrix = matrix4.projection(
  gl.canvas.width,
  gl.canvas.height,
  state.projectionZ,
);
const worldMatrix = matrix4.multiply(
  matrix4.translate(...state.translate),
  matrix4.xRotate(state.rotate[0]),
  matrix4.yRotate(state.rotate[1]),
  matrix4.zRotate(state.rotate[2]),
  matrix4.scale(...state.scale),
);

gl.uniformMatrix4fv(
  uniforms.matrix,
  false,
  matrix4.multiply(viewMatrix, worldMatrix),
);

gl.drawArrays(gl.TRIANGLES, 0, modelBufferArrays.numElements);
}
```

結果看起來像是這樣，是 3D 了，但是成像會如同圖 3-6，顯然怪怪的：

▲ 圖 3-6　稍做旋轉後的成像

3-13

3　3D & 物件

在上面已經啟用 `gl.CULL_FACE`，不面向觀看者的面確實不會被繪製，但是不夠，下圖箭頭指著的面是面向觀看者的，因為在 `a_position` 上排列在正面之後，導致正面繪製完成後，被之後的面覆蓋過去，如圖 3-7 箭頭指著的面就是在正面之後繪製，導致蓋過應該應在前面的成像：

▲ 圖 3-7　在正面之後繪製，導致蓋過應該應在前面成像

調換 `a_position` 或許可以解決，不過如果讓使用者可以旋轉，旋轉到背面時那麼又會露出破綻，因此需要另一個功能：`gl.DEPTH_TEST`，也就是深度測試，在 vertex shader 輸出的 `gl_Position.z` 除了給 clip space 之外，也可以作為深度資訊，如果準備要畫上的 pixel 比原本畫布上的來的更接近觀看者，顏色才會覆蓋上去，因此加入這行啟用這個功能：

```
function render(app) {
  // ...

  gl.enable(gl.CULL_FACE);
  gl.enable(gl.DEPTH_TEST);

  // ...

  gl.drawArrays(gl.TRIANGLES, 0, modelBufferArrays.numElements);
}
```

3-14

耶，一切就正確囉，如圖 3-8：

▲ 圖 3-8　啟用 gl.DEPTH_TEST 深度比較後，成像終於正確

最後把 HTML 上的控制項串到 *state* 上，就如同前面章節一樣使用 FormData：

```javascript
async function main() {
  // ...

  const controlsForm = document.getElementById('controls');
  controlsForm.addEventListener('input', () => {
    const formData = new FormData(controlsForm);

    app.state.projectionZ = parseFloat(formData.get('projection-z'));
    app.state.translate[0] = parseFloat(formData.get('translate-x'));
    app.state.translate[1] = parseFloat(formData.get('translate-y'));
    app.state.translate[2] = parseFloat(formData.get('translate-z'));
    app.state.rotate[0] = degToRad(parseFloat(formData.get('rotation-x')));
    app.state.rotate[1] = degToRad(parseFloat(formData.get('rotation-y')));
    app.state.rotate[2] = degToRad(parseFloat(formData.get('rotation-z')));
    app.state.scale[0] = parseFloat(formData.get('scale-x'));
    app.state.scale[1] = parseFloat(formData.get('scale-y'));
    app.state.scale[2] = parseFloat(formData.get('scale-z'));

    render(app);
```

```
});

// startLoop(app);
render(app);
}
```

更新狀態後,因為目前沒有啟用動畫,所以需要呼叫 `render(app)` 重畫畫面

一個可以根據使用者調整拉桿改變 transform 進行互動的 3D 物件就完成,讀者可以玩玩看這些拉桿,感受一下 3D transform 的效果,例如拉成如圖 3-9,整個轉到背面的樣子:

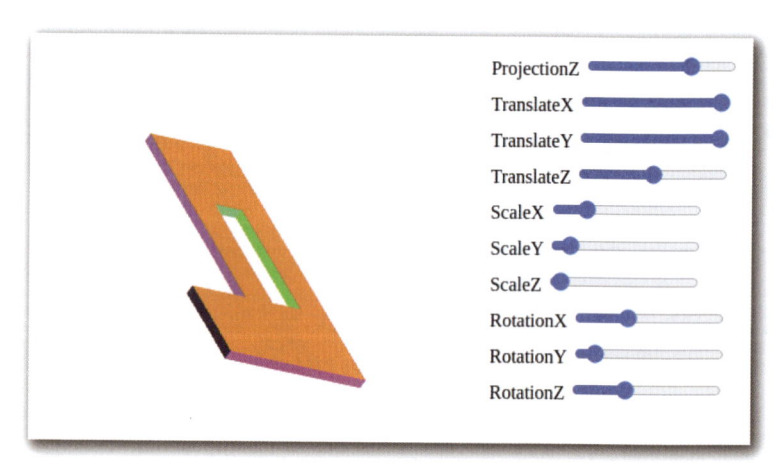

▲ 圖 3-9 調整各種 transform 參數,轉到背面的樣子

或者玩玩 ProjectionZ 以及 TranslateZ,感受 clip space 的深度,當初 `gl_Position` 的 x, y, z 必須在 +1 ~ -1 之間才會被畫在畫布上,x, y 的部份很好理解就是 2D 畫布上的位置,而現在開始有 z 軸,可以參考 `matrix4.projection` 內對於 z 軸的計算,使得 z 在投影前必須座落在 `+projectionZ/2` 至 `-projectionZ/2` 才會被畫進畫面上,像是圖 3-10 中的 3D 物件有部份已經在 clip space 以外而沒有進入成像:

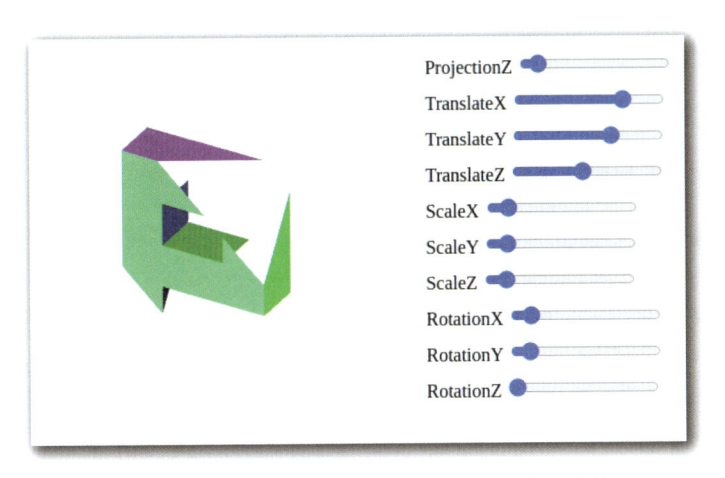

▲ 圖 3-10　在 clip space 以外而沒有進入成像

到此進度的完整程式碼可以在這邊找到：

</>	ch3/01/index.html
</>	ch3/01/main.js

也可以參考上線版本，透過瀏覽器打開此頁面：
https://webgl-book.pastleo.me/ch3/01/index.html

3-2　Perspective 3D 成像

　　使用 orthogonal 投影方法畫出來的畫面與我們在生活中從眼睛、相機看到的其實不同，這邊介紹更接近現實生活眼睛看到的成像方式：perspective[5] projection，使用 perspective 3D 投影渲染物件時，相當於模擬現實生活眼睛、相機捕捉的光線形成的投影，也是大多 3D 遊戲使用的投影方式，學會這種投影方式，才算是真正進入 3D 渲染的世界，那麼我們就開始吧！

5　https://zh.wikipedia.org/zh-tw/透視

3 3D & 物件

⌘ Orthogonal vs Perspective

先來回顧一下 orthogonal 投影,在這行產生的 transform 矩陣只是調整輸入座標使之落在 clip space 內

```
const viewMatrix = matrix4.projection(
  gl.canvas.width,
  gl.canvas.height,
  state.projectionZ,
);
```

為了讓輸入座標可以用螢幕長寬的 pixel 定位,這個轉換只是做拉伸(範圍由 `0 ~ 螢幕寬高` 轉換至 `0 ~ 2`)及平移(範圍由 `0 ~ 2` 轉換至 `-1 ~ +1`),orthogonal 投影視覺化看起來像是場景中如同圖 3-11 擺設,然後投影到畫面上如圖 3-12:

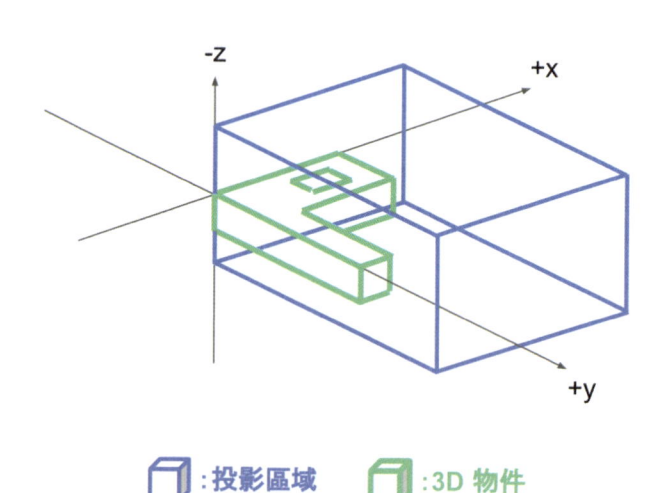

▲ 圖 3-11　orthogonal 投影的擺設

3-18

3-2 Perspective 3D 成像

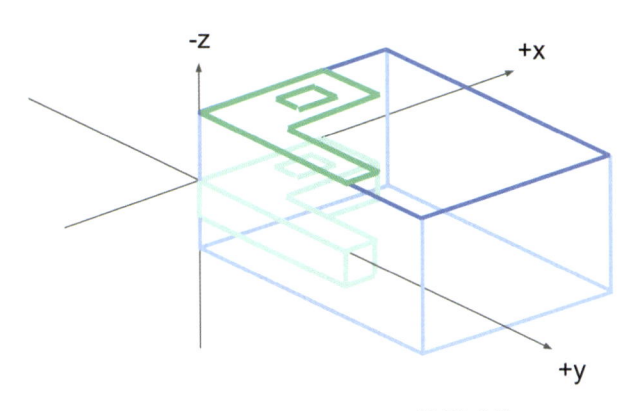

▲ 圖 3-12　orthogonal 投影成像

沒錯，orthogonal 投影其實是往 -z 的方向投影，也就是說在 clip space 是以 z = -1 的平面進行成像；還有另外一種講法：使用者看者螢幕是看向 clip space 的 +z 方向。

現實生活中比較可以找的到以 orthogonal 成像的範例就是影印機的掃描器，有一個大的面接收垂直於該面的光線；而眼睛、相機則是以一個小面積的感光元件，接收特定角度範圍內的光線，這樣的投影方式稱為 perspective projection（透視），兩者比較如圖 3-13：

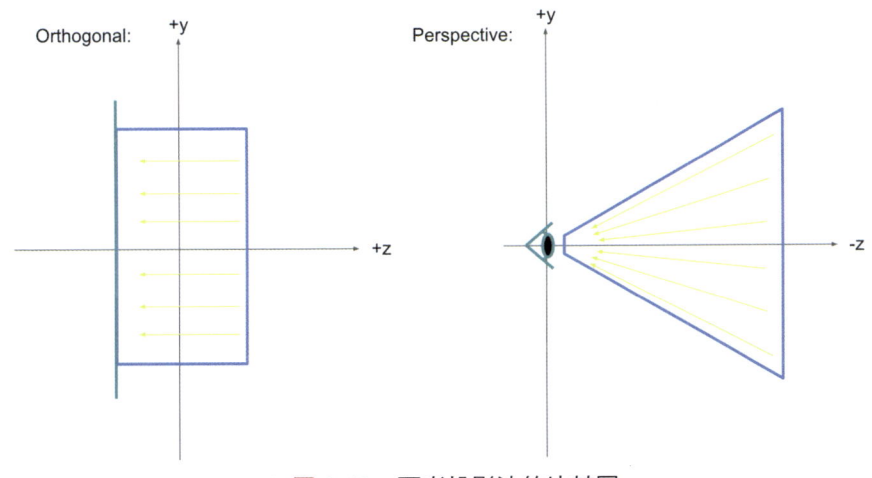

▲ 圖 3-13　兩者投影法的比較圖

3　3D & 物件

　　上面這張圖是從側面看的，x 軸方向與螢幕平面垂直，而藍色框起來表示可見、會 transform 到 clip space 的區域，在 orthogonal 是一個立方體；在 perspective 這個區域的形狀叫做錐台[6]（frustum），這個形狀 3D 的樣子像是圖 3-14 這樣：

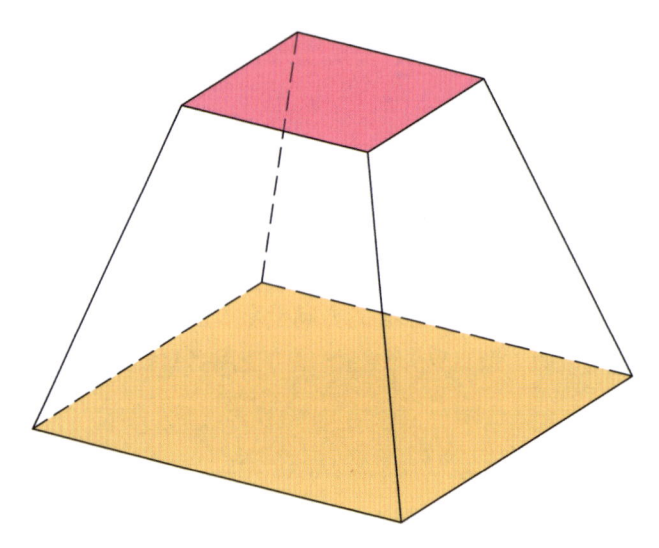

▲ 圖 3-14　perspective 投影會被 transform 到 clip space 的錐台 3D 圖

⌘ 產生 Perspective transform 矩陣

　　什麼樣的矩陣可以把 frustum 的區域轉換成 clip space 呢？很不幸的，其實這樣的矩陣不存在，因為這樣的轉換不是線性轉換（linear transformation），如果是線性轉換[7]，那麼轉換前後必須保持網格線平行並間隔均等，想像一下把上面 frustum 側邊的邊拉成 clip space 立方體的平行線，這個轉換（transform）就不是線性的。

6　https://zh.wikipedia.org/wiki/錐台
7　根據 3Blue1Brown 的 Youtube 影片—線性變換與矩陣：
　　https://youtu.be/kYB8IZa5AuE?t=152

3-20

幸好在 vertex shader 輸出 gl_Position 的第四個元素 `gl_Position.w` 有一個我們一直沒用到的功能：頂點位置在進入 clip space 之前，會把 `gl_Position.x`、`gl_Position.y`、`gl_Position.z` 都除以 `gl_Position.w`。有了這個功能，在距離相機越遠的地方輸出越大的 `gl_Position.w`，越遠的地方就能接受更寬廣的 xy 平面區域進入 clip space；也是因為會除以 `gl_Position.w`，所以我們在前面的章節必須要讓 `gl_Position.w` 輸出 1。

也就是說，這個 perspective transform 矩陣還要把遠近轉換到 .w 第四維度去，並且考慮轉換出來的 .x、.y、.z 會除以 .w，總體下來把場景錐台（frustum）區域轉換到 clip space，產生此轉換矩陣的 function 通常是這樣：

```
matrix4.perspective(
  fieldOfView,
  aspect,
  near,
  far,
)
```

`fieldOfView` 表示看出去的角度（上下），`aspect` 控制畫面寬高比（寬除以高），`near` 為靠近相機那面距離相機的距離，`far` 則為最遠相機能看到的距離。

那此轉換矩陣的實做是如何？因為透過透視呈現的 3D 遊戲或是應用程式基本上都使用到了這個轉換，因此 `matrix4.perspective()` 在各個 3D 函式庫中很常見，筆者從熱門的 WebGL 3D 渲染引擎 three.js 的 PerspectiveCamera[8]，從其原始碼找到產生 perspective transform 的 function[9]，

---

8 撰文當下 three.js new PerspectiveCamera(fov, aspect, near, far) 的原始碼：
 https://github.com/mrdoob/three.js/blob/6e897f/src/cameras/PerspectiveCamera.js#L6
9 撰文當下此 three.js function 為 Matrix4.makePerspective() 的原始碼：
 https://github.com/mrdoob/three.js/blob/6e897f/src/math/Matrix4.js#L773-L797，寫成數學式的過程有經過簡化、替換以符合 PerspectiveCamera 的 fov、aspect、near、far 參數

寫成數學式如下：

$$perspective(fov, aspect, near, far) = M$$

$$M = \begin{bmatrix} \frac{1}{aspect \cdot tan\left(\frac{fov}{2}\right)} & 0 & 0 & 0 \\ 0 & \frac{1}{tan\left(\frac{fov}{2}\right)} & 0 & 0 \\ 0 & 0 & \frac{near+far}{near-far} & \frac{2 \cdot near \cdot far}{near-far} \\ 0 & 0 & -1 & 0 \end{bmatrix}$$

注意接下來矩陣、向量將採用數學上的慣例，並非電腦程式的排列方式，並且以 M 表示 perspective transform 矩陣、fov 為 `fieldOfView` 的縮寫。

然而為什麼矩陣公式是這樣實做的呢？接下來我們花費一些篇幅把位置代入這個公式，觀察一下到 clip space 以及畫面上的過程。

令 A 表示 `a_position` 輸入的向量（正確來說，是與 perspective 矩陣 M 相乘的向量），P 表示與 M 相乘後輸出給 `gl_Position` 的向量，P' 表示 `gl_Position` 的 xyz 除以 w 的向量，也就是最終 clip space 中的位置：

$$let\ A = \begin{bmatrix} A_x \\ A_y \\ A_z \\ A_w \end{bmatrix} = \begin{bmatrix} A_x \\ A_y \\ A_z \\ 1 \end{bmatrix} \quad let\ P = MA = \begin{bmatrix} P_x \\ P_y \\ P_z \\ P_w \end{bmatrix} \quad let\ P' = \begin{bmatrix} P'_x \\ P'_y \\ P'_z \end{bmatrix} = \begin{bmatrix} \frac{P_x}{P_w} \\ \frac{P_y}{P_w} \\ \frac{P_z}{P_w} \end{bmatrix}$$

下圖 3-15 為場景入鏡相機的轉換示意圖，從 A 經過 M 轉換得到 P，並且除以 P_w 得到 P' 輸出到 clip space，圖中 (1)、(2) 綠色錐台區域表示 perspective 相機在場景內擷取的範圍，會轉換成圖中 (3)、(4) 的橘色 clip space 成像於畫

3-2　Perspective 3D 成像

面中，圖中 (2)、(3) 是從側面看的（看向 -x 方向），比較好標示長度以及角度：

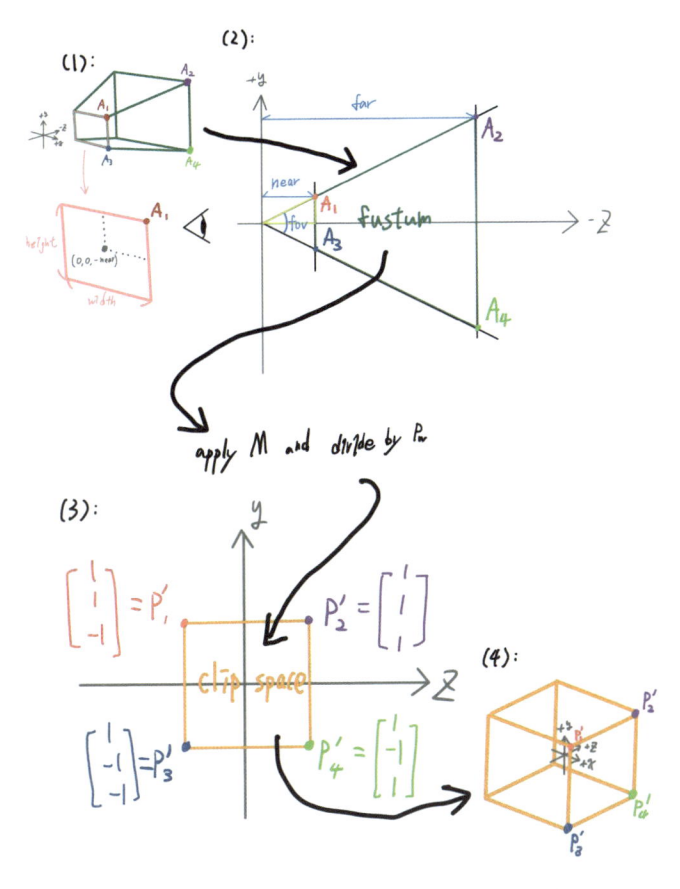

▲ 圖 3-15　perspective（透視）轉換示意圖

圖 3-15 中，四個點位於入鏡錐台的右邊四個角 A_1、A_2、A_3、A_4，經過 M 轉換後成為 P_1、P_2、P_3、P_4 除以各自的 P_w 得到 clip space 位置 P'_1、P'_2、P'_3、P'_4，圖中可以清楚看到這四個點轉換後在 clip space 中的座標位置（P'）。

以 A_1 為例，應該最後能轉換到 P'_1：[1, 1, -1]，我們來實際把 A_1 代入，用 A_{1x}、A_{1y}、A_{1z} 表示其 x、y、z 值，並嘗試把這些值展開來用 near、far、aspect 以及數學計算表示：

3-23

3 3D & 物件

$$MA_1 = \begin{bmatrix} \dfrac{1}{aspect \cdot tan\left(\dfrac{fov}{2}\right)} & 0 & 0 & 0 \\ 0 & \dfrac{1}{tan\left(\dfrac{fov}{2}\right)} & 0 & 0 \\ 0 & 0 & \dfrac{near + far}{near - far} & \dfrac{2 \cdot near \cdot far}{near - far} \\ 0 & 0 & -1 & 0 \end{bmatrix} \begin{bmatrix} A_{1x} \\ A_{1y} \\ A_{1z} \\ 1 \end{bmatrix}$$

看著圖 3-15 中 (2) 的黃色三角形，A_1 位於此三角型右上的角，可以知道黃色三角形下邊邊長是 near，同時也是 A_{1z} 的值，只不過是負的：

$$A_{1z} = -near$$

但是為什麼是負的呢？因為 near 本身只是長度，是正的，慣例上相機視角對著的方向為 -z，面向螢幕外（螢幕到使用者）的方向為 +z，圖 3-15 中 (2) 的 -z 軸方向就體現了這件事，因此可知 A_{1z} 是負的 near。

繼續看著圖 3-15 中 (2) 的黃色三角形，可以知道三角形右邊邊長即為 A_{1y}、下邊邊長是 near、左邊的角度是 fov 的一半，利用三角函數可以算出 A_{1y}：

$$A_{1y} = near \cdot tan\left(\dfrac{fov}{2}\right)$$

看著圖 3-15 中 (1) A_1 所在的粉紅色矩形，此矩形寬高比（寬除以高）即為 aspect，且正中心 x、y 值為 0，可知 A_{1x} 與 A_{1y} 的比例與寬高比相同，以這個比例關係算出 A_{1x}，並且將 A_{1y} 代入：

$$aspect = \dfrac{width}{height} = \dfrac{A_{1x}}{A_{1y}} \implies A_{1x} = aspect \cdot A_{1y} = aspect \cdot near \cdot tan\left(\dfrac{fov}{2}\right)$$

這麼一來 A_{1x}、A_{1y}、A_{1z} 的數值都展開用 near、far、aspect 以及數學計算表示了，帶入 MA_x 開始運算 P_1：

$$MA_1 = \begin{bmatrix} \dfrac{1}{aspect \cdot tan\left(\dfrac{fov}{2}\right)} & 0 & 0 & 0 \\ 0 & \dfrac{1}{tan\left(\dfrac{fov}{2}\right)} & 0 & 0 \\ 0 & 0 & \dfrac{near+far}{near-far} & \dfrac{2 \cdot near \cdot far}{near-far} \\ 0 & 0 & -1 & 0 \end{bmatrix} \begin{bmatrix} aspect \cdot near \cdot tan\left(\dfrac{fov}{2}\right) \\ near \cdot tan\left(\dfrac{fov}{2}\right) \\ -near \\ 1 \end{bmatrix}$$

$$= \begin{bmatrix} P_{1x} \\ P_{1y} \\ P_{1z} \\ P_{1w} \end{bmatrix}$$

根據 CH2-4 提到的矩陣運算，計算 P_{1x}，很快就會發現抵銷之後剩下 near：

$$P_{1x} = M_{(1,1)}A_{1x} + M_{(1,2)}\cancel{A_{1y}} + M_{(1,3)}\cancel{A_{1z}} + M_{(1,4)}\cancel{A_{1w}}$$

$$= \dfrac{1}{\cancel{aspect} \cdot \cancel{tan\left(\dfrac{fov}{2}\right)}} \cdot \cancel{aspect} \cdot near \cdot \cancel{tan\left(\dfrac{fov}{2}\right)} = near$$

$M_{(1,2)}$ 表示矩陣 M 的第一列、第二欄，$M_{(i,j)}$ 類似 CH2-4 圖 2-29、圖 2-30 的表示方式，矩陣乘法的完整計算也可以在 CH2-4 找到。P_{1y} 的計算與 P_{1x} 很相似：

$$P_{1y} = \cancel{M_{(2,1)}A_{1x}} + M_{(2,2)}A_{1y} + \cancel{M_{(2,3)}A_{1z}} + \cancel{M_{(2,4)}A_{1w}}$$

$$= \dfrac{1}{\cancel{tan\left(\dfrac{fov}{2}\right)}} \cdot near \cdot \cancel{tan\left(\dfrac{fov}{2}\right)} = near$$

P_{1z} 展開抵銷的過程稍微複雜點，但是最後也是剩下 -near：

$$P_{1z} = \cancel{M_{(3,1)}A_{1x}} + \cancel{M_{(3,2)}A_{1y}} + M_{(3,3)}A_{1z} + M_{(3,4)}A_{1w}$$

$$= -near \cdot \frac{near + far}{near - far} + \frac{2 \cdot near \cdot far}{near - far} = \frac{-near^2 - near \cdot far + 2 \cdot near \cdot far}{near - far}$$

$$= \frac{-near^2 + near \cdot far}{near - far} = \frac{near(far - near)}{near - far} = -near \cdot \frac{\cancel{near - far}}{\cancel{near - far}} = -near$$

P_{1w} 由田於 $M_{(4,3)}$ 有 -1，等同於從 A_{1z} 拿值過來乘以 -1：

$$P_{1w} = \cancel{M_{(4,1)}A_{1x}} + \cancel{M_{(4,2)}A_{1y}} + M_{(4,3)}A_{1z} + \cancel{M_{(4,4)}A_{1w}}$$

$$= -near \cdot (-1) = near$$

P_1 的所有元素都算出來了：

$$P_1 = \begin{bmatrix} near \\ near \\ -near \\ near \end{bmatrix}$$

P_1 全部的元素都是 near 或是 -near，並且再往下一步除以 P_{1w} 進入 clip space，其實就是除以 near：

$$P_1' = \begin{bmatrix} \frac{P_{1x}}{P_{1w}} \\ \frac{P_{1y}}{P_{1w}} \\ \frac{P_{1z}}{P_{1w}} \end{bmatrix} = \begin{bmatrix} \frac{near}{near} \\ \frac{near}{near} \\ \frac{-near}{near} \end{bmatrix} = \begin{bmatrix} 1 \\ 1 \\ -1 \end{bmatrix}$$

得到如圖 3-15 中 (3) 的 P'_1 在 clip space 所標示的位置 [1, 1, -1]，將成像到畫面右上角。

A_1、A_2、A_3、A_4 這些點都是入鏡錐台區的邊角，自然得轉換到 clip space 的邊角，在展開 A_{1x}、A_{1y} 的過程中就可以看到 aspect、$tan(\frac{fov}{2})$ 的出現，並且與矩陣的 scale 抵銷剩下 near，A_{1z} 也轉換成了 -near，最後除以 P_{1w} 的 near 抵銷成為 [1, 1, -1]；讀者有興趣可以把 A_2 或是 A_4 代入算算看，這兩個點因為位於 far 平面上，與矩陣運算完應該是剩下 far 值，最後除以 P_w 得到 1 或是 -1 位於 clip space 的對應邊角上。

為了推廣到更多點，把 A 純粹作為代數代入：

$$MA_1 = \begin{bmatrix} \frac{1}{aspect \cdot tan\left(\frac{fov}{2}\right)} & 0 & 0 & 0 \\ 0 & \frac{1}{tan\left(\frac{fov}{2}\right)} & 0 & 0 \\ 0 & 0 & \frac{near+far}{near-far} & \frac{2 \cdot near \cdot far}{near-far} \\ 0 & 0 & -1 & 0 \end{bmatrix} \begin{bmatrix} A_x \\ A_y \\ A_z \\ 1 \end{bmatrix}$$

$$= \begin{bmatrix} P_x \\ P_y \\ P_z \\ P_w \end{bmatrix} = \begin{bmatrix} \frac{A_x}{aspect \cdot tan\left(\frac{fov}{2}\right)} \\ \frac{A_y}{tan\left(\frac{fov}{2}\right)} \\ A_z \cdot \frac{near+far}{near-far} + \frac{2 \cdot near \cdot far}{near-far} \\ -A_z \end{bmatrix}$$

並且將 P 的 xyz 除以 w 算出 clip space 的 P'：

$$\begin{bmatrix} P'_x \\ P'_y \\ P'_z \end{bmatrix} = \begin{bmatrix} \frac{P_x}{P_w} \\ \frac{P_y}{P_w} \\ \frac{P_z}{P_w} \end{bmatrix} = \begin{bmatrix} \frac{-A_x}{aspect \cdot tan\left(\frac{fov}{2}\right) \cdot A_z} \\ \frac{-A_y}{tan\left(\frac{fov}{2}\right) \cdot A_z} \\ \frac{1}{A_z} \cdot \frac{2 \cdot near \cdot far}{(near - far)} - \frac{near + far}{near - far} \end{bmatrix}$$

從最後 P'_x、P'_y 的完整公式可以觀察到幾件事情：

- P'_x、P'_y 根據視角寬度（fov）越寬，其 $tan(\frac{fov}{2})$ 越大，使值縮小，越接近畫面正中央

- 對於 P'_x 來說，aspect 長寬比（寬除以長）越大，通常表示成像畫布更寬，這時 x 方向能容納更偏離 z 軸的物件入鏡

- P'_x、P'_y 會因為最後除以 P_w，等於除以 $-A_z$，有點像是形成入鏡錐台的 xy 平面切片，切片大小隨著 A 距離原點越遠（A_z 負值更多）而放大，這個切片內的點被縮放、轉換成 clip space 正方體的 xy 切片

Perspective 轉換矩陣 M 除了把 $-A_z$ 帶到 P_w 使用特殊規則使得整個錐台轉換成正方體 clip space 非線性轉換得以可行之外，本身只是把根據參數形成的入鏡錐台轉換成另一個正方形錐台。

最後來講講 P'_z 的情況，P_x 與 P_y 確實需要被 P_w 除，不過 P_z 應該可以不用被 P_w 除的，畢竟直接把 A_z 根據 near、far 平移縮放到 -1 ~ +1 即可，有可能因為其他應用可以用到或是硬體實做上比較容易，WebGL 在 P_z 在這邊也會除以 P_w，這導致 $M_{(3,3)}$ 所帶來的 z 數值會被自己抵銷掉；從這個 3D 業界慣用的 P'_z 公式來看，加入了平移到 $M_{(3,4)}$ 到 P_z，並且透過 P_z 會被 P_w（等於 $-A_z$）的

事實，讓 z（A_z）以倒數姿態存在在 P'_z 公式中，筆者把公式輸入視覺化工具來觀察 near、far 與 z 輸入輸出的影響：

https://www.desmos.com/calculator/umeabhlpjq，如圖 3-16：

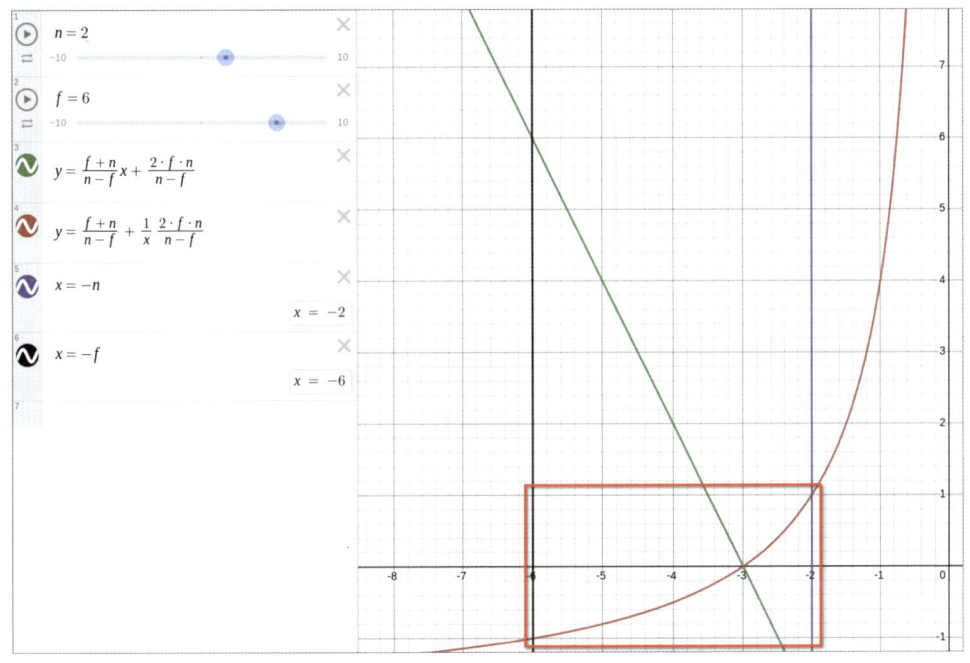

▲ 圖 3-16　z 軸透過 perspective 轉換的輸入輸出公式圖形

此圖中 x 軸為 A_z 做為輸入、`n = 2` 表示 near 為 2、`f = 6` 表示 far 為 6、綠色公式及畫出來的線是經過 M 轉換過的 P_z、紅色公式及畫出來的線則是再除以 w 轉的 P'_z，可以看到除了把 near、far 轉換到 -1 ~ +1 之外（紅色框框區域），還得到一個很重要的特性：接近 near 的物件，可以獲得較大範圍的深度範圍，意思是說接近鏡頭的物件前後分別會更精準！

走過這些運算流程以後，大概知道了 perspective 轉換矩陣的實做原理以及特性，依照矩陣 M 的產生公式實做到 `lib/matrix.js` 內：

```
export const matrix4 = {
  // ...
```

3　3D & 物件

```
perspective: (fieldOfView, aspect, near, far) => {
  const f = 1 / Math.tan(fieldOfView / 2);
  const rangeInv = 1.0 / (near - far);
  return [
    f / aspect, 0, 0, 0,
    0, f, 0, 0,
    0, 0, (near + far) * rangeInv, -1,
    0, 0, far * near * rangeInv * 2, 0,
  ]
},

// ...
}
```

把公式寫成程式時，除了行列要調換成符合電腦程式之外，可以看到公式中有重複、類似的計算，在 `perspective` function 內先計算好 f 以及 `rangeInv` 以重複利用計算結果。

⌘ 使用 perspective 轉換成像

來到主程式，把 `viewMatrix` 從原本 `matrix4.projection()` 改成 `matrix4.perspective()`，`fieldOfView` 先給上 45 度：

```
function render(app) {
  // ...

  const viewMatrix = matrix4.projection(
    gl.canvas.width,
    gl.canvas.height,
    state.projectionZ
  );
  const viewMatrix = matrix4.perspective(
    degToRad(45),
    gl.canvas.width / gl.canvas.height, // aspect
    0.1, // near
    2000, // far
  );
```

3-30

```
  // ...
}
```

存檔重整後會看到一片慘白，沒有錯誤，但是就是沒東西。以現在來說，因為 `matrix4.perspective()` 是從原點出發向著 -z 的方向看，而當初規劃 3D 模組的 z 軸是往 +z 的方向長的，更別說頂點時是用螢幕 pixel 為單位製作的，現在看不到東西其實很正常；要看到東西，就得『移動視角』讓物件在 frustum 內。

在這之前我們可以把之前的 `projectionZ` 控制改成對於 `fieldOfView` 的控制，一樣先建立初始狀態：

```javascript
async function setup() {
  // ...
  return {
    // ...
    state: {
      projectionZ: 400,
      fieldOfView: degToRad(45),
      translate: [150, 100, 0],
      rotate: [degToRad(30), degToRad(30), degToRad(0)],
      scale: [1, 1, 1],
    },
    time: 0,
  };
}
```

修改控制項 HTML，使得初始值一樣為 45 度，並且最大值設定為 180 度：

```html
<body>
  <canvas id="canvas"></canvas>
  <form id="controls">
    <div class="py-1">
      <label for="field-of-view">fieldOfView</label>
      <input
```

```html
        type="range" id="field-of-view" name="field-of-view"
        min="0" max="180" value="45"
      >

      <!-- ... -->
    </div>
  </form>
  <!-- ... -->
</body>
```

將控制項串到狀態上：

```javascript
async function main() {
  // ...

  const controlsForm = document.getElementById('controls');
  controlsForm.addEventListener('input', () => {
    const formData = new FormData(controlsForm);

    app.state.projectionZ = parseFloat(formData.get('projection-z'));
    app.state.fieldOfView =
      degToRad(parseFloat(formData.get('field-of-view')));

    // app.state.translate[0] = ...
    // ...

    render(app);
  });

  // startLoop(app);
  render(app);
}
```

最後在渲染階段使用狀態 state.fieldOfView：

```javascript
function render(app) {
  // ...

  const viewMatrix = matrix4.perspective(
```

3-32

```
  state.fieldOfView,
  gl.canvas.width / gl.canvas.height, // aspect
  0.1, // near
  2000, // far
);

// ...
}
```

串上後，在瀏覽器上把 fieldOfView 調整到接近約 160 度以上（最大為 180 度），就可以看到一些形狀的出現，如圖 3-17：

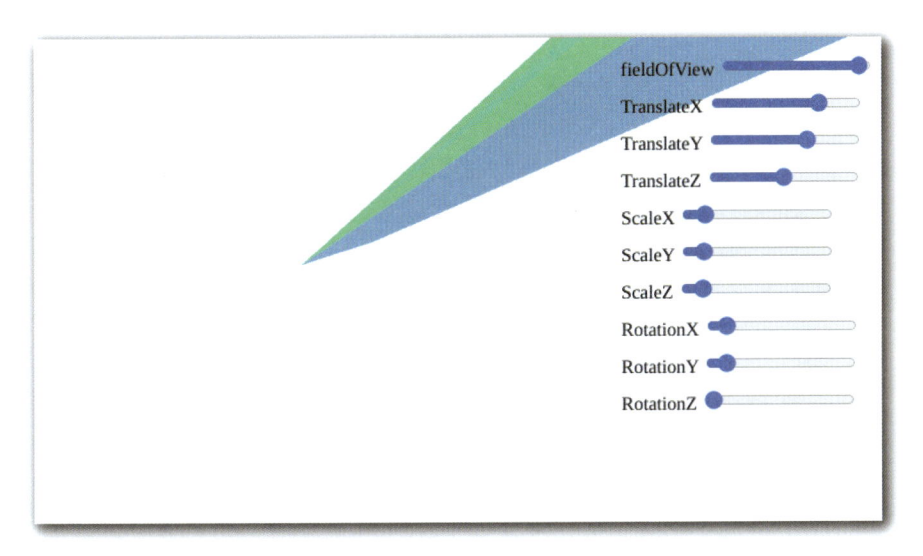

▲ 圖 3-17　改用 perspective 後要把 fieldOfView 拉到很大才能看到東西

到此進度的完整程式碼可以在這邊找到：

</>	ch3/02/index.html
</>	ch3/02/main.js

也可以參考上線版本，透過瀏覽器打開此頁面：
https://webgl-book.pastleo.me/ch3/02/index.html

3-3 視角 Transform

有了 `matrix4.perspective()` 使用眼睛 / 相機的方式進行成像，反而讓畫面變成一片慘白，假設把 3D 物件本身的 transform 取消，也就是 `worldMatrix` 設定成 `matrix4.identity()`，那麼 3D 物件與 frustum 區域的相對關係看起來像是圖 3-18 所示：

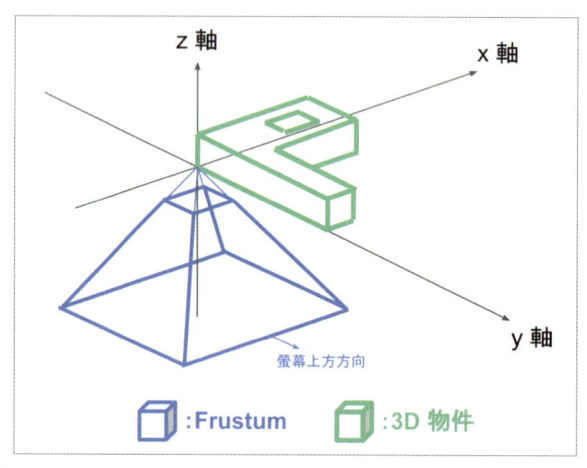

▲ 圖 3-18　clip space 回推形成的 frustum 與 P 物件相對位置

3D 物件不在 frustum 中，因此什麼都看不到，我們當然可以直接把 3D 物件做 translate 移動到 frustum 中，但是在現實生活中，如果架設好了一個場景，放了很多物件，而相機的位置不對，這時候我們會移動的是相機，而不是整個場景，這就是接下來要做的事情。

⌘ `viewMatrix` 與視角 transform

先前在製作作用在 `a_position` 上的 `u_matrix` 之前，會先產生兩個矩陣相乘：`viewMatrix` 與 `worldMatrix`，繪製多個物件時 `viewMatrix` 為同一顆相機 / 畫面下的所有物件共用，`worldMatrix` 則為物件本身的 transform，會因為不同物件而異；接下來要加入的視角 transform 想當然爾為同一顆相機或畫面下的

3-3 視角 Transform

所有物件共用,因此 viewMatrix 除了 clip space 的 transform 之外,也要開始包含視角相關的 transform,成為名符其實的 viewMatrix。

雖然筆者才剛說我們不應該移動整個場景來符合相機位置(就放在 viewMatrix 這部份的抽象來說確實像是移動相機本身),但是視角 transform 本身能做的事情就是移動場景,所有的 a_position、物件都會經過視角 transform:

```
clip space <= perspective <= 視角 transform <= worldMatrix <= a_position
              ^...viewMatrix transform...^
```

假設我們想要把相機放在如圖 3-19 的位置:

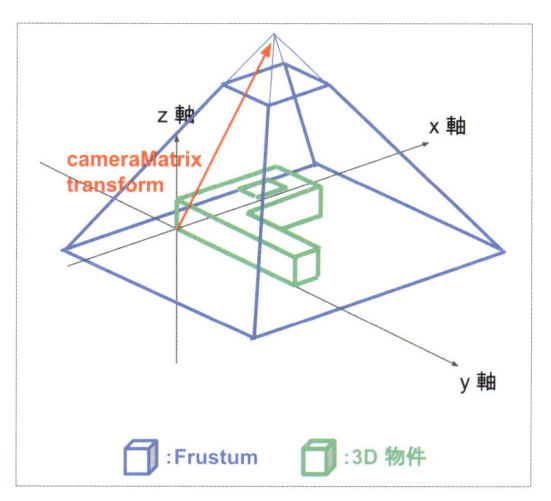

▲ 圖 3-19　相機 transform 示意圖

把移動相機的 transform 叫做 cameraMatrix ,因為視角 transform 只能移動整個場景,所以視角 transform 可以當成『反向做 cameraMatrix』,對整個場景做反向的 cameraMatrix,在定義 cameraMatrix 這件事情上,就真的抽象成移動相機了;反向這件事可以靠反矩陣(inverse matrix)[10] 來做到,什麼是

10　或稱為逆矩陣:https://zh.wikipedia.org/wiki/逆矩陣

3-35

3　3D & 物件

反矩陣？[11] 假設有一個轉換矩陣 M，一個向量 v，如果 M 的秩（Rank）與維度相同（也就是 Full Rank）[12]，那麼就可以找到一個 M^{-1} 表示 M 的反矩陣，這時 $M^{-1}Mv = v$，幾何意義上是說對於 v 做 M 轉換後，再做 M^{-1} 轉換會變成原本的 v，M^{-1} 也就是『反向做 M』，舉例來說，如果 M 是乘以 4 進行放大，那麼 M^{-1} 就是乘以 0.25 進行縮小、如果 M 是順時鐘轉 45 度，那麼 M^{-1} 就是逆時鐘轉 45 度。

以實際作用在 a_position 向量上的視角 transform（等於 `inverse(cameraMatrix)`）來說，看起來像是這樣：

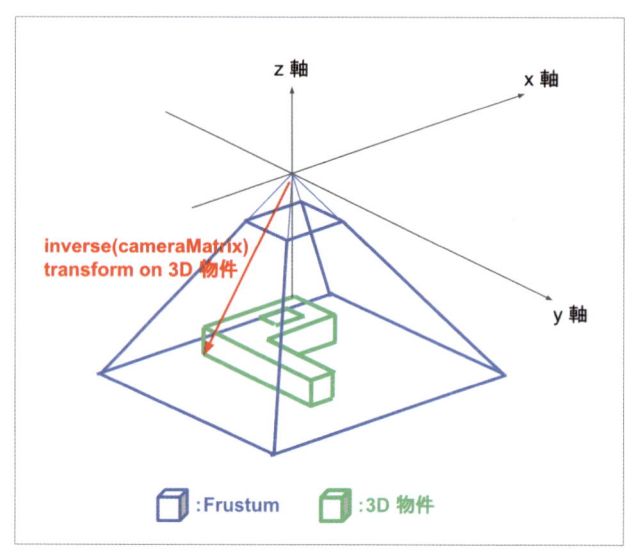

▲ 圖 3-20　反向的 cameraMatrix 示意圖

了解原理後，得在 `lib/matrix.js` 中加入 `matrix4.inverse()` 的實做，不過，有做過三階反矩陣運算就會知道計算量不小，更何況我們需要的是 4x4

11　可以參考 3Blue1Brown 的 Youtube 影片—反矩陣、行空間與零空間：
　　https://youtu.be/uQhTuRlWMxw
12　幾何意義上來說，這樣的矩陣可以轉換出來的空間維度與矩陣自己的維度相同，如果一個 3x3 矩陣的所有可能轉換結果維持為三維空間，不會塌陷成平面或是一個點，那麼此矩陣符合條件

3-3 視角 Transform

四階,寫成程式碼行數實在不少,筆者就不直接放在文章中了,有需要可以在完整程式碼中找到。

⌘ 實做視角 transform 到 viewMatrix 內

要實做圖 3-19 紅色箭頭的 `cameraMatrix`,用 `matrix4.translate()` 就足夠,因此在主程式 `render()` 內定義 viewMatrix 之前加上:

```javascript
function render(app) {
  // ...

  const cameraMatrix = matrix4.translate(250, 0, 400);

  // viewMatrix, worldMatrix...
}
```

並且像是上面說的,使用 `matrix4.inverse(cameraMatrix)` 加入 `viewMatrix`:

```javascript
function render(app) {
  // ...

  const cameraMatrix = matrix4.translate(250, 0, 400);
  const viewMatrix = matrix4.multiply(
    matrix4.perspective(
      state.fieldOfView,
      gl.canvas.width / gl.canvas.height, // aspect
      0.1, // near
      2000, // far
    ),
    matrix4.inverse(cameraMatrix),
  );

  // worldMatrix...
}
```

成像結果如圖 3-21:

3-37

3　3D & 物件

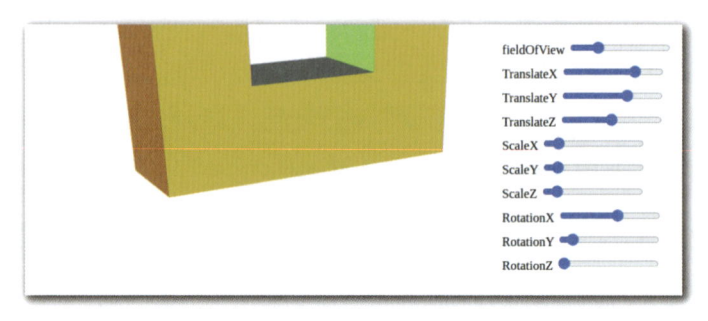

▲ 圖 3-21　成像結果

　　這個 P 上下顛倒了，而且現在現在看到的不是正面，在建構模型時，除了其背面往 +z 長之外，我們使用 2D 時慣用的 y 軸正向為螢幕下方方向，這些與 3D 中使用的慣例都是相反的，同時也可以看本篇第一張圖中標示的『螢幕上方方向』想像看到的畫面；如果要重新定位 P 形狀的 3D 模型實在是太累，因此筆者選擇修改 x 軸的旋轉的預設值轉 210 度過去：

```javascript
async function setup() {
  // ...
  return {
    gl,
    program, attributes, uniforms,
    buffers, modelBufferArrays,
    state: {
      fieldOfView: degToRad(45),
      translate: [150, 100, 0],
      rotate: [degToRad(210), degToRad(30), degToRad(0)],
      scale: [1, 1, 1],
    },
    time: 0,
  };
}
```

　　HTML 拉桿的初始值也改一下：

```html
<body>
  <canvas id="canvas"></canvas>
```

3-3 視角 Transform

```html
<form id="controls">
  <!-- ... -->
  <div class="py-1">
    <label for="rotation-x">RotationX</label>
    <input
      type="range" id="rotation-x" name="rotation-x"
      min="0" max="360" value="210"
    >
  </div>
  <!-- ... -->
</form>
<script type="module" src="main.js"></script>
</body>
```

此 P 形狀的 3D 模型就在 perspective 成像下完整呈現了，如圖 3-22：

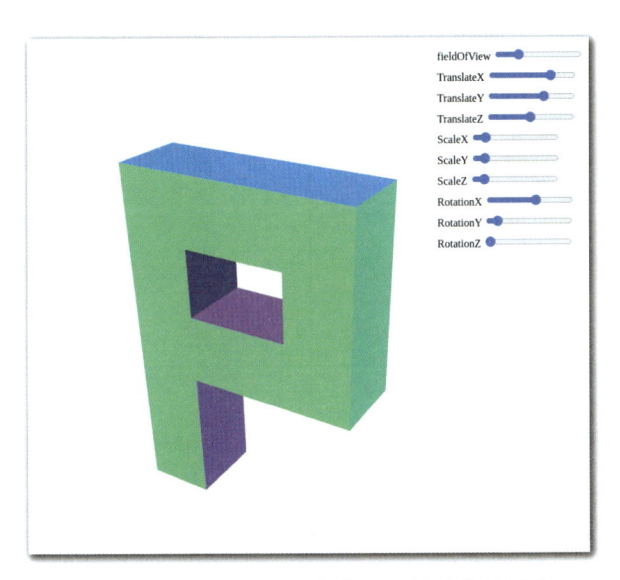

▲ 圖 3-22　旋轉後 P 完整呈現於成像結果中

到此進度的完整程式碼可以在這邊找到：

</>	ch3/03/index.html
</>	ch3/03/main.js

3-39

3 3D & 物件

也可以參考上線版本，透過瀏覽器打開此頁面：
https://webgl-book.pastleo.me/ch3/03/index.html

有了 perspective 投影以及加入反向 `cameraMatrix` 的 `viewMatrix`，我們擁有一套系統來模擬現實生活中眼睛、相機在想要的位置進行成像，方法也跟場景中的 3D 物件類似：在 `cameraMatrix` 中加入想要的 transform。

同時也可以來比較一下 orthogonal 與 perspective 投影的差別，最大的差別大概就是物件在不同 z 軸位置時成像的『遠近』了，在 orthogonal 投影下，距離投影面的遠近不會影響物件的大小，如圖組 3-23：

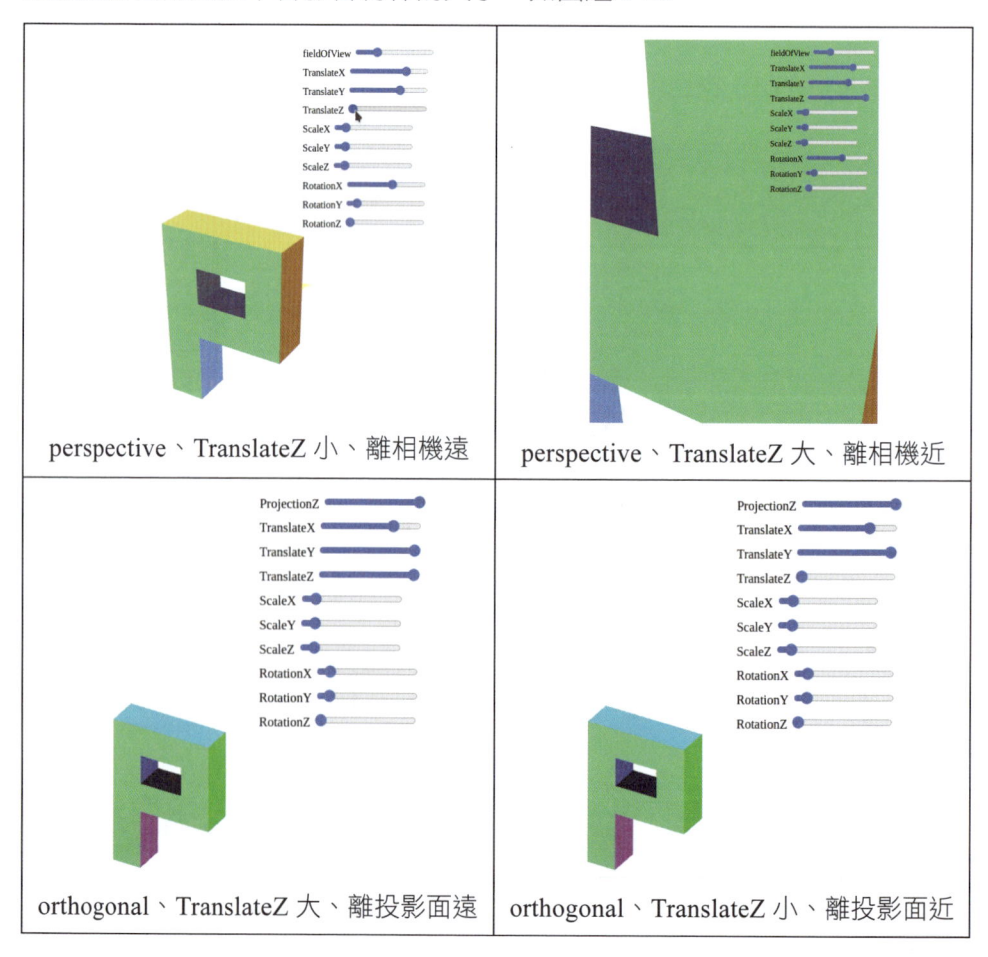

perspective、TranslateZ 小、離相機遠	perspective、TranslateZ 大、離相機近
orthogonal、TranslateZ 大、離投影面遠	orthogonal、TranslateZ 小、離投影面近

▲ 圖組 3-23　orthogonal 與 perspective 的投影差別

3-4 使相機看著目標

相機往往不會直直地往 -z 方向看,而且常常要對著某個目標,因此再介紹一個常用的 function:

⌘ `matrix4.lookAt()`

是產生一個矩陣的 function,且顧名思義做轉換使得相機『看著』一個目標,且也可以用於使物體的 -z 對著一個目標:

```
matrix4.lookAt(
  cameraPosition,
  target,
  up,
)
```

其前兩個參數意義蠻明顯的,分別是相機要放在什麼位置、看著的目標;接著不知道讀者在閱讀本書時,有沒有常常歪著頭看?對,`up` 就是控制這件事情,如果傳入 `[0, 1, 0]` 即表示正正的看,沒有歪著頭看。

關於 `matrix4.lookAt()` 的實做,可以想像到會有 `cameraPosition` 的平移,因此矩陣的一部分已經知道:

```
[
  ?, ?, ?, 0,
  ?, ?, ?, 0,
  ?, ?, ?, 0,
  cameraPosition.x, cameraPosition.y, cameraPosition.z, 1,
]
```

剩下的?部份則是相機的方向,首先需要知道從 `cameraPosition` 到 `target` 的方向向量 `kHat`,接著拿 `up` 與 `kHat` 向量做外積得到與兩者都垂直的向量 `iHat`,最後拿 `kHat` 與 `iHat` 做外積得到與兩者都垂直的向量 `jHat`,我們就得到變換矩陣的『基本向量』;當有一個 3x3 矩陣,其中會包含 3 個 3 維向量，分別稱為 i、j、k，這就是所謂的『基本向量』[13]，此 3x3 矩陣做轉換時就像是把一個空間的 x、y、z 軸 +1 方向轉變成 i、j、k，被轉換的向量放在空間中一起旋轉、縮放，結束之後產生的新向量即為轉換的結果，因此只要將上面算出互相垂直的 `iHat`、`jHat`、`kHat` 放好在矩陣中，就可以讓一開始看著 -z 的相機旋轉成 `kHat`、+y 旋轉成 `jHat`、+x 旋轉成 `iHat`。

因為我們只需要旋轉，不用縮放，要單位向量化（normalize）確保 i、j、k 都為單位向量[14]。

在上面這段提到三個在工具箱中尚未實做的運算：向量差異、外積、單位向量化，先根據公式在 `lib/matrix.js` 中實做這幾個 function。

```javascript
export const matrix4 = {
  // ...
  subtractVectors: (a, b) => ([
    a[0] - b[0], a[1] - b[1], a[2] - b[2]
  ]),
  cross: (a, b) => ([
    a[1] * b[2] - a[2] * b[1],
    a[2] * b[0] - a[0] * b[2],
    a[0] * b[1] - a[1] * b[0],
  ]),
  normalize: v => {
    const length = Math.sqrt(v[0] * v[0] + v[1] * v[1] + v[2] * v[2]);
    if (length > 0.00001) { // 確保不會除以零
      return [v[0] / length, v[1] / length, v[2] / length];
    } else {
      return [0, 0, 0];
    }
  },
```

13　關於三維線性轉換與基本向量（矢量），3Blue1Brown 的 Youtube 有更清楚明白的動畫做解說：https://youtu.be/rHLEWRxRGiM
14　向量長度為 1 的向量稱為單位向量：https://zh.wikipedia.org/zh-tw/單位向量

3-4 使相機看著目標

```
  // ...
};
```

最後把 `matrix4.lookAt()` 實做起來：

```
export const matrix4 = {
  // ...

  lookAt: (cameraPosition, target, up) => {
    // +z 將轉換為 kHat，計算由 cameraPosition 到 target 的『反』方向：
    const kHat = matrix4.normalize(
      matrix4.subtractVectors(cameraPosition, target)
    );
    // 拿 up 與 kHat 向量做外積得到與兩者都垂直的向量 iHat：
    const iHat = matrix4.normalize(matrix4.cross(up, kHat));
    // 拿 kHat 與 iHat 做外積得到與兩者都垂直的向量 jHat：
    const jHat = matrix4.normalize(matrix4.cross(kHat, iHat));

    // 擺放 iHat, jHat, kHat 以及相機平移到矩陣中：
    return [
      iHat[0], iHat[1], iHat[2], 0,
      jHat[0], jHat[1], jHat[2], 0,
      kHat[0], kHat[1], kHat[2], 0,
      cameraPosition[0],
      cameraPosition[1],
      cameraPosition[2],
      1,
    ];
  },

  // ...
};
```

讀者們可能會有一個疑問：最核心的相機方向 `kHat` 為何是由 `cameraPosition` 到 `target` 的『反』方向？因為 `matrix4.perspective()` 會讓相機看著的方向是『負』z 方向，而基本向量會從『正』z 轉換成 `kHat`，與其在 `matrix4.subtractVectors()` 之後乘上 -1，不如直接計算『反』方向以符合相機初始的『負』z 方向。

3-43

3 3D & 物件

⌘ 使用 matrix4.lookAt()

回到主程式，將 cameraMatrix 改用 matrix4.lookAt() 產生的矩陣，從同樣的位置 [250, 0, 400] 的位置看向 3D 物件，精確來說是看向 3D 模型平移後正面左上角的定點位置：

```
function render(app) {
  // ...
  const cameraMatrix = matrix4.translate(250, 0, 400);
  const cameraMatrix = matrix4.lookAt(
    [250, 0, 400],
    state.translate,
    [0, 1, 0],
  );
  // viewMatrix ...
}
```

存檔後，試著調整 TranslateX、TranslateY 以及 TranslateZ，確認相機會一直看著 3D 物件，如圖組 3-24 與 圖組 3-25，讀者可以嘗試感受一下物件與相機的相對位置與視角：

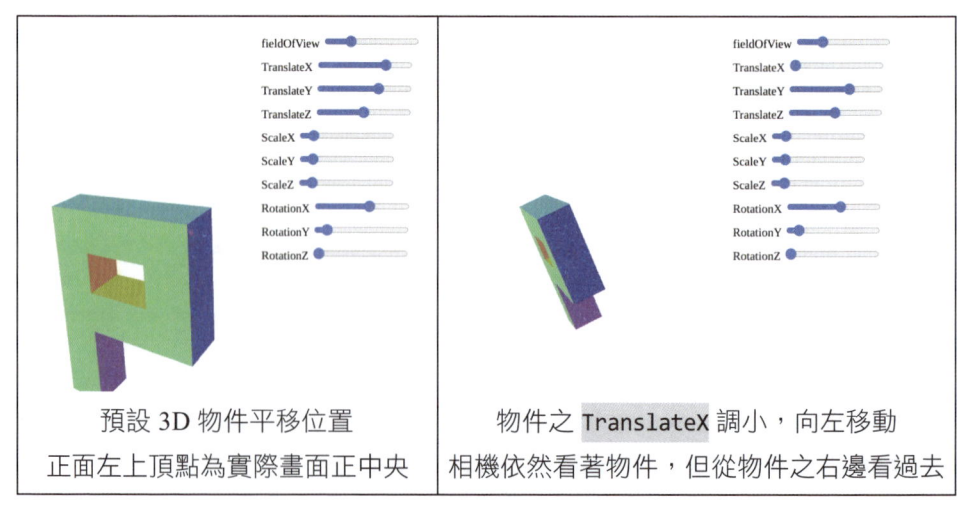

預設 3D 物件平移位置	物件之 TranslateX 調小，向左移動
正面左上頂點為實際畫面正中央	相機依然看著物件，但從物件之右邊看過去

▲ 圖組 3-24　測試 matrix4.lookAt() 看著物件之效果（一）

3-44

3-4　使相機看著目標

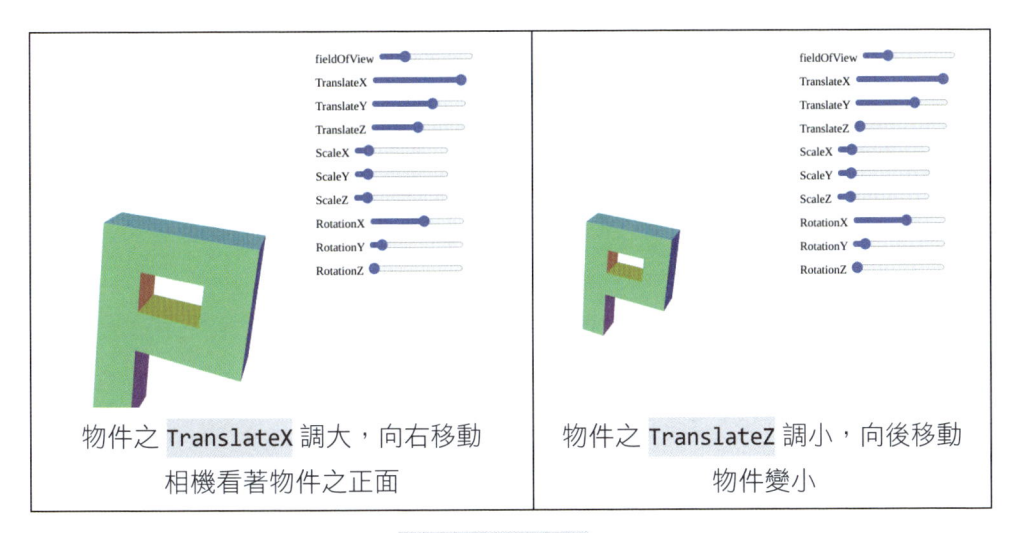

物件之 `TranslateX` 調大，向右移動
相機看著物件之正面

物件之 `TranslateZ` 調小，向後移動
物件變小

▲ 圖組 3-25　測試 `matrix4.lookAt()` 看著物件之效果（二）

⌘ 移動相機

接下來加入相機位置的控制，不過這次不要再用 `<input type='range' />` 的滑桿了，我們來使用鍵盤 WASD 或上下左右、用滑鼠或觸控按住畫面上下左右半部來移動相機。

因為要接入的事件很多，而且這些事件都是按下開始移動，放開時候停止，因此在接收到事件時設定相機的速度，再由 `requestAnimationFrame` 的迴圈來進行相機位置的更新以及重新渲染，重拾前面章節所提到的動畫功能，我們加上這兩個狀態：

```
async function setup() {
  // ...
  return {
    gl,
    program, attributes, uniforms,
    buffers, modelBufferArrays,
    state: {
      fieldOfView: degToRad(45),
      translate: [150, 100, 0],
```

3-45

3　3D & 物件

```
      rotate: [degToRad(210), degToRad(30), degToRad(0)],
      scale: [1, 1, 1],
      cameraPosition: [250, 0, 400],
      cameraVelocity: [0, 0, 0],
    },
    time: 0,
  };
}
```

在 `render()` 中讓 `matrix4.lookAt()` 串上剛建立的狀態，相機位置從寫死的 `[250, 0, 400]` 改成可變動的狀態：

```
function render(app) {
  // ...

  const cameraMatrix = matrix4.lookAt(
    state.cameraPosition,
    state.translate,
    [0, 1, 0],
  );
  // viewMatrix ...
}
```

啟用、取消註解 startLoop，使用 cameraVelocity 來更新 cameraPosition：

```
function startLoop(app, now = 0) {
  const timeDiff = now - app.time;
  app.time = now;

  app.state.cameraPosition[0] += app.state.cameraVelocity[0] * timeDiff;
  app.state.cameraPosition[1] += app.state.cameraVelocity[1] * timeDiff;
  app.state.cameraPosition[2] += app.state.cameraVelocity[2] * timeDiff;

  render(app, timeDiff);
  requestAnimationFrame(now => startLoop(app, now));
}
```

移除原本的在初始時以及有 form 事件更新狀態時才重新渲染，改呼叫

3-46

startLoop() 啟動動畫：

```javascript
async function main() {
  // ...

  const controlsForm = document.getElementById('controls');
  controlsForm.addEventListener('input', () => {
    // update app.state...

    render(app);
  });

  startLoop(app);
  render(app);
}
```

動畫設定完成，再把事件接起來就可以了，我們要接上的事件有鍵盤的 `keydown`、`keyup`；滑鼠的 `mousedown`、`mouseup`；觸控螢幕的 `touchstart`、`touchend`，在 main() 初始化時進行事件監聽：

```javascript
async function main() {
  // setup and controlsForm ...
  app.gl.canvas.addEventListener('mousedown', event => {
    handlePointerDown(app, event);
  });
  app.gl.canvas.addEventListener('mouseup', () => {
    handlePointerUp(app);
  });
  app.gl.canvas.addEventListener('touchstart', event => {
    handlePointerDown(app, event.touches[0]);
  });
  app.gl.canvas.addEventListener('touchend', () => {
    handlePointerUp(app);
  });
  document.addEventListener('keydown', event => {
    handleKeyDown(app, event);
  });
  document.addEventListener('keyup', event => {
```

3-47

```
    handleKeyUp(app, event);
  });

  startLoop(app);
}
```

由於滑鼠以及觸控事件都是指標事件，因此 mousedown、touchstart 共用同一個 function handlePointerDown 來處裡事件：

```
function handlePointerDown(app, touchOrMouseEvent) {
  // 使 x, y 以畫面中央為原點
  const x = touchOrMouseEvent.pageX - app.gl.canvas.width / 2;
  const y = touchOrMouseEvent.pageY - app.gl.canvas.height / 2;
  if (x * x > y * y) {
    // x 絕對值大於 y，向左或向右移動相機
    if (x > 0) {
      app.state.cameraVelocity[0] = 0.5;
    } else {
      app.state.cameraVelocity[0] = -0.5;
    }
  } else {
    // y 絕對值大於 x，向上或向下移動相機
    if (y < 0) {
      app.state.cameraVelocity[1] = 0.5;
    } else {
      app.state.cameraVelocity[1] = -0.5;
    }
  }
}
```

而 mouseup、touchend 共用同一個 function handlePointerUp，收到這些事件時重設相機速度：

```
function handlePointerUp(app) {
  app.state.cameraVelocity[0] = 0;
  app.state.cameraVelocity[1] = 0;
  app.state.cameraVelocity[2] = 0;
}
```

用來處裡按下鍵盤按鍵事件的 function `handleKeyDown` 需要查看按下、放開的按鍵是什麼去修改相機移動速度，來達到讓使用者用鍵盤 WASD 或上下左右控制相機之位置：

```javascript
function handleKeyDown(app, event) {
 switch (event.code) {
   case 'KeyA':
   case 'ArrowLeft':
     app.state.cameraVelocity[0] = -0.5;
     break;
   case 'KeyD':
   case 'ArrowRight':
     app.state.cameraVelocity[0] = 0.5;
     break;
   case 'KeyW':
   case 'ArrowUp':
     app.state.cameraVelocity[1] = 0.5;
     break;
   case 'KeyS':
   case 'ArrowDown':
     app.state.cameraVelocity[1] = -0.5;
     break;
 }
}
```

同樣地需要 `handleKeyUp` function 看放開的按鈕來決定重設 x 還是 y 方向的移動：

```javascript
function handleKeyUp(app, event) {
  switch (event.code) {
    case 'KeyA':
    case 'ArrowLeft':
    case 'KeyD':
    case 'ArrowRight':
      app.state.cameraVelocity[0] = 0;
      break;
    case 'KeyW':
    case 'ArrowUp':
```

```
      case 'KeyS':
      case 'ArrowDown':
        app.state.cameraVelocity[1] = 0;
        break;
    }
}
```

這麼一來鍵盤、滑鼠、觸控螢幕的事件控制就都接起來了，為了方便觀察相機位置，最後簡單地用 HTML 文字顯示出相機的位置，先在 HTML 加上顯示用的文字元素：

```
<body>
  <canvas id="canvas"></canvas>
  <form id="controls">
    <p>cameraPosition: [
      <span id="cameraPositionX">0</span>,
      <span id="cameraPositionY">0</span>,
      <span id="cameraPositionZ">0</span>]
    </p>
    <!-- ... -->
  </form>
</body>
```

並且在更新 `app.state.cameraPosition` 之後將其值透過 `.toFixed()` 把浮點數轉換成字串指定到文字元素內容中：

```javascript
function startLoop(app, now = 0) {
  // update cameraPosition...
  document.getElementById("cameraPositionX").textContent =
    app.state.cameraPosition[0].toFixed(2);
  document.getElementById("cameraPositionY").textContent =
    app.state.cameraPosition[1].toFixed(2);
  document.getElementById("cameraPositionZ").textContent =
    app.state.cameraPosition[2].toFixed(2);

  // ...
}
```

整個串完，回到瀏覽器就可以用使用鍵盤 WASD 或上下左右、用滑鼠或觸控按住畫面上下左右半部直覺地在 xy 平面上移動相機囉，用起來如圖 3-26、圖 3-27：

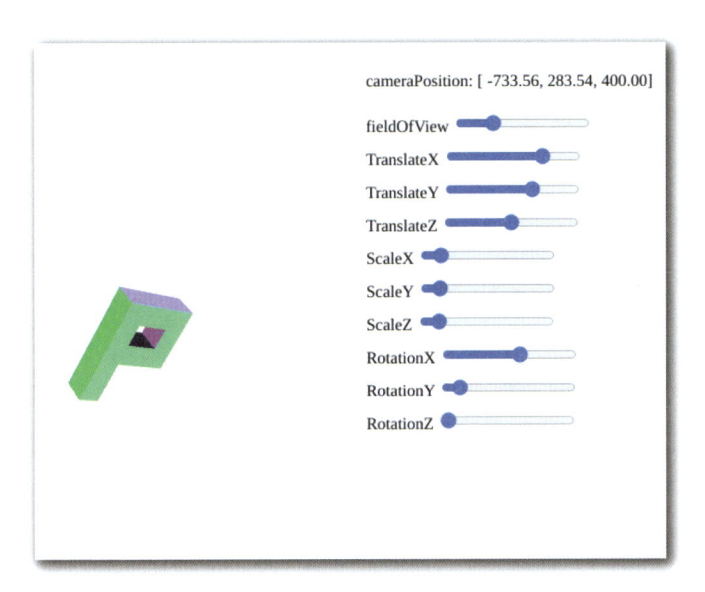

▲ 圖 3-26　將相機往左上方移動

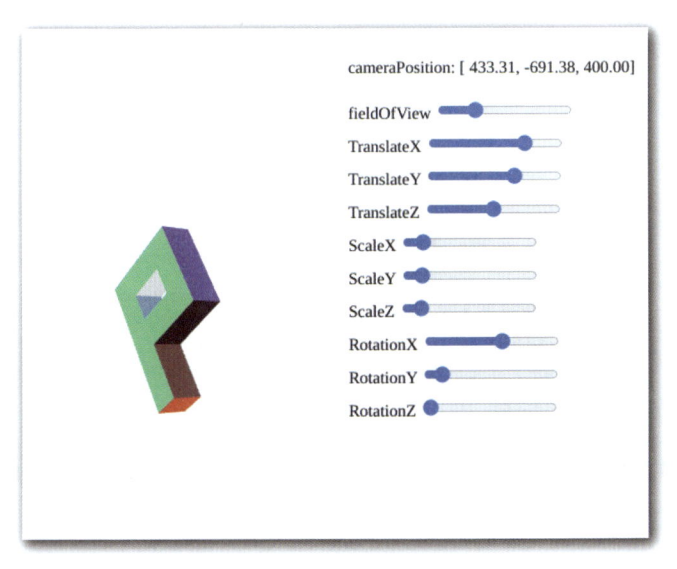

▲ 圖 3-27　將相機往右下方移動

3-51

3 3D & 物件

到此進度的完整程式碼可以在這邊找到：

| </> | ch3/04/index.html |
| </> | ch3/04/main.js |

也可以參考上線版本，透過瀏覽器打開此頁面：
https://webgl-book.pastleo.me/ch3/04/index.html

3-5 渲染多個物件

本書到目前的範例，畫面上都只有一個物件，既然已經介紹完 3D 物件的產生、在空間中的 transform、相機控制以及 perspective 投影到畫布上，接下來來讓所謂『場景』比較有場景的感覺，加入一顆球體以及地板，如圖 3-28：

▲ 圖 3-28　接下來的目標：在畫面中渲染多個物件

3-52

⌘ 重構程式碼使得加入多物件變得容易

之前在準備 P 字母 3D 模型資料準備的時候，分別對 `a_position`、`a_color` 製作了 attributes 資料、vertexAttribArray 以及 buffer，這些都屬於 P 形狀這個『物件』的內容，接下來要加入其他物件，因此建立一個 objects 變數來存這些物件『們』，在 `objects` 下每個物件自己再有一個 Javascript object 來存放 attributes、vertexAttribArray 以及 buffer 等資訊：

```javascript
async function setup() {
  // program, attributes, uniforms ...

  // 存放 3D 物件『們』的容器
  const objects = {};

  // ...
}
```

對於原本 P 形狀 3D 物件，我們將建立 `objects.pModel`，首先把原本 `modelBufferArrays.vertexDataArrays`、`modelBufferArrays.numElements`、`buffers` 的建立流程放置到一個花括弧 {} 內，形成一個明顯的區塊[15]，同時對於 let、const 這類現今較為推薦的變數建立語法有變數區域的功能：

```javascript
async function setup() {
  // program, attributes, uniforms ...

  // 存放 3D 物件『們』的容器
  const objects = {};

  { // pModel, P 形狀的物件 初始化
    // 對 modelBufferArrays 解構，取出 vertexDataArrays 以及 numElements
    const { vertexDataArrays, numElements } = createModelBufferArrays();
    const buffers = {}; // pModel 專用的 buffers

    // a_position
```

---

15 https://developer.mozilla.org/zh-TW/docs/Web/JavaScript/Reference/Statements/block

```
    buffers.position = gl.createBuffer();
    gl.bindBuffer(gl.ARRAY_BUFFER, buffers.position);

    gl.enableVertexAttribArray(attributes.position);
    gl.vertexAttribPointer(
      attributes.position,
      // size, type, normalize, stride, offset ...
    );

    gl.bufferData(
      gl.ARRAY_BUFFER,
      new Float32Array(vertexDataArrays.a_position),
      gl.STATIC_DRAW,
    );

    // a_color
    buffers.color = gl.createBuffer();
    gl.bindBuffer(gl.ARRAY_BUFFER, buffers.color);

    gl.enableVertexAttribArray(attributes.color);
    gl.vertexAttribPointer(
      attributes.color,
      // size, type, normalize, stride, offset ...
    );

    gl.bufferData(
      gl.ARRAY_BUFFER,
      new Float32Array(vertexDataArrays.a_color),
      gl.STATIC_DRAW,
    );
  }

  // ...
}
```

　　這邊基本上是把之前『畫什麼』的程式移動整理到專屬於 P 形狀物件的區域，其中 `gl.vertexAttribPointer` 設定 attribute 拿資料的方法有進行省略，整理的過程中讀者們可以順便複習一下，如果覺得有需要可以參考第一章『畫什麼』的詳細說明；`attributes`、`buffer` 建立綁定完成之後，我們把在此建立

3-5 渲染多個物件

對於此 P 形狀物件的 `vertexDataArrays`（頂點資料陣列，包含 `a_postion`、`a_color`）、`numElements` 以及 `buffers` 存放到 `objects.pModel` 中：

```javascript
async function setup() {
  // ...
  // 存放 3D 物件『們』的容器
  const objects = {};

  { // pModel, P 形狀的物件 初始化
    // 對 modelBufferArrays 解構，取出 vertexDataArrays 以及 numElements
    const { vertexDataArrays, numElements } = createModelBufferArrays();
    const buffers = {}; // pModel 專用的 buffers

    // a_position, a_color 與 buffer 的綁定 ...

    objects.pModel = {
      vertexDataArrays, numElements,
      buffers,
    };
  }

  // ...
}
```

整理到 P 形狀物件的花括弧 `{}` 區域內之後，記得把花括弧 `{}` 區域外原本用來建立 `modelBufferArrays`、`buffers` 以及 attribute `a_position`、`a_color` 之綁定程式移除。

在 `setup()` 回傳之 `app` 內的資料結構改放 `objects`：

```javascript
async function setup() {
  // objects, objects.pModel …

  return {
    gl,
    program, attributes, uniforms,
```

3-55

3 3D & 物件

```
    buffers, modelBufferArrays,
    objects,
    state: {
      // fieldOfView, translate ...
    },
    time: 0,
  };
}
```

對應的 `setup()` 改接收 `objects`：

```
function render(app) {
  const {
    gl,
    program, uniforms,
    modelBufferArrays,
    objects,
    state,
  } = app;

  // ...
}
```

最後修改讓原本使用 `modelBufferArrays` 的程式改從 `objects` 取用，並像是 `setup()` 一樣建立專屬於 P 物件的程式區塊，這時 `worldMatrix` 也應該放進來，此 transform 表示在 P 物件在世界中位置的轉換，相較之下 `viewMatrix` 屬於多個物件共用的 transform：

```
function render(app) {
  // cameraMatrix, viewMatrix ...

  { // pModel, P 形狀的物件 轉換、繪製
    const worldMatrix = matrix4.multiply(
      matrix4.translate(...state.translate),
      matrix4.xRotate(state.rotate[0]),
      matrix4.yRotate(state.rotate[1]),
      matrix4.zRotate(state.rotate[2]),
      matrix4.scale(...state.scale),
```

3-56

```
  );
  gl.uniformMatrix4fv(
    uniforms.matrix,
    false,
    matrix4.multiply(viewMatrix, worldMatrix),
  );

  gl.drawArrays(gl.TRIANGLES, 0, objects.pModel.numElements);
  }
}
```

這時存檔回到瀏覽器檢查是否一切運作正常，如果是那麼就表示重構完成，可以繼續進行往下一步囉！

⌘ 加入純色物件繪製功能

因為待會要加入的物件都是純色，因此不需要傳送 a_color 進去指定每個 vertex / 三角形的顏色，我們可以讓 fragment shader 接收 uniform u_color 來指定整個物件的顏色：

```
#version 300 es
precision highp float;

in vec3 v_color;
uniform vec3 u_color;
out vec4 outColor;

void main() {
  outColor = vec4(v_color + u_color, 1);
}
```

這邊直接讓兩種來源相加：v_color + u_color，因為在 attribute a_color 沒輸入的時候 RGB 三個 channel 都會是 0，因此 v_color 就會是 [0, 0, 0]，對於 a_color 有值的 P 物件來說，我們要做的就是在繪製 P 物件時把 u_color 設定成 [0, 0, 0]；同時要記得取得 uniform 的位置：

3　3D & 物件

```
async function setup() {
  // program ...

  const attributes = {
    position: gl.getAttribLocation(program, 'a_position'),
    color: gl.getAttribLocation(program, 'a_color'),
  };
  const uniforms = {
    matrix: gl.getUniformLocation(program, 'u_matrix'),
    color: gl.getUniformLocation(program, 'u_color'),
  };

  // objects ...
}
```

最後在 `render()` 繪製 P 物件時設定 `u_color` 設定成 `[0, 0, 0]`，如同上段所說：

```
function render(app) {
  { // pModel, P 形狀的物件 轉換、繪製
    // worldMatrix, gl.uniformMatrix4fv ...

    gl.uniform3f(uniforms.color, 0, 0, 0);

    gl.drawArrays(gl.TRIANGLES, 0, objects.pModel.numElements);
  }
}
```

存檔回到瀏覽器，此時 P 物件應該還是一樣沒有變化，到此進度的完整程式碼可以在這邊找到：

</>	ch3/05/index.html
</>	ch3/05/main.js

也可以參考上線版本，透過瀏覽器打開此頁面：

https://webgl-book.pastleo.me/ch3/05/index.html

雖然 P 物件沒有變化，不過讀者可以嘗試修改 u_color 輸入數值觀察整個 P 物件的顏色變化，例如輸入 [1, 0, 0]：

```
function render(app) {
  { // pModel, P 形狀的物件 轉換、繪製
    // worldMatrix, gl.uniformMatrix4fv ...

    gl.uniform3f(uniforms.color, 1, 0, 0);

    gl.drawArrays(gl.TRIANGLES, 0, objects.pModel.numElements);
  }
}
```

這樣會使得在 fragment shader 中輸出之紅色值 `v_color + u_color` 大於等於浮點數 `1.0`，也就是紅色全滿，看起來如圖 3-29，原本正面的綠色變成黃橘色：

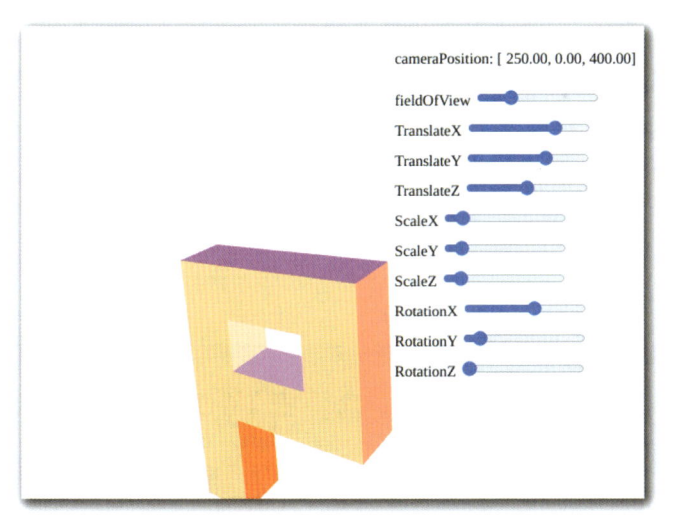

▲ 圖 3-29　u_color 輸入 [1, 0, 0]

3-59

3　3D & 物件

⌘ TWGL: WebGL 小幫手

我們接下來要產生球體以及地板所需的 `a_position`，也就是每個三角形各個頂點的位置，難道我們又要寫一長串程式來產生這些資料了嗎？如果是常用的形狀如球體、長方形、平面等網路上已經有許多寫好的程式可以用了，接下來我們要使用套件叫做：TWGL ─ A Tiny WebGL helper Library。

TWGL 的文件首頁：https://twgljs.org/docs/index.html

這個套件裡面不僅可以產生球體、平面等物件所需的資料，同時他也是一層對 WebGL 的薄薄包裝，讀者們應該也有感受到 WebGL API 的冗長，像是從 CH2 把 `createShader`、`createProgram` 放到 `lib/utils.js` 就是把細節做一層包裝，TWGL 提供了許多這樣的 function，從 program 到 vertex attribute、buffer 等操作都有，使得程式碼可以減少不少，在套件文件首頁上就有不少 WebGL API 以及 TWGL 的比較。

不過本篇就先只用到 `twgl.primitives` 來產生球體、平面物件的資料，引入這個套件有很多方法 [16]，因為我們已經在 ES module 中，筆者已經將其 v4.24.0 的 ES module 版本 [17] 放在範例程式碼包的這個位置讓各位讀者可以直接引入：

</> lib/vendor/twgl-full.module.js

使用 `import * as twgl from '...'` 把 TWGL 提供的所有 function 都引入並放在 `twgl` 變數上：

---

[16] 例如使用 npm、bower 或是下載放置 TWGL 到專案資料夾內：https://github.com/greggman/twgl.js#download

[17] https://github.com/greggman/twgl.js/blob/5360d4977ed33618878bf993f7e52ea105b956ad/dist/4.x/twgl-full.module.js

```
import { createShader, createProgram, degToRad } from '../../lib/utils.js';
import { matrix4 } from '../../lib/matrix.js';

import * as twgl from '../../lib/vendor/twgl-full.module.js';

window.twgl = twgl;
```

把引入的 `twgl` 變數綁定到 `window.twgl` 上，回到瀏覽器打開 Chrome 開發人員工具之 Console，輸入 `twgl` 可以透過自動提示看到 twgl 內有許多工具可以使用，如圖 3-30；而用來產生 3D 物件頂點資料的工具則是放在 `twgl.primitives` 中，在 Console 上可以看到琳瑯滿目的 function 可產生各式各樣形狀的 3D 物件，如圖 3-31。

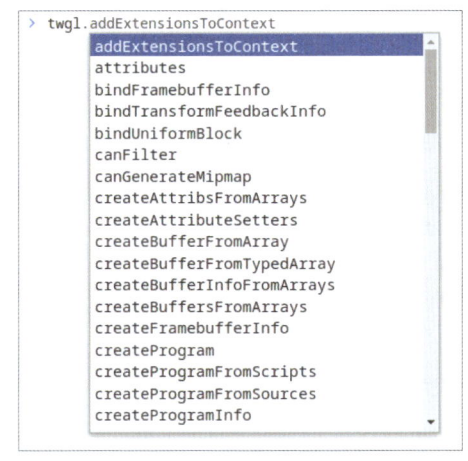

▲ 圖 3-30　自動提示列出之 twgl 內容　　▲ 圖 3-31　twgl.primitives 內容

⌘ 建立球體物件

球體的英文是 sphere，在 `twgl.primitives` 中可以找到 createSphereVertices[18] 這個 function：

`twgl.primitives.createSphereVertices(`

18　https://twgljs.org/docs/module-twgl_primitives.html#.createSphereVertices

3 3D & 物件

```
  radius,
  subdivisionsAxis,
  subdivisionsHeight,
)
```

第一個參數表示半徑、第二三個參數表示要分成多少個區段產生頂點，分越多這個球體就越精緻，我們先用這樣的設定印出來看看：

```
console.log(
  twgl.primitives.createSphereVertices(
    10, // radius
    32, // subdivisionsAxis
    32, // subdivisionsHeight
  )
);
```

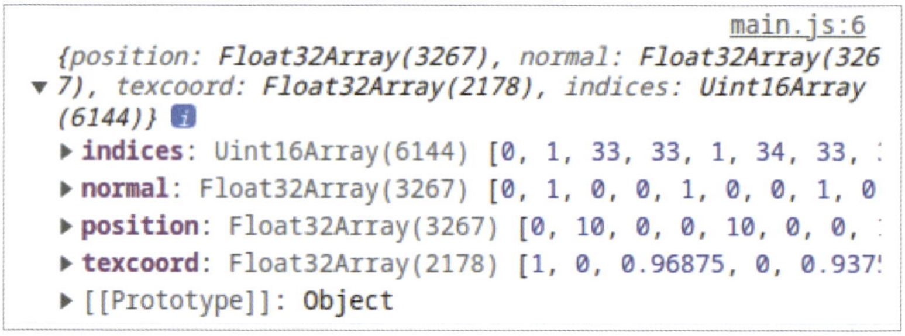

```
                                                              main.js:6
{position: Float32Array(3267), normal: Float32Array(326
7), texcoord: Float32Array(2178), indices: Uint16Array
(6144)}
  ▶ indices: Uint16Array(6144) [0, 1, 33, 33, 1, 34, 33, 
  ▶ normal: Float32Array(3267) [0, 1, 0, 0, 1, 0, 0, 1, 0
  ▶ position: Float32Array(3267) [0, 10, 0, 0, 10, 0, 0,
  ▶ texcoord: Float32Array(2178) [1, 0, 0.96875, 0, 0.937
  ▶ [[Prototype]]: Object
```

▲ 圖 3-32　twgl.primitives.createSphereVertices() 產生之球體資料

在圖 3-32 中可以看到產生的資料中有 `position`，這就是球體的點位置資料，是一個有 3267 個元素的浮點數陣列；同時還有 `texcoord` 取用 texture 時對應圖片中的位置、`normal` 法向量，那麼 `indices` 是什麼？在 WebGL 中還有一種 indexed element 模式使得繪製時透過 `indices` 類似指標去取得其指向之 vertex attribute 的值，可以避免重複的頂點資料，不過也因為這樣，這邊的 `position`、`texcoord`、`normal` 的資料排列方式必須使用 indexed element 模式透過 indices 取得頂點資料來渲染，與目前我們實做的讀取方式不同，然而我們目前也不打算使用 indexed element 模式進行繪製，幸好 TWGL 很好心地提供 twgl.primitives.deindexVertices() 可以轉換為直接以頂點為單位的

3-62

`position`、`texcoord`、`normal` 資料，使用起來如下：

```
console.log(
  twgl.primitives.deindexVertices(
    twgl.primitives.createSphereVertices(10, 32, 32)
  )
);
```

```
                                                         main.js:6
▼ {position: Float32Array(18432), normal: Float32Array(18
  432), texcoord: Float32Array(12288)} ⓘ
  ▶ normal: Float32Array(18432) [0, 1, 0, 0, 1, 0, 0.0980
  ▶ position: Float32Array(18432) [0, 10, 0, 0, 10, 0, 0.
  ▶ texcoord: Float32Array(12288) [1, 0, 0.96875, 0, 1, 0
  ▶ [[Prototype]]: Object
```

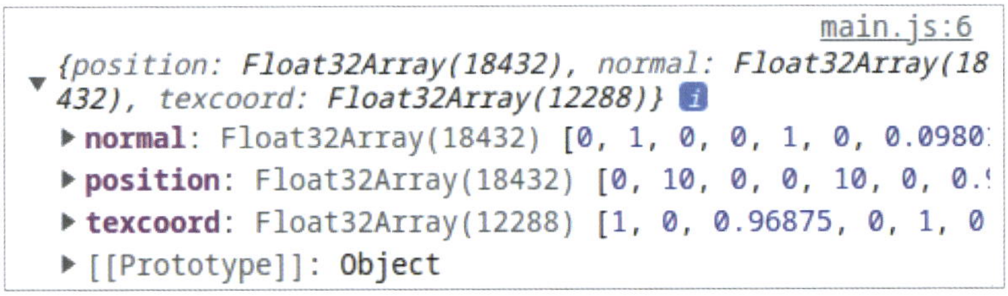

▲ 圖 3-33　twgl.primitives.deindexVertices() 展開之球體資料

經過 `twgl.primitives.deindexVertices()` 轉換後可以看到 `position` 經過 `indices` 展開之後長度變成 18432，可見重複的資料有多少，關於透過這層 `indices` 類似指標去取得 attribute 資料的 indexed element 模式待後面章節再來詳談。

球體的頂點位置資料有著落了，而且球體也將是一個場景中的『物件』，如同 P 形狀的 3D 物件一樣，因此在 `setup()` 中的區塊建立一個專屬於球體資料準備的程式區塊，把 TWGL 產生球體 `vertexDataArrays` 資料之程式實做進去：

```
async function setup() {
  // objects, pModel ...

  { // sphere, 球體
    // 產生、轉換成球體 vertexDataArrays 資料
    const vertexDataArrays = twgl.primitives.deindexVertices(
      // 產生球體 indexed element vertexDataArrays 資料
      twgl.primitives.createSphereVertices(10, 32, 32)
    );
  }
```

3-63

```
  // return ...
}
```

對於此球體除了需要 `position` 之外，還需要 `numElements`，也就是到時候繪製時在 `gl.drawArrays()` 請 GPU 繪製的頂點數量；事實上 `vertexDataArrays.position` 除了頂點位置資料之外，其身上有個 `numComponents` 表示『每個頂點（Element）有多少個數字』，因此將 `vertexDataArrays.position` 的陣列長度除以 `vertexDataArrays.position.numComponents` 即可：

```
async function setup() {
  // objects、pModel ...

  { // sphere, 球體
    // 產生、轉換成球體 vertexDataArrays 資料
    const vertexDataArrays = twgl.primitives.deindexVertices(
      // 產生球體 indexed element vertexDataArrays 資料
      twgl.primitives.createSphereVertices(10, 32, 32)
    );

    // 計算頂點（element）數量：
    const numElements = (
      vertexDataArrays.position.length /
      vertexDataArrays.position.numComponents
    );
  }

  // return ...
}
```

以 3D 狀況來說，`vertexDataArrays.position` 理所當然地每三個元素（x、y、z）為一個頂點，如果嘗試把 `vertexDataArrays.position.numComponents` 印在 Console 上就可以看到 3 這個數字；同時此 `numElements` 也應該跟 indexed element 版本之 `indices` 長度相同。

剩下的球體『畫什麼』資料準備就跟之前一樣，替 `vertexDataArrays.position` 建立 buffer 並與 vertex shader 的 attribute 建立關聯：

```javascript
async function setup() {
  // objects, pModel ...

  { // sphere, 球體
    // vertexDataArrays, numElements ...

    const buffers = {};

    // a_position
    buffers.position = gl.createBuffer();
    gl.bindBuffer(gl.ARRAY_BUFFER, buffers.position);

    gl.enableVertexAttribArray(attributes.position);
    gl.vertexAttribPointer(
      attributes.position,
      vertexDataArrays.position.numComponents, // size
      gl.FLOAT, // type
      false, // normalize
      0, // stride
      0, // offset
    );

    gl.bufferData(
      gl.ARRAY_BUFFER,
      new Float32Array(vertexDataArrays.position),
      gl.STATIC_DRAW,
    );
  }

  // return ...
}
```

其中比較值得注意的是 `gl.vertexAttribPointer()` 在 `size` 參數上我們可以使用 `vertexDataArrays.position.numComponents`，雖然都是數字 3，但是這樣我們就利用了這個 `numComponents` 抽象出來的資料結構，而非手寫寫死數字

3。

最後在 `objects` 物件清單中建立 `sphere` 球體物件：

```js
async function setup() {
  // objects, pModel ...

  { // sphere, 球體
    // vertexDataArrays, numElements, buffers, a_position ...

    objects.sphere = {
      vertexDataArrays, numElements,
      buffers,
    };
  }

  // return ...
}
```

球體的『畫什麼』交待給 GPU 了，在 `render()` 中比照 P 形狀之 3D 模型加入程式碼區塊進行繪製，筆者將球體放置在 [300, -80, 0] 的位置，並且放大 3 倍，顏色使用前面加入之 uniform u_color 設定此球體為藍色（#437bd0）：

```js
function render(app) {
  // cameraMatrix, viewMatrix, pModel ...

  { // sphere, 球體
    const worldMatrix = matrix4.multiply(
      // 將球體放置在 [300, -80, 0] 的位置：
      matrix4.translate(300, -80, 0),
      // 放大 3 倍：
      matrix4.scale(3, 3, 3),
    );

    gl.uniformMatrix4fv(
      uniforms.matrix,
      false,
      matrix4.multiply(viewMatrix, worldMatrix),
```

```
  );

  // 設定此球體為藍色（#437bd0）純色物件：
  gl.uniform3f(uniforms.color, 67/255, 123/255, 208/255);

  gl.drawArrays(gl.TRIANGLES, 0, objects.sphere.numElements);
 }
}
```

原本相機經過 `matrix4.lookAt()` 會一直看著 P 形狀之 3D 模型，為了方便看到完整的場景，改成看著 `[250, 0, 0]` 的位置：

```
function render(app) {
  // app, useProgram ...

  const cameraMatrix = matrix4.lookAt(
    state.cameraPosition,
    state.translate,
    [250, 0, 0],
    [0, 1, 0],
  );

  // viewMatrix ...
  // pModel, sphere ...
}
```

存檔重整後，球體有出現了，但是原本的 P 物體消失，畫面中還有一條不知道是什麼的東西，如圖 3-34：

3 3D & 物件

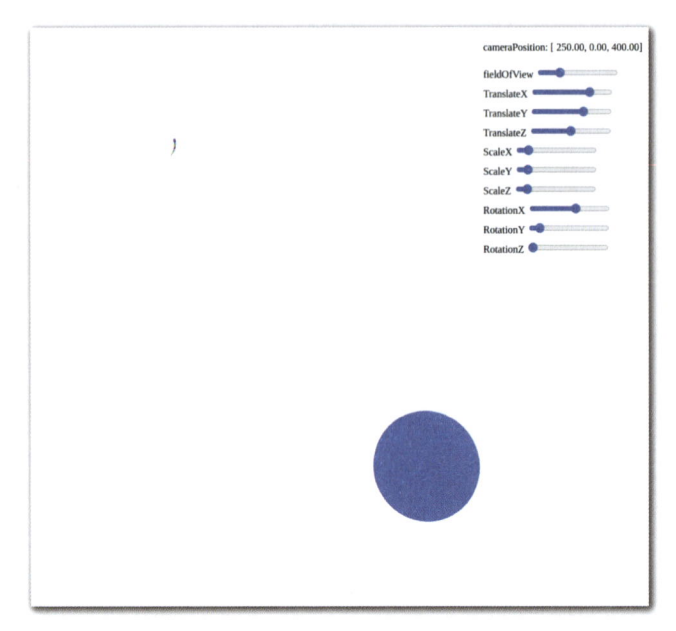

▲ 圖 3-34　加入球體渲染之結果

為什麼呢？在 `render()` 分別繪製 P 物件以及球體時，只有切換了 uniform，而在 `setup()` 設定好的 vertex attribute 與 buffer 之間的關係顯然在 `render()` 這邊沒有進行切換，事實上，在 setup() 中第二次呼叫的 `gl.vertexAttribPointer()` 就把 position attribute 改成與球體的 position buffer 綁定，因此最後繪製時兩次 gl.drawArrays() 都是繪製球體，只是第一次繪製時候 `objects.pModel.numElements` 比球體頂點數少很多所以只有一小條出現。

要解決這個問題，我們需要 Vertex Attribute Object 這個功能，接下來會繼續介紹，到此進度的完整程式碼可以在這邊找到：

</>	ch3/06/index.html
</>	ch3/06/main.js

也可以參考上線版本，透過瀏覽器打開此頁面：
https://webgl-book.pastleo.me/ch3/06/index.html

3-68

⌘ Vertex Attribute Object（VAO）

回想一下把 buffer 與 vertex attribute 建立關係的部份，也就是圖 3-35 紅色框起來的區域：

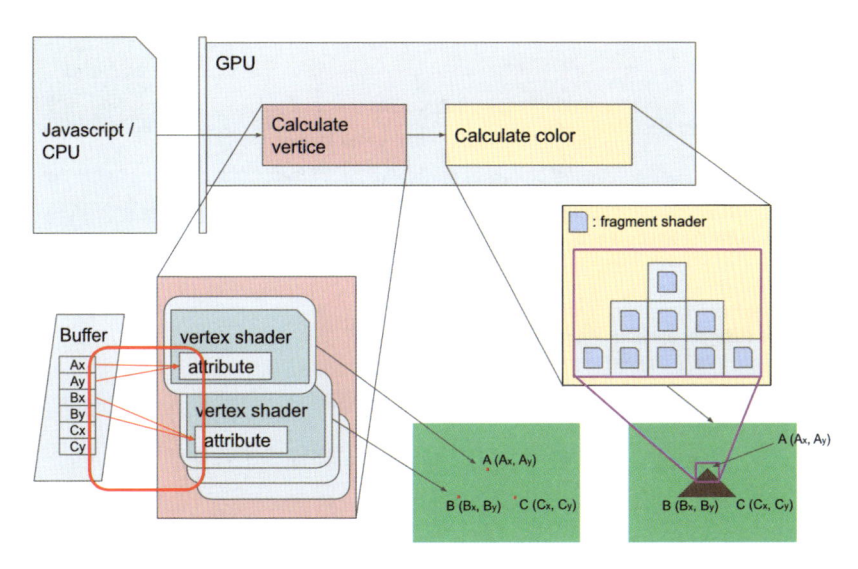

▲ 圖 3-35　『畫』這個動作中 buffer 與 attribute 之關聯

經過 `gl.bindBuffer()` 指定對準的 buffer，接著呼叫 `gl.vertexAttribPointer()` 來指定 vertex 的 attribute 使用對準好的 buffer，也就是說在不使用 vertex attribute object（接下來簡稱 VAO）的情況下，我們其實也可以在每次 `gl.drawArrays()` 之前更換 vertex attribute 使用的 buffer，但是對於每一個 attribute 就要執行一次 `gl.bindBuffer()` 以及 `gl.vertexAttribPointer()`，如果我們有兩個或甚至更多 attribute 的時候，除了程式碼更複雜之外，多餘的 GPU call 也會讓性能下降。

因此，VAO 就來拯救我們了，透過建立一個 VAO，我們會獲得一個『工作空間』，在這個工作空間建立好 vertex attribute 與 buffer 的關聯，接著切換到其他 VAO 指定 attribute-buffer 時，不會影響到原本 VAO/ 工作空間 attribute-buffer 的關聯，要執行繪製時再切換回原本的 VAO 進行繪製；假設

3 3D & 物件

我們有兩個物件分別叫做 `pModel` 與 `sphere`，以及兩個 attribute `a_position` 與 `a_color`，那麼使用 VAO 的狀況下會像是圖 3-36 所示：

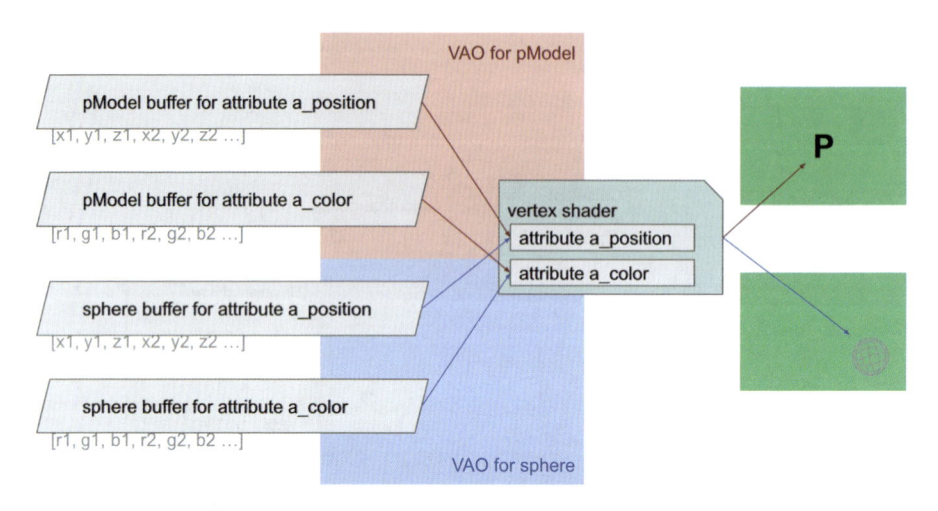

▲ 圖 3-36　使用 VAO 功能產生、切換『工作空間』

⌘ 使用 VAO 功能建立並使用『工作空間』

在 `setup()` 時，要分別為各個物件建立並切換到各自的『工作空間』，再進行 buffer 與 attribute 的綁定；先從 P 形狀的物件開始改，呼叫 `gl.createVertexArray` 建立專屬於 P 物件的 `vao`；緊接著使用 `gl.bindVertexArray` 切換到此 `vao` 所代表的工作空間，接著才進行 buffer 的建立以及 buffer 與 attribute 的綁定：

```
async function setup() {
  // ...

  { // pModel, P 形狀的物件
    // vertexDataArrays, numElements ...

    // 產生 pModel 專屬的 VAO『工作空間』
    const vao = gl.createVertexArray();
    // 切換到 pModel 專屬的 VAO『工作空間』
```

3-70

```
  gl.bindVertexArray(vao);

  // a_position, a_color ...
  //   gl.createBuffer(), gl.bindBuffer()
  //   gl.enableVertexAttribArray()
  //   gl.vertexAttribPointer()
  //   gl.bufferData()

  // ...
  }

  // ...
}
```

同時也把建立的 vao 放入該物件的 Javascript object 中，在渲染時期得以使用此 vao 進行工作空間的切換：

```
async function setup() {
  // ...

  { // pModel, P 形狀的物件
    // vertexDataArrays, numElements, vao ...
    // a_position, a_color ...

    objects.pModel = {
      vertexDataArrays, numElements,
      vao, buffers,
    };
  }

  // ...
}
```

球體也一樣，建立、切換至其專屬的 vao，並放到對應的 Javascript object：

```
async function setup() {
  // ...
```

3-71

```
{ // sphere, 球體
  // vertexDataArrays, numElements ...

  // 產生 sphere 專屬的 VAO 『工作空間』
  const vao = gl.createVertexArray();
  // 切換到 sphere 專屬的 VAO 『工作空間』
  gl.bindVertexArray(vao);

  // a_position ...
  //   gl.createBuffer(), gl.bindBuffer()
  //   gl.enableVertexAttribArray()
  //   gl.vertexAttribPointer()
  //   gl.bufferData()

  objects.sphere = {
    vertexDataArrays, numElements,
    vao, buffers,
  };
}

// ...
}
```

畫龍點睛的時刻來了，在 render() 中繪製物件之前只要加上 `gl.bindVertexArray()` 先切到對應的 VAO，那麼就會回到之前設定好的 attribute-buffer 關聯狀態，筆者習慣在專屬於物件程式區塊的一開始就切過去：

```
function render(app) {
  // app, cameraMatrix, viewMatrix ...

  { // pModel, P 形狀的物件
    gl.bindVertexArray(objects.pModel.vao);

    // worldMatrix, gl.uniformMatrix4fv(uniforms.matrix) ...
    // gl.uniform3f(uniforms.color, 0, 0, 0);

    gl.drawArrays(gl.TRIANGLES, 0, objects.pModel.numElements);
```

3-5 渲染多個物件

```
  }
  { // sphere, 球體
    gl.bindVertexArray(objects.sphere.vao);

    // worldMatrix, gl.uniformMatrix4fv(uniforms.matrix) ...
    // gl.uniform3f(uniforms.color, 67/255, 123/255, 208/255);

    gl.drawArrays(gl.TRIANGLES, 0, objects.sphere.numElements);
  }
}
```

存檔重整，P 物件與球體都能正常的繪製出來，如圖 3-37 所示：

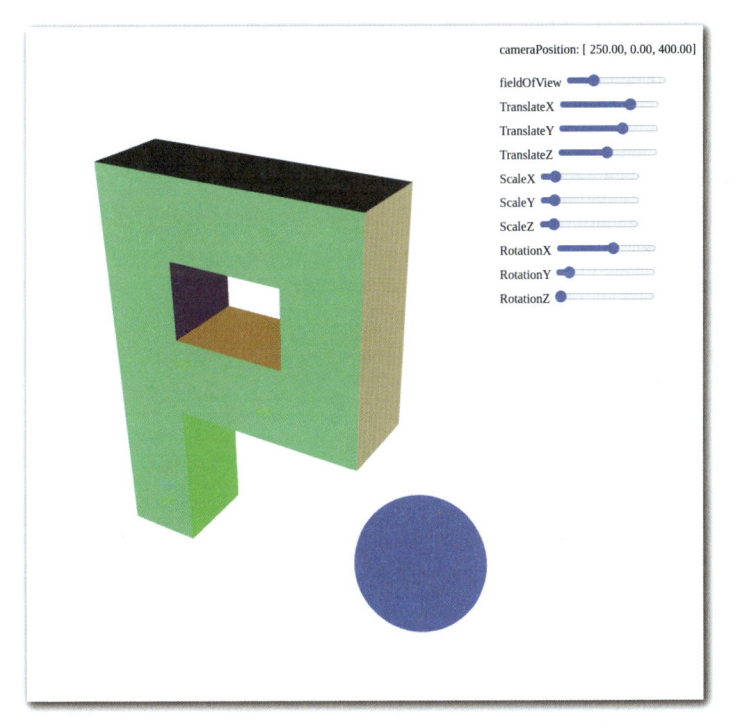

▲圖 3-37　使用 VAO 後，P 物件與球體皆能正常繪製

⌘ 補充：在 WebGL1 中啟用 VAO 功能

在 WebGL1 的 spec（標準）中，VAO 並沒有被包含在內，幸好大多數瀏覽器在 WebGL1 中都透過 WebGL extension 來支援了，這個 extension 不是指要從 Chrome web store 或是 Firefox Add-ons 下載的瀏覽器擴充套件，比較像是 WebGL spec 上沒有指定要支援，但是各家瀏覽器可以自行加入的功能，所以得看各家瀏覽器的臉色來決定特定功能能不能用；以 VAO 來說各家瀏覽器都以 `OES_vertex_array_object` 這個 WebGL extension 來提供，為了啟用此 WebGL extension，在 `canvas.getContext('webgl')`; 之後放上這些程式：

```javascript
async function setup() {
  // after canvas.getContext('webgl');
  const oesVaoExt = gl.getExtension('OES_vertex_array_object');
  if (oesVaoExt) {
    gl.createVertexArray =
      (...args) => oesVaoExt.createVertexArrayOES(...args);
    gl.deleteVertexArray =
      (...args) => oesVaoExt.deleteVertexArrayOES(...args);
    gl.isVertexArray =
      (...args) => oesVaoExt.isVertexArrayOES(...args);
    gl.bindVertexArray =
      (...args) => oesVaoExt.bindVertexArrayOES(...args);
  } else {
    throw new Error(
      'Your browser does not support WebGL ext: OES_vertex_array_object'
    );
  }

  // ...
}
```

經過 `gl.getExtension()` 取得 WebGL extension，經過 `if` 檢查沒問題有東西的話，在 `gl` 物件上建立 `createVertexArray`、`deleteVertexArray` 等 function 來模仿 WebGL2 VAO 相關功能的 interface（界面）。

⌘ 補上地板

地板只是一個 plane，也就是 2 個三角形、6 個頂點即可做出來，手刻 a_position 的 buffer 資料並不是難事，不過筆者這邊透過上篇 import 進來的 TWGL 幫忙，使用 `twgl.primitives.createPlaneVertices(width, depth)`[19] 建立 xz 平面，長深（寬）都給 1，大小再透過 transform 調整，並且記得加上 VAO 的建立與切換，剩下的程式碼就跟球體那邊差不多依樣畫葫蘆，讀者可以試著自己實做看看，而非第一時間就直接看答案：

```javascript
async function setup() {
  // gl, objects.pModel, objects.sphere ...

  { // ground
    const vertexDataArrays = twgl.primitives.deindexVertices(
      twgl.primitives.createPlaneVertices(1, 1)
    );
    const numElements = (
      vertexDataArrays.position.length /
      vertexDataArrays.position.numComponents
    );
    const vao = gl.createVertexArray();
    gl.bindVertexArray(vao);

    const buffers = {};

    // a_position
    buffers.position = gl.createBuffer();
    gl.bindBuffer(gl.ARRAY_BUFFER, buffers.position);

    gl.enableVertexAttribArray(attributes.position);
    gl.vertexAttribPointer(
      attributes.position,
      vertexDataArrays.position.numComponents, // size
      gl.FLOAT, // type
      false, // normalize
```

---

19 https://twgljs.org/docs/module-twgl_primitives.html#.createPlaneVertices

3　3D & 物件

```
    0, // stride
    0, // offset
  );

  gl.bufferData(
    gl.ARRAY_BUFFER,
    new Float32Array(vertexDataArrays.position),
    gl.STATIC_DRAW,
  );

  objects.ground = {
    vertexDataArrays, numElements,
    vao, buffers,
  };
 }

 // return { ...app }
}
```

地板渲染部份，筆者放大成 500 長深（寬）、放置在 `[250, -100, -50]` 的位置，並且純色顏色設定成灰色 #808080：

```
function render(app) {
  // app, viewMatrix ...
  // pModel, sphere ...

  { // ground
    gl.bindVertexArray(objects.ground.vao);

    const worldMatrix = matrix4.multiply(
      // 放大成 500 長深（寬）
      matrix4.translate(250, -100, -50),
      // 放置在 [250, -100, -50] 的位置
      matrix4.scale(500, 1, 500),
    );

    gl.uniformMatrix4fv(
      uniforms.matrix,
      false,
```

3-76

```
    matrix4.multiply(viewMatrix, worldMatrix),
  );

  // 純色顏色設定成灰色 #808080
  gl.uniform3f(uniforms.color, 0.5, 0.5, 0.5);

  gl.drawArrays(gl.TRIANGLES, 0, objects.ground.numElements);
  }
}
```

存檔重整，就得到 P 物件、球體加上地板，如圖 3-38 這樣，不覺得開始有場景的感覺了嗎？

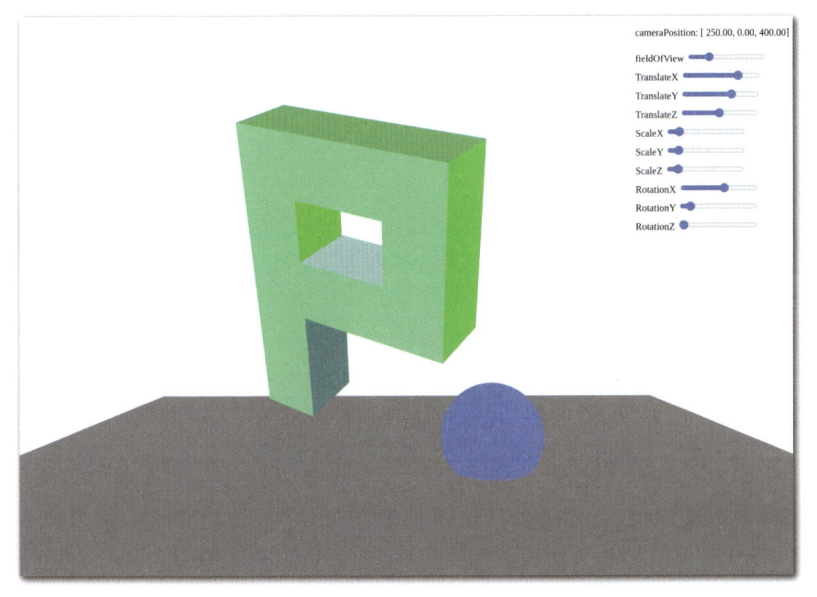

▲ 圖 3-38　P 形狀、球體與地板組成的場景

到此進度的完整程式碼可以在這邊找到：

</>	ch3/07/index.html
</>	ch3/07/main.js

also 可以參考上線版本，透過瀏覽器打開此頁面：
https://webgl-book.pastleo.me/ch3/07/index.html

『CH3：3D & 物件』就到這邊，讀者可以嘗試左右移動視角感受一下這個 3D 場景，有了地板加上視角的移動可以讓人類對於這個場景開始有『空間感』，本章節一開始從 orthogonal 投影至平面開始，學會 perspective 成像、控制相機視角，最後渲染多個物件形成 3D 場景，是不是很有成就感呢？

不過不知道有沒有覺得球體感覺很不立體，因為我們在這個物件渲染時不論從哪邊看，每個面都是 uniform 指定的統一純色，接下來將加入光對於物體表面顏色的影響，使物體看起來更立體。

Lighting

在這個章節中將會加入『光』的元素，使得物體在有光照射的時候才會有顏色，並利用上個章節提到的 TWGL 讓程式碼可以寫的比較愉快，最後加入反射光以及更細緻的表面細節，使得物體的渲染栩栩如生！

4 Lighting

新的章節使用新的一組 `index.html`、`main.js` 作為開始，完整程式碼可以在這邊找到：

</>	ch4/00/index.html
</>	ch4/00/main.js

也可以參考上線版本，透過瀏覽器打開此頁面：

https://webgl-book.pastleo.me/ch4/00/index.html

這次的起始點跑起來就有一個木質地板跟一顆球，並且延續 CH3 的相機位置控制方式，如下圖 4-1：

▲ 圖 4-1　程式碼起始點即渲染出木質地板以及球體

還記得 CH2 中介紹的 texture 嗎？此起始點已經直接將之整合在 3D 中囉，需要的技巧並沒有什麼不同，同樣是讓 GPU 對圖片取樣讀取出一個 pixel

所要顯示的顏色，取樣時所需要的圖片位置（texcoord）在前一章 TWGL 產生的 `vertexDataArrays` 就有了，這邊的球體以及地板 attribute 資料使用 TWGL 產生，因此 `a_texcoord` 所需的資料可以直接使用 `vertexDataArrays.texcoord`，讀者可以試著閱讀 ch4/00/main.js，不知道是否能找到對應的串接程式呢？從 fragment shader 中進行圖片顏色取樣的 `texture(u_texture, v_texcoord)` 兩個參數 `u_texture`、`v_texcoord` 開始追朔看看，敘述整個 3D 物件的 texture 是怎麼『貼』圖到物件表面的，如果覺得有困難，可以回去參考 CH2 的內容。

至於 texture 的圖片筆者從遊戲素材分享網站 https://opengameart.org/ 找到這兩張 CC0[1] 的圖來使用：

- 木質地板：https://opengameart.org/content/2048%C2%B2-wooden-texture
  - 選用壓縮檔中的 wooden-floor/woodfloor_c.jpg
- 球體：https://opengameart.org/content/commission-medieval
  - 選用 steel，於壓縮檔中的 Medieval/steel/steel_diffuse.tga

筆者將素材包下載並稍做轉換以及壓縮，放置在範例資料夾中的這兩個位置：

</>	assets/woodfloor.webp
</>	assets/steel.webp

除此之外，這個起始點的場景的『尺度』與上個章節的不一樣，之前的因為是從 2D 符合螢幕 pixel 寬高的狀態修改而來，導致物件的大小、位置距離動輒好幾百，從這邊開始調整至接近一般 3D 系統的大小，使得場景內的『1 單位』約為人們現實生活中的 1 公尺：

---

[1] CC0 表示作者拋棄該著作依著作權法所享有之權利，製作應用程式需要素材時可以放心使用：https://creativecommons.org/publicdomain/zero/1.0/deed.zh_TW

4 Lighting

- 相機初始位置在 `[0, 0, 8]`，看著 `[0, 0, 0]`，可視角度為 45 度，擷取 0.1 至 200 之間深度的成像
- 球體（sphere）放置在 `[0, 0, 0]`，半徑為 1
- 地板（ground）放置在 `[0, -1, 0]`，原本大小為 1x1 放大 10 倍
  - 也就是 y = -1、10x10 大小的地板

4-1 法向量（Normals）與散射光（Diffuse）

目前木質地板與球只是按照原本 texture 上的顏色進行繪製，本篇的目標是加入一個從無限遠的地方照射過來的白色平行光（directional light），並且在物體表面計算『散射（diffuse）』[2] 之後從任意角度觀察到的顏色，現實物理來說，散射光表示光線照射到粗糙表面導致光線會往任意角度反射，如圖 4-2 中紅色箭頭 diffuse reflection 表示了散射光的方向[3]：

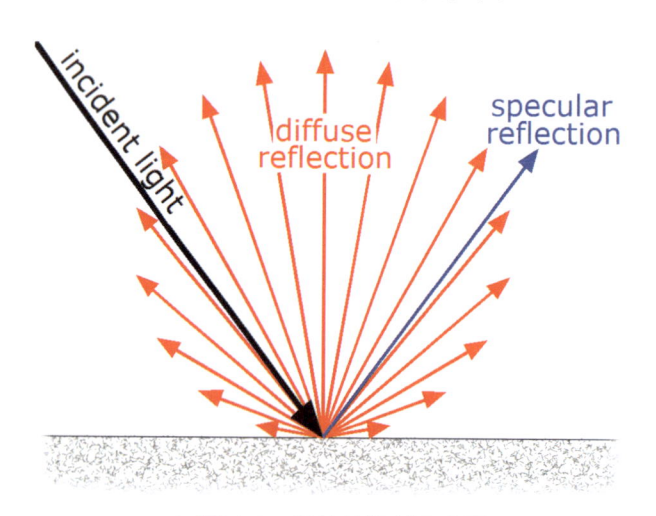

▲ 圖 4-2　光線的散射示意圖

2　https://zh.wikipedia.org/wiki/漫反射
3　圖 4-2 由 GianniG46 以 CC BY-SA 3.0 發布在
　　https://en.wikipedia.org/wiki/Diffuse_reflection#/media/File:Lambert2.gif

4-1 法向量（Normals）與散射光（Diffuse）

⌘ 散射光形成的亮度與法向量（Normal）

因為是白光，所以散射之後的顏色其實就是物體表面顏色經過一個明暗度的處理，而明暗度要怎麼計算呢？筆者畫了下方的意示圖圖 4-3 嘗試解釋，首先，如果是在光照不到的區域，像是紅色面，與光平行或是背對著光，那麼就會是全黑；被照射到的區域，如綠色與藍色面，因為一個單位的光通量在與垂直的面上可以形成較小的區域（在綠色面上的橘色線段較藍色面短），一個單位的面積獲得的光通量就比較高，因此綠色面比藍色面來的更亮：

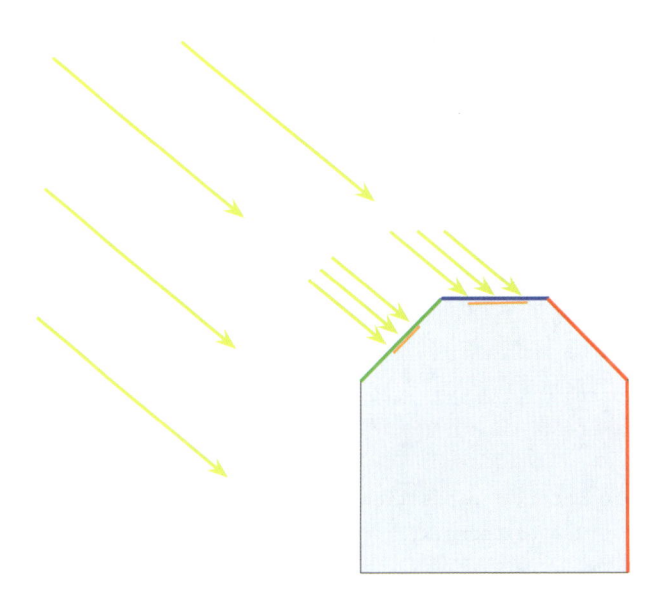

▲ 圖 4-3　平行光在各個面形成不同入射角，產生光通量的不同

總和以上，在一個面上接收到光線的入射角越垂直於該面，明暗度越高，而『越垂直於一個面』這件事，就是（入射角的反向）越接近該面的法向量[4]，面的法向量英文稱為 normal，在 3D 系統中一個面的 normal 表示了該面面向的方向；在 fragment shader 內，我們可以取光線方向的反向，再與

4　https://zh.wikipedia.org/wiki/法線

4 Lighting

normal 做內積[5]（兩者要確保皆為單位向量[6]）來計算出『光線入射角垂直於該面的程度』，得到 -1 ~ +1 之間的值，小於零表示該面背對著入射光，應該為全黑，+1 表示該面正對著入射光、介於 0 ~ +1 之間就是『光線入射角垂直於該面的程度』，到時候可以直接乘以 rgb 調整明暗度。

之前使用 `twgl.primitives` 產生的頂點資料，不只有 `position`、`texcoord`，還有 `normal`，也就是這邊可以與光線方向進行運算的法向量，場景中的球以及地板都是使用 TWGL 生成的，首先要做的事情就是把這些 normal 傳入 shader。

在 vertex shader 內加入 `a_normal` attribute：

```
#version 300 es
in vec4 a_position;
in vec2 a_texcoord;
in vec3 a_normal;

uniform mat4 u_matrix;

out vec2 v_texcoord;

void main() {
  gl_Position = u_matrix * a_position;
  v_texcoord = vec2(a_texcoord.x, 1.0 - a_texcoord.y);
}
```

取得 attribute 的位置：

```
async function setup() {
  // canvas, gl, program ...

  const attributes = {
```

5　https://zh.wikipedia.org/wiki/內積
6　https://zh.wikipedia.org/wiki/單位向量

4-6

4-1 法向量（Normals）與散射光（Diffuse）

```js
    position: gl.getAttribLocation(program, 'a_position'),
    texcoord: gl.getAttribLocation(program, 'a_texcoord'),
    normal: gl.getAttribLocation(program, 'a_normal'),
  };

  // uniforms, textures, objects ...
}
```

對於球體（sphere）建立 attribute 與 buffer 之間的關聯：

```js
async function setup() {
  // canvas, gl, program, uniforms, textures, attributes, objects ...

  { // sphere
    // const vertexDataArrays = ... twgl.primitives.createSphereVertices ...
    // vao, a_position, a_texcoord ...

    // a_normal
    buffers.normal = gl.createBuffer();
    gl.bindBuffer(gl.ARRAY_BUFFER, buffers.normal);

    gl.enableVertexAttribArray(attributes.normal);
    gl.vertexAttribPointer(
      attributes.normal,
      vertexDataArrays.normal.numComponents, // size
      gl.FLOAT, // type
      false, // normalize
      0, // stride
      0, // offset
    );

    gl.bufferData(
      gl.ARRAY_BUFFER,
      new Float32Array(vertexDataArrays.normal),
      gl.STATIC_DRAW,
    );

    // objects.sphere = { ... }
  }
}
```

4-7

4　Lighting

對於地板（ground）也是一樣，建立 attribute 與 buffer 之間的關聯：

```
async function setup() {
  // objects, sphere ...

  { // ground
    // const vertexDataArrays = ... twgl.primitives.createPlaneVertices ...
    // vao, a_position, a_texcoord ...

    // a_normal
    buffers.normal = gl.createBuffer();
    gl.bindBuffer(gl.ARRAY_BUFFER, buffers.normal);

    gl.enableVertexAttribArray(attributes.normal);
    gl.vertexAttribPointer(
      attributes.normal,
      vertexDataArrays.normal.numComponents, // size
      gl.FLOAT, // type
      false, // normalize
      0, // stride
      0, // offset
    );

    gl.bufferData(
      gl.ARRAY_BUFFER,
      new Float32Array(vertexDataArrays.normal),
      gl.STATIC_DRAW,
    );

    // objects.ground = { ... }
  }
}
```

⌘ 法向量的 Transform

計算明暗度會在面投影到螢幕的每個 pixel 上進行，也就是 fragment shader，透過 varying 接收到法向量，而 vertex shader 主要的工作是把法向量

4-1 法向量（Normals）與散射光（Diffuse）

pass 到 fragment shader，但是有一個問題：物體會旋轉，我們讓頂點的位置透過 `u_matrix` 做 transform，假設有一個物體轉了 90 度，那麼法向量也應該一起轉 90 度才對。

所以我們就直接讓法向量 `a_normal` 與 `u_matrix` 相乘來得到旋轉後的結果嗎？想想看 `u_matrix` 可能不只有旋轉、縮放、平移（`worldMatrix`），還有包含視角、投影到螢幕上的 `viewMatrix`，因此不能直接對 `a_normal` 進行 `u_matrix` 轉換，勢必得要多傳送另一個矩陣，這個用來轉換 `a_normal` 的矩陣可以從 `worldMatrix` 出發，畢竟至少不會包含視角以及螢幕投影的轉換，但是 `worldMatrix` 除了旋轉之外還會包含平移以及縮放。

平移的部份倒是還好，到時候在計算之時本來就只會抽取 3x3 的部份進行運算，也就自然不包含平移；縮放的部份就比較複雜了，如果一個物體放大到 2 倍，使用 `worldMatrix` 轉換後 `a_normal` 向量長度也就會跟著放大到 2 倍，到時候與光線方向做內積之前為了得先把 normal 以及光線向量皆調整成長度為 1 的單位向量來運算，才能確保內積結果的變因只有兩者的角度、最大值為 1，不過還不只如此，如果 `worldMatrix` 只有沿著 x 軸放大到 2 倍呢？接下來需要用一個 2D 的狀況來舉例，有一個等腰直角三角形如下圖 4-4：

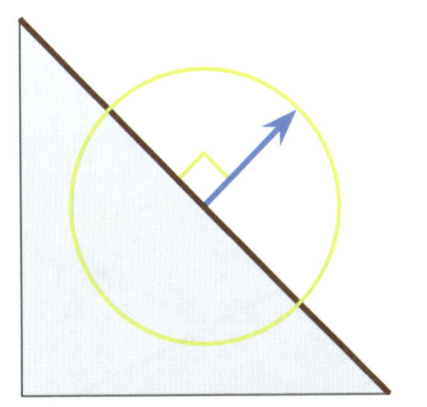

▲ 圖 4-4　2D 三角形斜面與法向量作為範例

4-9

4　Lighting

我們接下來關注褐色那面，以及藍色箭頭表示的法向量，黃色正方形用來對標直角、黃色圓形假設半徑為 1 用來表示單位向量應有的長度，此時 normal 法向量為單位向量。

接著沿著 x 軸放大到 2 倍，就像是用變形工具直接進行左右拉伸，結果如圖 4-5：

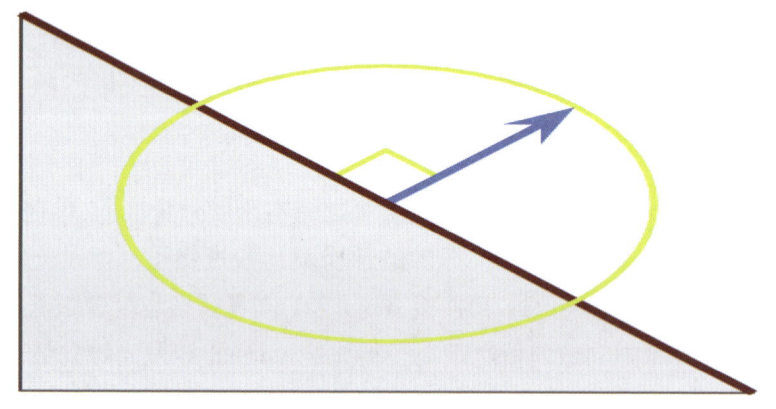

▲ 圖 4-5　2D 三角形沿著 x 方向放大到 2 倍

除了三角形本身左右拉伸之外，同樣的 tranform 也作用在法向量上，顯然圖 4-5 的藍色箭頭已經不再跟褐色面垂直，這時直接再把法向量取單位向量成為圖 4-6 的橘色箭頭：

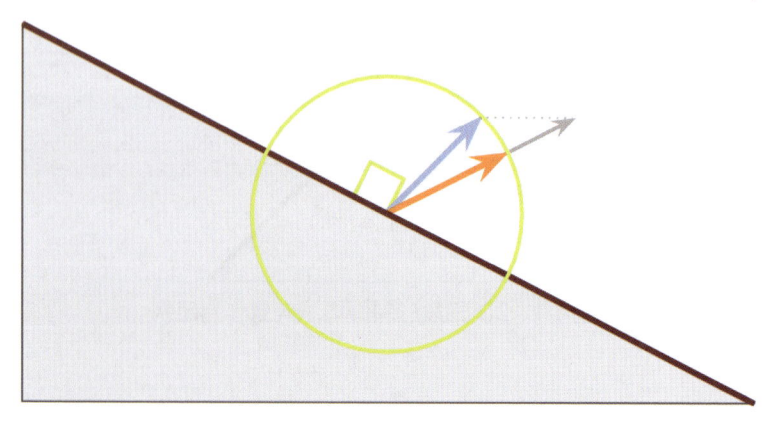

▲ 圖 4-6　將法向量沿著 x 方向放大到 2 倍後轉換為單位向量

4-1　法向量（Normals）與散射光（Diffuse）

圖 4-6 中，先讓黃色正方形、圓型回歸測量功能，淡藍色箭頭為原本的法向量，經過『沿著 x 軸放大到 2 倍』的 transform 後成為灰色箭頭（也就等同圖 4-5 的藍色箭頭），最後轉換成單位向量成為橘色箭頭，如果我們拿這個橘色箭頭跟光線方向做內積很明顯沒辦法得到正確的亮度數值，不能正確反應出褐色面對於光線產生的明暗程度。

看著圖 4-6 中表示正確直角角度的黃色正方形就會發現，在沿著 x 方向放大物體之時，對於法向量的轉換應該反而是沿著 x 方向進行某種程度的『縮小』，這麼一來角度就可以貼回垂直於褐色面的狀態，事實上這個縮小的程度就是作用於物體的**倒數**，以目前對物體『沿著 x 軸放大到 2 倍』的範例來說，就是對 normal 法向量沿著 x 軸縮小到 $\frac{1}{2}$ 倍，這麼一來就可以得到圖 4-7 的灰色箭頭：

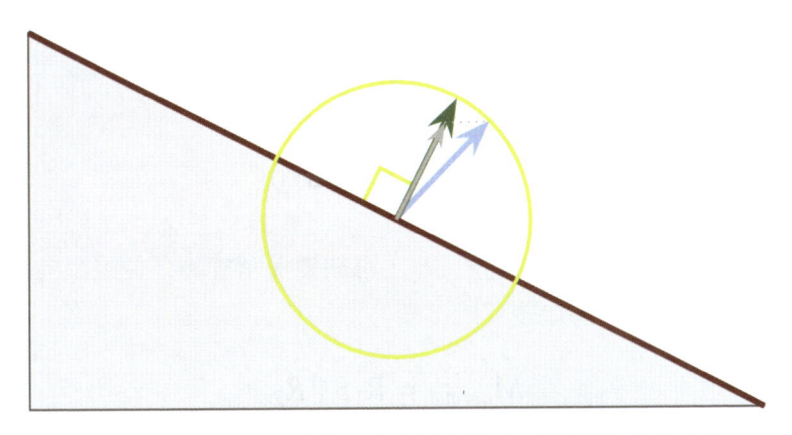

▲ 圖 4-7　將法向量沿著 x 方向縮小到 倍後轉換為單位向量

沿著 x 軸縮小到 $\frac{1}{2}$ 倍後，一樣轉換成單位向量成為綠色箭頭，這時這個綠色箭頭就能表示『沿著 x 軸放大到 2 倍』三角形褐色面的法向量。

總和以上，一個要做用在 3D 物件的 `worldMatrix`，其對應用來轉換 normal 的矩陣應該包含 `worldMatrix` 的旋轉以及『縮放的相反』，要得到這個矩陣其實比想像中容易，只要**取 `worldMatrix` 的反矩陣，再取轉置矩陣**[7]即可。

7　將矩陣中的行與列對調，稱為轉置：https://zh.wikipedia.org/zh-tw/轉置矩陣

4 Lighting

　　筆者當時學到的時候對於如此簡單的解決方法很驚訝，原本以為之後在計算 `worldMatrix` 的過程都要額外計算一份，看到旋轉時加入、縮放時加入縮放的相反，像這樣可以拿『已經算好』的 `worldMatrix` 多做點加工就得到轉換 normal 之矩陣的方法，可說是免除不少程式碼的複雜度；至於為什麼取反矩陣再取轉置矩陣，這邊用幾個反矩陣、轉置矩陣的特性做個簡單的證明：

　　假設有個矩陣 M 是用來轉換物體的 `worldMatrix`，其對應用來轉換 normal 法向量的矩陣設為 M_{normal}，我們要證明取 M 的反矩陣，再取轉置矩陣，即為 M_{normal}：

$$M_{normal} = \left(M^{-1}\right)^T$$

　　平移到時候可以簡單地把多餘的維度去除即可，也就是說現在只需要考慮旋轉、縮放所組成的線性轉換，且所有線性轉換都可以拆解成先做一次旋轉、做一次縮放接著再做一次旋轉；將 M 拆解成旋轉 R_1、縮放 S 接著再做旋轉 R_2（做為式 ①）如下：

$$M = R_1 S R_2 \implies ①$$

　　對於轉換法向量用的矩陣 M_{normal}，應該要保留旋轉，並且反向縮放的部份，使用 R_1、S、R_2 拆解之後寫成這樣：

$$M_{normal} = R_1 S^{-1} R_2$$

　　開始運算之前，先說明幾個重要的特性，首先純粹的旋轉矩陣其反矩陣等同於轉置矩陣 [8]，一個矩陣的反反矩陣為自己，且一個旋轉矩陣的反矩陣也是純旋轉，因此一個旋轉矩陣的反矩陣再轉置等於自己（作為式 ②）：

$$R^{-1} = R^T,\ R = \left(R^{-1}\right)^{-1} = \left(R^{-1}\right)^T \implies ②$$

8　https://zh.wikipedia.org/zh-tw/旋轉矩陣#性質

4-1 法向量（Normals）與散射光（Diffuse）

另外，一個純縮放的矩陣進行轉置之後不會有變化，因為縮放矩陣[9]中除了中間斜排之外應該都為 0，行列對調不會有作用，且縮放矩陣的反矩陣也同樣是縮放矩陣，因此一個縮放矩陣的反矩陣等於反矩陣再轉置（作為式③）：

$$S = S^T, \ (S^{-1}) = (S^{-1})^T \implies ③$$

同時，對於做矩陣的轉置或反矩陣，先相乘，做轉置或反矩陣後，與展開後再做的順序顛倒（分別作為式④、式⑤）：

$$(AB)^T = B^T A^T \implies ④, \quad (AB)^{-1} = B^{-1} A^{-1} \implies ⑤$$

開始導公式，我們可以把 M_{normal} 的旋轉、縮放矩陣利用式②、式③換成各自的反矩陣再轉置：

$$M_{normal} = R_1 S^{-1} R_2 \xrightarrow{using\ ②} \left(R_1^{-1}\right)^T S^{-1} \left(R_2^{-1}\right)^T \xrightarrow{using\ ③} \left(R_1^{-1}\right)^T \left(S^{-1}\right)^T \left(R_2^{-1}\right)^T$$

接著利用式④、式⑤把反矩陣、轉置推到最外的括弧，過程中 R_1 與 R_2 剛好會對調兩次：

$$M_{normal} = \left(R_1^{-1}\right)^T \left(S^{-1}\right)^T \left(R_2^{-1}\right)^T \xrightarrow{using\ ④} \left(R_2^{-1} S^{-1} R_1^{-1}\right)^T \xrightarrow{using\ ⑤} \left(\left(R_1 S\ R_2\right)^{-1}\right)^T$$

兩層括弧內的旋轉、縮放、旋轉是不是很眼熟？沒錯，就是 M，使用式①取代：

$$M_{normal} = \left(\left(R_1 S\ R_2\right)^{-1}\right)^T \xrightarrow{using\ ①} \left(M^{-1}\right)^T$$

9　https://zh.wikipedia.org/zh-tw/變換矩陣#縮放

4　Lighting

這麼一來證明就完成了，轉換 normal 法向量的矩陣可以透過取 `worldMatrix` 的反矩陣、再取轉置矩陣來取得。

關於這方面的證明讀者也可以參考維基百科古典幾何的證明法[10]或是當初筆者學習時閱讀的原文：

https://paroj.github.io/gltut/Illumination/Tut09%20Normal%20Transformation.html

⌘ 實做法向量的 Transform

在上面討論法向量的旋轉時，有提到一個新的矩陣運算叫做轉置矩陣，這樣的矩陣數學運算與 `matrix4.inverse()` 一樣應放在 `lib/matrix.js` 工具箱中，我們將命名為 `matrix4.transpose()`，根據其定義『將矩陣中的行與列對調』，實做並不難：

```javascript
// lib/matrix.js

export const matrix4 = {
  // ...

  transpose: m => {
    return [
      m[0], m[4], m[8], m[12],
      m[1], m[5], m[9], m[13],
      m[2], m[6], m[10], m[14],
      m[3], m[7], m[11], m[15],
    ];
  },
  // inverse ...
}
```

講解法向量轉換的原理時花了一長串篇幅，程式碼實做部份在計算物件的 `worldMatrix` 後，用這麼一小段程式碼便能『取 `worldMatrix` 的反矩陣、再

10　https://en.wikipedia.org/wiki/Normal_(geometry)#Transforming_normals

4-1 法向量（Normals）與散射光（Diffuse）

取轉置矩陣』獲得轉換 `a_normal` 法向量的矩陣：

```
matrix4.transpose(matrix4.inverse(worldMatrix))
```

算出矩陣程式碼已經水落石出，我們現在需要一個傳送這個新矩陣的 uniform，於 `u_matrix` 類似，這邊命名為 `u_normalMatrix`，於 vertex shader 中宣告：

```glsl
#version 300 es
in vec4 a_position;
in vec2 a_texcoord;
in vec3 a_normal;

uniform mat4 u_matrix;
uniform mat4 u_normalMatrix;

out vec2 v_texcoord;

void main() {
  gl_Position = u_matrix * a_position;
  v_texcoord = vec2(a_texcoord.x, 1.0 - a_texcoord.y);
}
```

同樣地，在 `setup()` 中先取得 uniform 變數位置為 `uniforms.normalMatrix`：

```js
async function setup() {
  // gl, program, attributes ...
  const uniforms = {
    matrix: gl.getUniformLocation(program, 'u_matrix'),
    normalMatrix: gl.getUniformLocation(program, 'u_normalMatrix'),
    color: gl.getUniformLocation(program, 'u_color'),
    texture: gl.getUniformLocation(program, 'u_texture'),
  };

  // ...
}
```

4 Lighting

把這段程式直接寫在 `render()` 中渲染球體（sphere）與地板（ground）時輸入給 `u_normalMatrix` 矩陣內容的位置：

```javascript
function render(app) {
  { // both sphere and ground
    // ...
    const worldMatrix = matrix4.multiply(/* ... */);

    gl.uniformMatrix4fv(uniforms.matrix, /* ... */);

    gl.uniformMatrix4fv(
      uniforms.normalMatrix,
      false,
      matrix4.transpose(matrix4.inverse(worldMatrix)),
    );

    // ... gl.drawArrays
  }
}
```

在 vertex shader 內將 `a_normal` 使用 `u_normalMatrix` 做轉換，並透過 varying `v_normal` 傳送到 fragment shader：

```glsl
#version 300 es
in vec4 a_position;
in vec2 a_texcoord;
in vec3 a_normal;

uniform mat4 u_matrix;
uniform mat4 u_normalMatrix;

out vec2 v_texcoord;
out vec3 v_normal;

void main() {
  gl_Position = u_matrix * a_position;
  v_texcoord = vec2(a_texcoord.x, 1.0 - a_texcoord.y);
  v_normal = mat3(u_normalMatrix) * a_normal;
}
```

4-1 法向量（Normals）與散射光（Diffuse）

在這邊使用 GLSL 的 `mat3()` 把 `mat4` `u_normalMatrix` 多餘的維度去除，同時平移效果也一起被移除，而且 `a_normal` 本來就是 `vec3` 需要用 `mat3` 進行運算。

⌘ 計算散射亮度 Diffuse

為了方便調整光線方向觀察不同方向的成像，在 fragment shader 這邊加入 uniform `u_lightDir`：

```glsl
#version 300 es
precision highp float;

in vec2 v_texcoord;

uniform vec3 u_color;
uniform sampler2D u_texture;
uniform vec3 u_lightDir;

out vec4 outColor;

void main() {
  vec3 color = u_color + texture(u_texture, v_texcoord).rgb;
  outColor = vec4(color, 1);
}
```

以及對應的 uniform 位置、狀態，並且一開始直直向下（-y 方向）照射：

```js
async function setup() {
  // ... gl, canvas, attributes
  const uniforms = {
    matrix: gl.getUniformLocation(program, 'u_matrix'),
    normalMatrix: gl.getUniformLocation(program, 'u_normalMatrix'),
    color: gl.getUniformLocation(program, 'u_color'),
    texture: gl.getUniformLocation(program, 'u_texture'),
    lightDir: gl.getUniformLocation(program, 'u_lightDir'),
  };
```

4-17

4 Lighting

```
// ...

return {
  // ...
  state: {
    fieldOfView: degToRad(45),
    cameraPosition: [0, 0, 8],
    cameraVelocity: [0, 0, 0],
    lightDir: [0, -1, 0],
  },
  time: 0,
};
}
```

因為整個場景的光線方向都是固定的，因此在 `render()` 中繪製物件之前設定好 `u_lightDir`：

```
function render(app) {
  // app, useProgram ...

  gl.uniform3f(uniforms.lightDir, ...state.lightDir);

  { // sphere and ground ...
    // ... gl.drawArrays( ... );
  }
}
```

回到 fragment shader，接收 vertex shader 傳送過來的 varying `v_normal`：

```
#version 300 es
precision highp float;

in vec2 v_texcoord;
in vec3 v_normal;

uniform vec3 u_color;
uniform sampler2D u_texture;
uniform vec3 u_lightDir;
```

4-18

4-1 法向量（Normals）與散射光（Diffuse）

```
void main() {
  // ...
}
```

這麼一來在 fragment shader 就有全部所需要的輸入資料，實做計算明暗度計算：

```
1  void main() {
2    vec3 color = u_color + texture(u_texture, v_texcoord).rgb;
3
4    // 轉換 normal, 光線反向成單位向量：
5    vec3 normal = normalize(v_normal);
6    vec3 surfaceToLightDir = normalize(-u_lightDir);
7
8    // 算出明暗度：
9    float colorLight = clamp(dot(surfaceToLightDir, normal), 0.0, 1.0);
10
11   // 輸出顏色，改成乘以數值為 0 至 1 之間的明暗度
12   outColor = vec4(color * colorLight, 1);
13 }
```

`main()` 的第 5 行使用 GLSL 內建的 `normalize` function[11] 算出 `v_normal` 的單位向量，完成『法向量轉換』的最後一步，這時 `vec3 normal` 變數就是真的可以用來與光線方向做比較的法向量；第 6 行計算『表面到光源』的方向，同樣使之為單位向量。

`main()` 的第 9 行大概就是本篇最關鍵的一行，如同上方講的使用計算明暗度：`dot(surfaceToLightDir, normal)`，`dot()` 為 GLSL 內建的內積 function[12]，不過為了避免數值跑到負的，再套上 GLSL 另一個內建 function `clamp()`[13] 把數值限制一個範圍上，後面兩個參數指定最小最大值，第一個參

---

11 https://www.khronos.org/registry/OpenGL-Refpages/es3.0/html/normalize.xhtml
12 https://www.khronos.org/registry/OpenGL-Refpages/es3.0/html/dot.xhtml
13 https://www.khronos.org/registry/OpenGL-Refpages/es3.0/html/clamp.xhtml

4　Lighting

數大於最大值時 `clamp()` 會回傳最大值、小於最小值時回傳最小值、介於範圍時原封不動回傳，在這邊也就是限制在 `0.0` 到 `1.0` 之間，產生出明暗度數值，最後在輸出顏色的地方為 rgb 原本的 color 乘上這個明暗度，存檔重整後渲染如圖 4-8 所示：

▲ 圖 4-8　套用光線照射 diffuse 亮度後之成像

可以從球形的明暗看得出光線對於一個面的照射角度的成像，正對著光線來源的面才有顏色，傾斜角度為大就越暗；為了方便觀察及測試，我們來加入幾個使用者控制來調整平行光方向以及球體 x 軸的縮放，先在 html 上加入幾個 range input：

```html
<body>
  <canvas id="canvas"></canvas>
  <form id="controls">
    <div class="py-1">
      <label for="light-rot-x">lightRotX</label>
      <input
        type="range" id="light-rot-x" name="light-rot-x"
        min="-90" max="90" value="0"
      >
    </div>
    <div class="py-1">
      <label for="light-rot-z">lightRotZ</label>
      <input
        type="range" id="light-rot-z" name="light-rot-z"
```

4-1 法向量（Normals）與散射光（Diffuse）

```html
      min="-90" max="90" value="0"
    >
  </div>
  <div class="py-1">
    <label for="sphere-scale-x">sphereScaleX</label>
    <input
      type="range" id="sphere-scale-x" name="sphere-scale-x"
      min="0" max="10" value="1" step="0.01"
    >
  </div>
</form>
<script type="module" src="main.js"></script>
</body>
```

這邊加入的 `lightRotX`、`lightRotZ` 筆者定義為『平行光方向用 x 軸、z 軸旋轉的角度』，一開始皆為 0 度，可調整範圍從 -90~+90 度，到時候更新到 `app.state` 中的 `lightDir`，同時補上 `sphereScaleX` 作為球體 x 軸的縮放：

```js
async function setup() {
  // ...
  return {
    // ...
    state: {
      fieldOfView: degToRad(45),
      cameraPosition: [0, 0, 8],
      cameraVelocity: [0, 0, 0],
      sphereScaleX: 1,
      lightDir: [0, -1, 0],
    },
    time: 0,
  };
}
```

渲染時把 sphereScaleX 加入到球體的 transform 矩陣中：

```js
function render(app) {
  // ...

  { // sphere
```

4-21

4　Lighting

```javascript
  gl.bindVertexArray(objects.sphere.vao);

  const worldMatrix = matrix4.multiply(
    matrix4.translate(0, 0, 0),
    matrix4.scale(state.sphereScaleX, 1, 1),
  );
  // ...
}
```

最後把使用者控制項事件串接到 app.state 上：

```javascript
async function main() {
  // ...

  const controlsForm = document.getElementById('controls');
  controlsForm.addEventListener('input', () => {
    const formData = new FormData(controlsForm);

    const lightRotXRad =
      degToRad(parseFloat(formData.get('light-rot-x')));
    const lightRotZRad =
      degToRad(parseFloat(formData.get('light-rot-z')));

    // 對 [0, -1, 0] 旋轉 x 軸 lightRotXRad、旋轉 z 軸 lightRotZRad
    app.state.lightDir[0] = (
      -1 * Math.cos(lightRotXRad) * Math.sin(lightRotZRad)
    );
    app.state.lightDir[1] = (
      -1 * Math.cos(lightRotXRad) * Math.cos(lightRotZRad)
    );
    app.state.lightDir[2] = -1 * Math.sin(lightRotXRad);

    app.state.sphereScaleX = (
      parseFloat(formData.get('sphere-scale-x'))
    );
  });

  // ...
}
```

4-1　法向量（Normals）與散射光（Diffuse）

這邊從 `light-rot-x`、`light-rot-z` input 到 `app.state.lightDir` 的串接稍微說明一下：8~11 行把原始 input 角度（字串）轉成浮點數、再從角度轉成弧度，而平行光方向從原本往正下向方照射：`[0, -1, 0]`，經過 x 軸、z 軸旋轉之後 x 可以從 `y = -1` 中得到 `-1 *cos(lightRotXRad) * sin(lightRotZRad)`、z 得到 `-1 * sin(lightRotXRad)`，y 則成為 `-1 * cos(lightRotXRad) * cos(lightRotZRad)`，把這些公式寫成程式如 14~20 行。

存檔回到瀏覽器，這時就可以調整平行光的方向來觀察物體明暗的變化，例如把 lightRotX 調大，使得光線往 -z 方向偏移，如圖 4-9 所示：

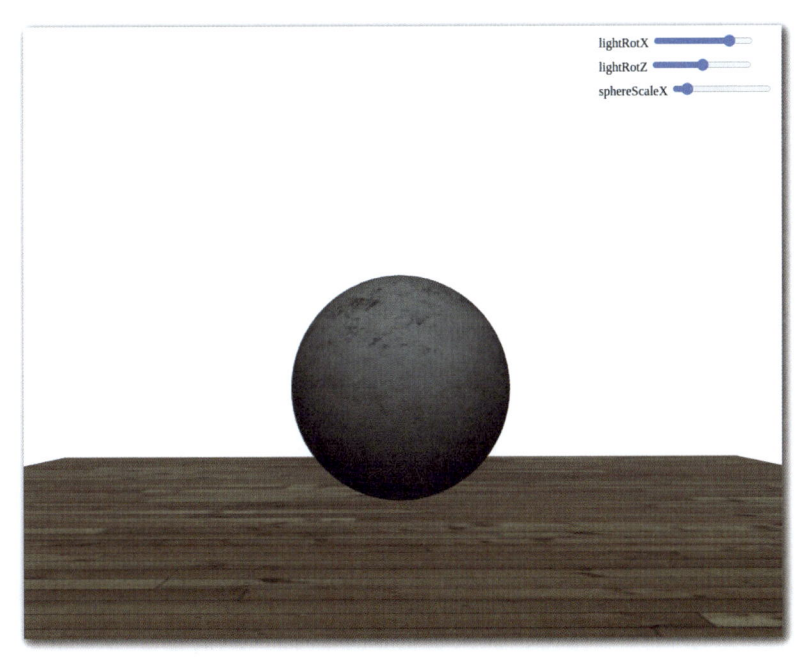

▲ 圖 4-9　經使用者調整光線往 -z 方向偏移之成像

把球體拉長，觀察亮度是否根據該面面對光線的方向有正確的亮度，如圖 4-10：

4-23

4　Lighting

▲ 圖 4-10　拉長球體後之成像

到此進度的完整程式碼可以在這邊找到：

</>	ch4/01/index.html
</>	ch4/01/main.js

也可以參考上線版本，透過瀏覽器打開此頁面：

https://webgl-book.pastleo.me/ch4/01/index.html

如果有讀者好奇，可以試試看把法向量的 transform 改成直接用 `worldMatrix`，也就是不做反矩陣以及轉置矩陣，這麼一來旋轉、縮放會照著物體做轉換，而非旋轉以及『縮放的相反』：

```javascript
function render(app) {
  // ...
  { // sphere and ground
    // ...
    gl.uniformMatrix4fv(
      uniforms.normalMatrix,
```

4-24

```
      false,
      worldMatrix,
      // matrix4.transpose(matrix4.inverse(worldMatrix)),
    );
    // ...
  }
}
```

這時如果把球體拉伸,就可以看到因為法向量不正確的轉換,導致亮部不正確,如圖 4-11:

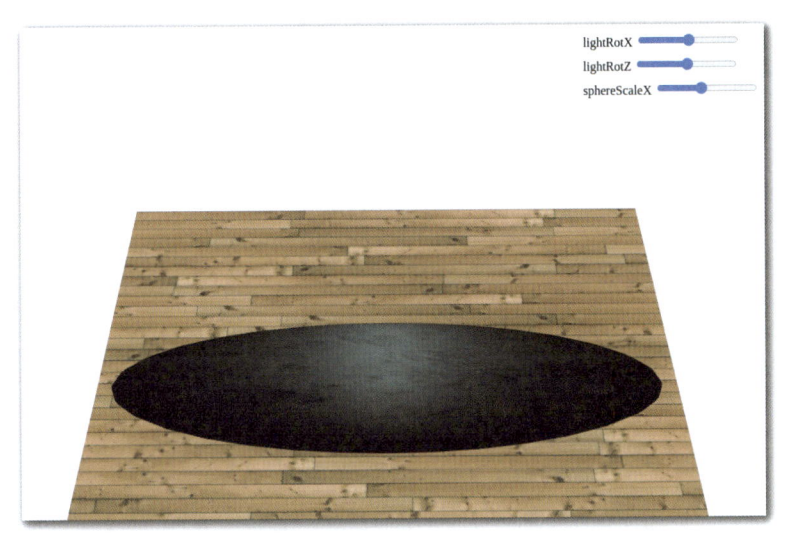

▲ 圖 4-11　不對法向量做『縮放相反』轉換時拉長球體之成像

4-2 Indexed Element

　　回想 CH3 那個 P 形狀的 3D 模型,傳入 `a_position` buffer 的資料中,因為必須以三角形頂點為單位,輸入的資料會有不少重複部份,顯然有些浪費記憶體,這件事情也已經可以在 CH3 使用 TWGL 產生頂點資料時,經過 `deindexVertices()`『展開』資料之前與之後的陣列長度看出重複的資料量,以當時球體 position 為例,原本有 3267 個數值,展開後變成 18432 個。

4-25

4 Lighting

⌘ 什麼是 Indexed Element？舉個正方形範例來說明：

本書到目前為止的繪製模式都是使用 `gl.drawArrays()` 進行繪製，這個模式下 vertex shader 的 attribute 依序取用 buffer 中的資料計算出頂點位置以及對應的 varying，除此之外還有一種 indexed element 模式使得繪製時透過 indices 類似指標去取得其指向之 vertex attribute 的值；這邊筆者舉一個範例，如果我們要繪製這樣的正方形，各個點的座標以及分成的三角形如圖 4-12 所示：

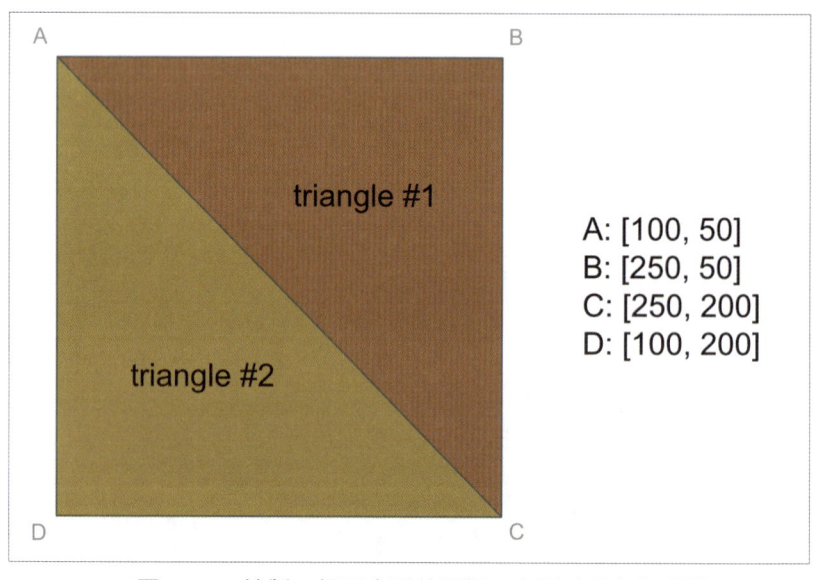

▲ 圖 4-12　繪製一個正方形的兩個三角形以及各個頂點

若使用 `gl.drawArrays()` 的繪製模式，必須要讓輸入的每個 attribute buffer 都以三角形頂點為單位輸入，雖然這邊只有 4 個點，但是我們必須輸入 6 個頂點，像是圖 4-13 這樣，圖中 position buffer 之平行四邊形為我們實際要輸入到 buffer 中的資料：

4-2　Indexed Element

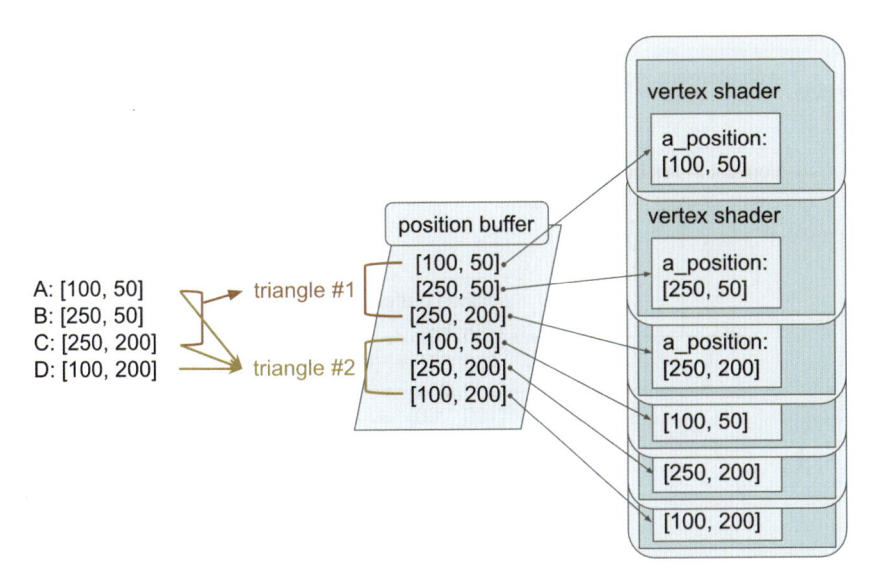

▲ 圖 4-12　gl.drawArrays() 繪製模式下 buffer 與 shader attribute 取用關係

vertex shader 第一次執行的 `a_position` 從 position buffer 中使用第一組資料、第二次第二筆、第三次第三筆，以此類推，可以明顯看到輸入的 position buffer 有兩組資料是完全重複的（`[100, 50]` 以及 `[250, 50]` 這兩組），在複雜的 3D 物件中很可能會有更多重複的資料造成記憶體的浪費。

為解決此問題，WebGL 提供另一種 indexed element 的繪製模式：`gl.drawElements()`，透過一個叫做 `ELEMENT_ARRAY_BUFFER`（圖 4-13 中的 element index buffer）的 buffer 當成指標，每次 vertex shader 執行時取得的所有 attribute 將變成指標所指向的那組資料，也因此『畫什麼』的部份需要多傳送一整組『指標』buffer，但是其他 attribute buffer 如 position、texcoord、normal 就可以省去重複的資料。

以這個正方形舉例的話，所需要傳送的 buffer 以及最後 vertex shader 各次執行所得到 attribute 之內容如圖 4-13 所示：

4-27

4　Lighting

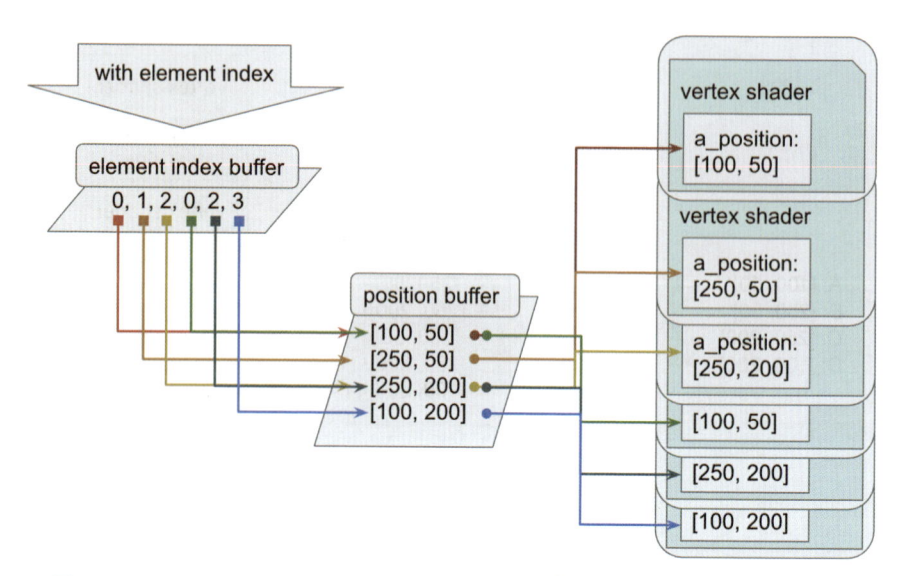

▲ 圖 4-13　gl.drawElements() 繪製模式下 buffer 與 shader attribute 取用關係

可以到圖 4-13 中 position buffer 就不會有重複的資料了，vertex shader 第一次執行的 `a_position` 由 position buffer 中根據 element index buffer 左到右第一個數字 0 取用第一筆資料得到 `[100, 50]`、第二次根據 element index buffer 第二個數字 1 取第二筆資料得到 `[250, 50]`、第五次根據 element index buffer 第五個數字 2 取取用第三筆得到 `[250, 200]`、第六次 index 為 3 取用第四筆得到 `[100, 200]`。

⌘ 改用 Indexed Element

總結來說 indexed element 繪製模式就是多輸入一組 indices 陣列到 element index buffer 來描述每次 vertex shader 之每個 attribute 要從 buffer 中拿出第幾筆資料，接下來我們就來把原本使用 `gl.drawArrays()` 的繪製模式改成 indexed element 使用 `gl.drawElements()` 來進行繪製。

之前不想要用 indexed element 功能，因此使用 `twgl.primitives.deindexVertices()` 展開 `twgl.primitives` 產生的頂點資料，現在可

4-28

4-2　Indexed Element

以不用展開了，`vertexDataArrays` 可以直接接收 `twgl.primitives.createXXXVertices()` 的回傳值，同時 `numElements` 作為頂點數量可以直接使用 `vertexDataArrays.indices` 陣列的長度 `vertexDataArrays.indices`：

```javascript
async function setup() {
  // ...
  { // sphere
    const vertexDataArrays = twgl.primitives.deindexVertices(
      twgl.primitives.createSphereVertices(1, 32, 32)
    );
    const vertexDataArrays = (
      twgl.primitives.createSphereVertices(1, 32, 32)
    );
    const numElements = (
      vertexDataArrays.position.length /
      vertexDataArrays.position.numComponents
    );
    const numElements = vertexDataArrays.indices.length;

    // vao, buffers, a_position, a_texcoord ...
  }

  { // ground
    const vertexDataArrays = twgl.primitives.deindexVertices(
      twgl.primitives.createPlaneVertices()
    );
    const vertexDataArrays = twgl.primitives.createPlaneVertices();
    const numElements = (
      vertexDataArrays.position.length /
      vertexDataArrays.position.numComponents
    );
    const numElements = vertexDataArrays.indices.length;

    // vao, buffers, a_position, a_texcoord ...
  }
  // ...
}
```

4 Lighting

接著建立 element index buffer 並把 `vertexDataArrays.indices` 輸入，記得 sphere 和 ground 各要做一次：

```
async function setup() {
  // ...
  { // both sphere and ground
    // a_position, a_texcoord, a_normal ...

    // indexed element indices
    buffers.indices = gl.createBuffer();
    gl.bindBuffer(gl.ELEMENT_ARRAY_BUFFER, buffers.indices);
    gl.bufferData(
      gl.ELEMENT_ARRAY_BUFFER,
      vertexDataArrays.indices,
      gl.STATIC_DRAW,
    );

    // objects.sphere / objects.ground = { ... }
  }
  // ...
}
```

這邊可以看到 `gl.bindBuffer()` 與 `gl.bufferData()` 的第一個參數 target 是 `gl.ELEMENT_ARRAY_BUFFER`，與 position、texcoord 等一般 attribute 用的 `gl.ARRAY_BUFFER` 不同。

最後把 `gl.drawArrays()` 改用 `gl.drawElements()`，使用多輸入的這組 indices 陣列來描述每次 vertex shader 之每個 attribute 要從 buffer 中拿出第幾筆資料：

```
function render(app) {
  // ...
  { // sphere
    // ...
    gl.drawArrays(gl.TRIANGLES, 0, objects.sphere.numElements);
    gl.drawElements(
```

4-30

```
    gl.TRIANGLES,
    objects.sphere.numElements,
    gl.UNSIGNED_SHORT,
    0,
  );
}

{ // ground
  // ...
  gl.drawArrays(gl.TRIANGLES, 0, objects.ground.numElements);
  gl.drawElements(
    gl.TRIANGLES,
    objects.ground.numElements,
    gl.UNSIGNED_SHORT,
    0,
  );
}
// ...
}
```

`gl.drawElements()` 的第一個參數與 `gl.drawArrays()` 相同表示要繪製的形狀 `gl.TRIANGLES`；第二個參數為頂點數量，也就是 `gl.drawArrays()` 的第三個參數；而 `gl.drawElements()` 的第三個參數為 element index buffer 的資料格式，給 `gl.UNSIGNED_SHORT` 以符合 `vertexDataArrays.indices` 的格式：`Uint16Array`。

改完重整後不會有變化，到此進度的完整程式碼可以在這邊找到：

</>	ch4/02/index.html
</>	ch4/02/main.js

也可以參考上線版本，透過瀏覽器打開此頁面：

https://webgl-book.pastleo.me/ch4/02/index.html

4-31

4　Lighting

因為我們的資料全部靠 TWGL 產生，讀者或許對於 indexed element 狀況下輸入的實際資料不是很有感，可以嘗試把使用 Web 開發工具之 Console 輸入 `app.objects.ground.vertexDataArrays` 查看一個平面（兩個三角形）的 position、texcoord、normal 以及 indices 資料，如圖 4-14 所示：

```
> app.objects.ground.vertexDataArrays
<· {position: Float32Array(12), normal: Float32Array(12), texcoord: Float32Array(8), indices: Uint16Array(6)}
   ▼ indices: Uint16Array(6)
       0: 0
       1: 2
       2: 1
       3: 2
       4: 3
       5: 1
       numComponents: 3
     ▶ push: f ()
     ▶ reset: f (opt_index)
       numElements: (...)
     ▶ buffer: ArrayBuffer(12)
       byteLength: 12
       byteOffset: 0
       length: 6
       Symbol(Symbol.toStringTag): "Uint16Array"
     ▶ get numElements: f ()
     ▶ [[Prototype]]: TypedArray
   ▶ normal: Float32Array(12) [0, 1, 0, 0, 1, 0, 0, 1, 0, 0, 1, 0, numComponents: 3, num
   ▶ position: Float32Array(12) [-0.5, 0, -0.5, 0.5, 0, -0.5, -0.5, 0, 0.5, 0.5, 0, 0.5,
   ▶ texcoord: Float32Array(8) [0, 0, 1, 0, 0, 1, 1, 1, numComponents: 2, numElements: (
   ▶ [[Prototype]]: Object
```

▲ 圖 4-14　`app.objects.ground.vertexDataArrays` 在 indexed element 繪製模式之內容

讀者可以試著依照這邊的資料，繪製一個類似圖 4-12 由兩個三角形組成的正方形，標示出頂點位置以及這兩個三角形分別使用了哪些點，藉此對於資料的擺放、取用更加了解。

4-3　請 TWGL 替程式碼減肥

如果讀者稍有實務軟體工程經驗，那麼肯定會覺得目前實做出的程式碼中有許多重複的部份，最明顯的應該屬 setup() 中對於每個物件、每個 attribute 進行 `gl.createBuffer()`、`gl.vertexAttribPointer()` 到 `gl.bufferData()`。在之前 attribute 還不算太多時還可以接受，但是為了與光線運算加入的 normal 後，筆者開始覺得是時候正視並處理這兩個問題，現在某些程式碼區塊很明顯可以拆成可重複呼叫的 function，例如各個物件從 `vertexDataArrays`、VAO、attribute 與 buffer 建立綁定到對 buffer 輸入資料的流程，而這些『把繁瑣的流程拆成可重複呼叫之 function』的工作已經由 TWGL 做好，因此本篇將把這些流程改用 TWGL 進行重構以對程式碼減肥。

⌘ 使用 TWGL 建立 `programInfo`、設定 uniforms

老實說，WebGL 有提供 API 來列舉、取得 GLSL program 中的 attribute[14] 以及 uniform[15] 資訊，因此先前寫的 `gl.getAttribLocation()`、`gl.getUniformLocation()` 是可以被自動化的，這個自動化在 TWGL 中已經幫我們實做於 `twgl.createProgramInfo()`，而且同時也包含了 shader 的編譯、program 的連結（Link），也就是我們在 CH1『怎麼畫』、後來搬到 `lib/utils.js` 的 `createShader()`、`createProgram()`，甚至再往上包裝成一個 function 可以直接傳入 vertex、fragment shader 的原始碼，最後回傳整個稱為 programInfo 的資料結構，這邊嘗試呼叫並用 `console.log()` 顯示這個 `programInfo`：

```
async function setup() {
  // canvas, gl ...

  console.log(
    twgl.createProgramInfo(gl, [vertexShaderSource, fragmentShaderSource])
  );
}
```

---

14 列舉、取得 attribute 的 WebGL API 為 `gl.getActiveAttrib()`：
https://developer.mozilla.org/en-US/docs/Web/API/WebGLRenderingContext/getActiveAttrib

15 列舉、取得 attribute 的 WebGL API 為 `gl.getActiveUniform()`：
https://developer.mozilla.org/en-US/docs/Web/API/WebGLRenderingContext/getActiveUniform

4 Lighting

`programInfo` 之內容如圖 4-15 所示：

```
{program: WebGLProgram, uniformSetters: {…}, attribSetters: {…}, uniformBlockSpec: {…},
 transformFeedbackInfo: {…}}
  ▼ attribSetters:
    ▶ a_normal: f (b)
    ▼ a_position: f (b)
        location: 0
        length: 1
        name: ""
      ▶ prototype: {constructor: f}
        arguments: (...)
        caller: (...)
        [[FunctionLocation]]: twgl-full.module.js:7110
      ▶ [[Prototype]]: f ()
      ▶ [[Scopes]]: Scopes[3]
    ▶ a_texcoord: f (b)
    ▶ [[Prototype]]: Object
  ▶ program: WebGLProgram {}
  ▶ transformFeedbackInfo: {}
  ▶ uniformBlockSpec: {blockSpecs: {…}, uniformData: Array(5)}
  ▼ uniformSetters:
    ▶ u_color: f (v)
    ▶ u_lightDir: f (v)
    ▶ u_matrix: f (v)
    ▶ u_normalMatrix: f (v)
    ▶ u_texture: f (textureOrPair)
    ▶ [[Prototype]]: Object
  ▶ [[Prototype]]: Object
```

▲ 圖 4-15　app.objects.ground.vertexDataArrays 在 indexed element 繪製模式之內容

可以在 `programInfo` 中看到 attributes 跟 uniforms 都已經被偵測好分別放在 `.attribSetters` 與 `.uniformSetters` 之中，編譯、連結完成的 shader 程式放置在 `.program`，同時 shader 中變數位置也可以透過 `.attribSetters.a_xxx.location`、`.uniformSetters.a_xxx.location` 來取得，因此原本寫的一長串編譯、連結 shader 程式到取得 attributes、uniforms 位置的程式碼整個可以用一個 `twgl.createProgramInfo()` 取代掉：

```js
async function setup() {
  // canvas, gl ...

  const vertexShader = createShader(
    gl,
    gl.VERTEX_SHADER,
    vertexShaderSource,
```

4-3 請 TWGL 替程式碼減肥

```javascript
);
const fragmentShader = createShader(
  gl,
  gl.FRAGMENT_SHADER,
  fragmentShaderSource,
);
const program = createProgram(gl, vertexShader, fragmentShader);

const attributes = {
  position: gl.getAttribLocation(program, 'a_position'),
  texcoord: gl.getAttribLocation(program, 'a_texcoord'),
  normal: gl.getAttribLocation(program, 'a_normal'),
};
const uniforms = {
  matrix: gl.getUniformLocation(program, 'u_matrix'),
  normalMatrix: gl.getUniformLocation(program, 'u_normalMatrix'),
  color: gl.getUniformLocation(program, 'u_color'),
  texture: gl.getUniformLocation(program, 'u_texture'),
  lightDir: gl.getUniformLocation(program, 'u_lightDir'),
};
const programInfo = twgl.createProgramInfo(
  gl,
  [vertexShaderSource, fragmentShaderSource],
);

// ...
}
```

但是因為 `programInfo` 跟我們原本存放 attribute、uniform 位置的方式不同，要去修改對於變數位置的使用方式，先從 attribute 開始：

```javascript
async function setup() {
  // programInfo, objects ...
  { // both sphere and ground
    const vertexDataArrays = twgl.primitives.createSphereVertices(
      1, 32, 32
    );
    const numElements = vertexDataArrays.indices.length;
    const vao = gl.createVertexArray();
```

4 Lighting

```
  gl.bindVertexArray(vao);

  const buffers = {};

  // a_position, a_texcoord, a_normal
  buffers.xxx = gl.createBuffer();
  gl.bindBuffer(gl.ARRAY_BUFFER, buffers.xxx);

  gl.enableVertexAttribArray(attributes.xxx)
  gl.enableVertexAttribArray(programInfo.attribSetters.a_xxx.location);
  gl.vertexAttribPointer(
    attributes.xxx,
    programInfo.attribSetters.a_xxx.location,
    vertexDataArrays.xxx.numComponents, // size
    gl.FLOAT, // type
    false, // normalize
    0, // stride
    0, // offset
  );
  // ...
  }
  // ...
}
```

原本有個 `attributes` 物件,使用 `attributes.xxx` 來取用變數位置(`xxx` 是 `position`、`texcoord`、`normal`,請自行代入),現在將改成 `programInfo.attribSetters.a_xxx.location`,記得 sphere 跟 ground 的 position、texcoord、normal,總共 6 個 attribute-buffer 都要修改。

在 `setup()` 與 `render()` 之間傳送之 shader 程式、變數資訊現在改為 `programInfo`,因此做出對應的修改:

```
async function setup() {
  // ...

  return {
    gl,
```

```
    program, attributes, uniforms,
    programInfo,
    textures, objects,
    state: {
      // ...
    },
    time: 0,
  };
}

function render(app) {
  const {
    gl,
    program, uniforms,
    programInfo,
    textures, objects,
    state,
  } = app;

  // ...
}
```

`render()` 內首先在設定使用之 shader 程式的地方改用 programInfo：

```
function render(app) {
  // app, gl.canvas, gl.viewport ...

  gl.useProgram(programInfo.program);

  // cameraMatrix, sphere, ground ...
}
```

uniform 的部份，TWGL 有包好一個功能強大的 `twgl.setUniforms()`，使用上把 `programInfo` 傳入，第二個參數是一個 key-value 的物件，key 為 uniform 變數名稱、value 即為 uniform 要輸入的值，先修改光線方向 `lightDir` 的設定，使用起來如下：

4-37

4 Lighting

```javascript
function render(app) {
  // cameraMatrix, viewMatrix ...

  const textureUnit = 0;
  gl.uniform3f(uniforms.lightDir, ...state.lightDir);
  twgl.setUniforms(programInfo, {
    u_lightDir: state.lightDir,
  });

  // sphere, ground ...
}
```

而且 `twgl.setUniforms()` 不只是如此，原本設定 uniform 的 `gl.uniformXX()` 每次呼叫只能設定一個 uniform，而且如果是設定 texture，則還要多呼叫 `gl.bindTexture()`、`gl.activeTexture()` 等，這部份在 `twgl.createProgramInfo()` 時因為有偵測型別，發現設定 uniform 是 texture 便會自動執行 texture 設定該做的動作，因此只要使用 `twgl.setUniforms()`，就能一次設定許多 uniform 並且根據變數型別自動做對應的設定，這邊對球體以及地板渲染之各個 uniform 改用 `twgl.setUniforms()` 進行設定：

```javascript
function render(app) {
  // ...

  { // both sphere and ground
    // gl.bindVertexArray(), worldMatrix ...

    gl.uniformMatrix4fv(
      uniforms.matrix,
      false,
      matrix4.multiply(viewMatrix, worldMatrix),
    );

    gl.uniformMatrix4fv(
      uniforms.normalMatrix,
      false,
      matrix4.transpose(matrix4.inverse(worldMatrix)),
    );
```

```
    gl.uniform3f(uniforms.color, 0, 0, 0);

    gl.bindTexture(gl.TEXTURE_2D, textures.steel);
    gl.activeTexture(gl.TEXTURE0 + textureUnit);
    gl.uniform1i(uniforms.texture, textureUnit);
    twgl.setUniforms(programInfo, {
      u_matrix: matrix4.multiply(viewMatrix, worldMatrix),
      u_normalMatrix: matrix4.transpose(matrix4.inverse(worldMatrix)),
      u_color: [0, 0, 0],
      u_texture: textures.steel, // or textures.wood for ground
    });

    gl.drawElements(/* ... */);
  }
}
```

改用 `twgl.createProgramInfo()` 以及 `twgl.setUniforms()` 重構後，功能不變，但是程式碼已經減少了不少，到此進度的完整程式碼可以在這邊找到：

</>	ch4/03/index.html
</>	ch4/03/main.js

也可以參考上線版本，透過瀏覽器打開此頁面：
https://webgl-book.pastleo.me/ch4/03/index.html

⌘ 使用 TWGL 的 bufferInfo 取代繁瑣的 attribute-buffer 設定

在每個物件裡頭的每個 attribute，都要分別 `gl.createBuffer()`、`gl.bindBuffer()`、`gl.enableVertexAttribArray()`、`gl.vertexAttribPointer()` 並且傳送資料 `gl.bufferData()`，有經驗的開發者應該很快可以看得出來這邊可以用某種資料結構描述這些 attribute 的設定值以及資料，老實說，透過 `twgl.primitives.createXXXVertices()` 所建立的 `vertexDataArrays` 其實就是這樣的資料結構，我們可以整組傳給 `twgl.createBufferInfoFromArrays()` 把所

4 Lighting

有的 buffer 一次建立好並包裝成 `bufferInfo` 這個資料結構，並且透過 `twgl.createVAOFromBufferInfo()` 傳入 `programInfo` 與 `bufferInfo` 建立 buffer-attribute 關聯與 VAO，甚至 TWGL 也在『畫』這個動作做了包裝可以直接使用 `bufferInfo`，也就是說從 `gl.createVertexArray()` 到 `gl.bufferData()` 全部都可以用 TWGL 簡單幾行取代，要刪除的行數實在太多，這邊直接寫改完後 `setup()` 內準備物件資料的樣子：

```
async function setup() {
  // programInfo, textures ...

  const objects = {};

  { // sphere
    const vertexDataArrays = (
      twgl.primitives.createSphereVertices(1, 32, 32)
    );
    const bufferInfo = twgl.createBufferInfoFromArrays(
      gl, vertexDataArrays,
    );
    const vao = twgl.createVAOFromBufferInfo(gl, programInfo, bufferInfo);

    objects.sphere = {
      vertexDataArrays,
      vao, bufferInfo,
    };
  }

  { // ground
    const vertexDataArrays = twgl.primitives.createPlaneVertices();
    const bufferInfo = twgl.createBufferInfoFromArrays(
      gl, vertexDataArrays,
    );
    const vao = twgl.createVAOFromBufferInfo(gl, programInfo, bufferInfo);

    objects.ground = {
      vertexDataArrays,
      vao, bufferInfo,
    };
```

```
}

  // gl.enable(), return { ... objects ... }
}
```

沒錯，這邊大概刪除了一兩百行程式碼，這就是引用套件自動化的強大之處，在繼續之前，可以打開 Console 輸入 `app.objects` 看看各個物件由 `twgl.createBufferInfoFromArrays()` 產生的 `bufferInfo` 資料結構，如圖 4-16 所示，在 `bufferInfo` 內有 `numElements` 表示 `gl.drawElement()` 或 `gl.drawArrays()` 時要畫的頂點數量、`elementType` 也先幫我們填好 `gl.UNSIGNED_SHORT`，同時還有 indexed element 的 `indices`、`attribs` 描述各個 attribute 以及根據 `vertexDataArrays` 建立的 buffer：

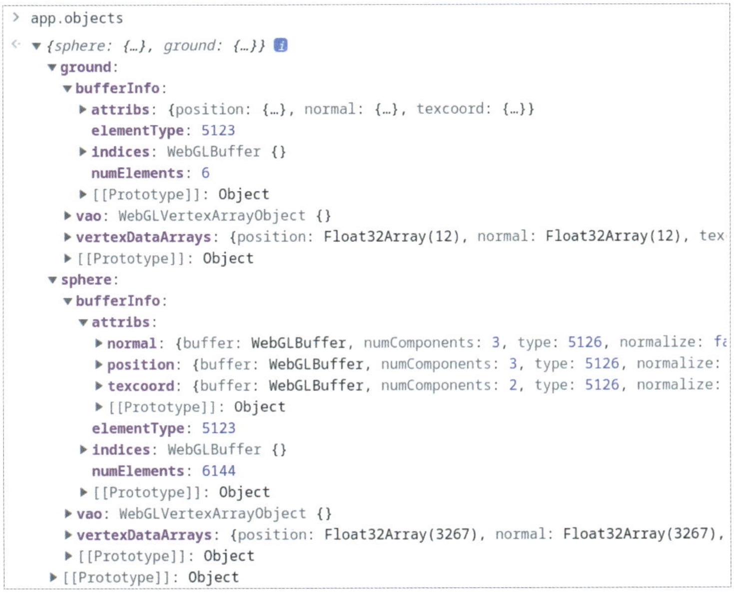

▲ 圖 4-16　app.objects 球體與地板的 bufferInfo 內容

這邊有一個問題，在 vertex shader 中 attribute 的變數名稱都有 `a_` 開頭方便我們知道這是一個 attribute，導致 `twgl.primitives.createXXXVertices()` 所回傳的資料無法直接跟 vertex shader 的 attribute 變數名稱對起來，幸好 TWGL 有提供 `twgl.setAttributePrefix()` 設定 attribute 的 prefix（前綴），像

4 Lighting

這樣執行於 `setup()` 一開始即可：

```javascript
async function setup() {
  const canvas = document.getElementById('canvas');
  const gl = canvas.getContext('webgl2');

  twgl.setAttributePrefix('a_');

  // programInfo, textures, sphere, ground ...
}
```

加上這一行之後，回到 Console 輸入 `app.objects` 能看到現在 `attribs` 的各個變數名稱都能如圖 4-17 這樣與我們的 shader 對起來囉：

```
> app.objects
< ▼ {sphere: {…}, ground: {…}} ⓘ
    ▼ ground:
      ▼ bufferInfo:
        ▼ attribs:
          ▶ a_normal: {buffer: WebGLBuffer, numComponents: 3, type: 5126, normalize
          ▶ a_position: {buffer: WebGLBuffer, numComponents: 3, type: 5126, normali
          ▶ a_texcoord: {buffer: WebGLBuffer, numComponents: 2, type: 5126, normali
          ▶ [[Prototype]]: Object
          elementType: 5123
        ▶ indices: WebGLBuffer {}
          numElements: 6
        ▶ [[Prototype]]: Object
      ▶ vao: WebGLVertexArrayObject {}
      ▶ vertexDataArrays: {position: Float32Array(12), normal: Float32Array(12), te
      ▶ [[Prototype]]: Object
    ▼ sphere:
      ▼ bufferInfo:
        ▼ attribs:
          ▶ a_normal: {buffer: WebGLBuffer, numComponents: 3, type: 5126, normalize
          ▶ a_position: {buffer: WebGLBuffer, numComponents: 3, type: 5126, normali
          ▶ a_texcoord: {buffer: WebGLBuffer, numComponents: 2, type: 5126, normali
          ▶ [[Prototype]]: Object
```

▲ 圖 4-17 bufferInfo.attribs 各個 attribute 加上 a_ 的 prefix

最後 TWGL 也在『畫』這個動作也做的包裝為 `twgl.drawBufferInfo()`，讓我們可以直接傳入 `bufferInfo` 進行繪製，取代 `gl.drawElement()` 或 `gl.drawArrays()`：

```javascript
function render(app) {
  // ...
```

4-42

```
{ // sphere
  // ...

  gl.drawElements(
    gl.TRIANGLES, objects.sphere.numElements, gl.UNSIGNED_SHORT, 0,
  );
  twgl.drawBufferInfo(gl, objects.sphere.bufferInfo);
}

{ // ground
  // ...

  gl.drawElements(
    gl.TRIANGLES, objects.ground.numElements, gl.UNSIGNED_SHORT, 0,
  );
  twgl.drawBufferInfo(gl, objects.ground.bufferInfo);
}
}
```

在渲染時期關於 VAO 的切換還是使用 WebGL API `gl.bindVertexArray()` 來切換，要留意 buffer-attribute 的關聯是存放在 VAO『工作區域』的，`twgl.createBufferInfoFromArrays()` 跟 `twgl.drawBufferInfo()` 是不會幫忙處理的，因此要記得使用 `twgl.createVAOFromBufferInfo()` 建立好 VAO 並且透過 `gl.bindVertexArray()` 好好切換工作區域。

存檔回到瀏覽器進行測試，功能運作起來依然是一個球體、地板以及散射光效果，但是程式碼簡短了非常多，到此進度的完整程式碼可以在這邊找到：

</>	ch4/04/index.html
</>	ch4/04/main.js

也可以參考上線版本，透過瀏覽器打開此頁面：
https://webgl-book.pastleo.me/ch4/04/index.html

4 Lighting

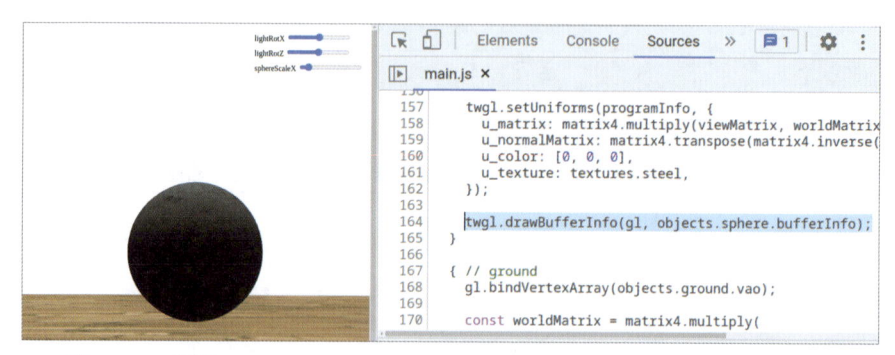

▲ 圖 4-17　採用 TWGL bufferInfo、twgl.drawBufferInfo() 來重構後功能不變

4-4　Specular 反射光

有了散射光的計算，物體的表面根據有沒有被光照射到而顯示；而本篇將介紹計算上較為複雜的 specular 反射光 [16]，筆者覺得加上這個效果之後，物體就可以呈現金屬、或是光滑表面的質感，開始跳脫死板的顏色，接下來以此畫面為目標：

▲ 圖 4-18　反射光效果

16　https://zh.wikipedia.org/wiki/光的反射定律、
　　https://en.wikipedia.org/wiki/Specular_reflection

4-4 Specular 反射光

在散射光時我們關注的是物體表面是否有光線照射，如果沒有或是入射角沒有直射時降低表面顏色的亮度；反射光關注的是是否有光線從光源經過物體表面反射後到達相機，有的話『提高』亮度（正確來說是加上反射光顏色）。

反射光的計算方法與所需要的資料

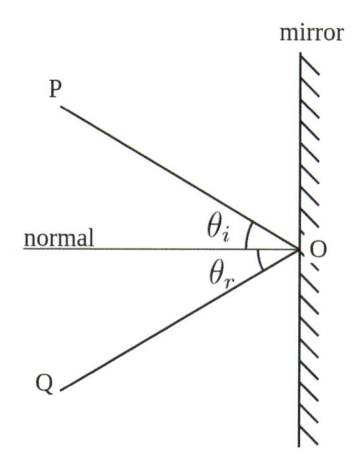

▲ 圖 4-19　光線反射示意圖

在反射光的維基百科頁面中，反射光光線的方向如圖 4-19 所示，入射角與反射角的角度相同，也就是 θ_i 與 θ_r 相同，在本篇實做目標圖 4-18 中，其中球體上的白色反光區域，就是光線入射角與反射角角度很接近的地方；而在 fragment shader 內，與計算散射時一樣，與其計算角度，不如利用單位向量的內積，先計算光線方向反向 `surfaceToLightDirection` 與表面到相機方向 `surfaceToViewerDirection` 的『中間向量』，也就是 `surfaceToLightDirection` 與 `surfaceToViewerDirection` 兩個向量箭頭頂點的中間位置延伸而成的單位向量 `halfVector`，再拿 `halfVector` 與法向量做內積得到反射光的明暗度，如圖 4-20 所示：

4 Lighting

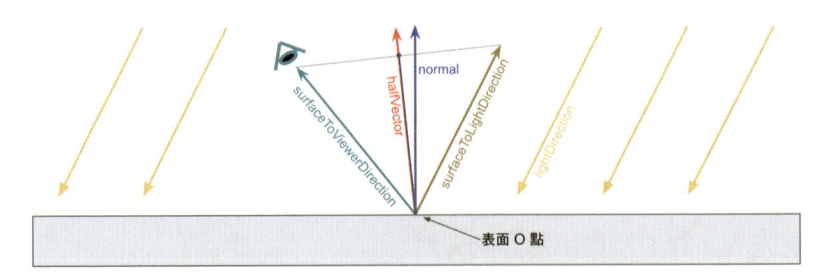

▲ 圖 4-20　反射光亮度計算方式示意圖

　　為了知道表面 O 點到相機方向，我們要在 shader 內計算出表面的座標位置，也就是只有經過 `worldMatrix` 做 transform、沒有經過 `viewMatrix` 轉換的 position，因此除了同時包含 `worldMatrix` 與 `viewMatrix` 的 `u_matrix` 之外，也得傳 `worldMatrix`，我們就叫這個 uniform `u_worldMatrix`；另外也需要傳送相機的位置進去 `u_worldViewerPosition`，才能計算表面到相機的向量，在上一篇改用 TWGL 改寫之後增加 uniform 只需要直接在 `render()` 的 `twgl.setUniforms()` 增加傳送的資料即可：

```js
function render(app) {
  // viewMatrix ...

  twgl.setUniforms(programInfo, {
    u_worldViewerPosition: state.cameraPosition,
    u_lightDir: state.lightDir,
  });

  { // both sphere and ground
    // bindVertexArray, worldMatrix ...

    twgl.setUniforms(programInfo, {
      u_matrix: matrix4.multiply(viewMatrix, worldMatrix),
      u_worldMatrix: worldMatrix,
      u_normalMatrix: matrix4.transpose(matrix4.inverse(worldMatrix)),
      // u_color, u_texture ...
    });
  }
}
```

⌘ Vertex Shader：計算表面到相機方向

應該不難想像，面上某個點的『表面到相機方向』是三角形頂點到相機方向的中間值，符合 varying 特性，也就是說我們可以讓這個方向由 vertex shader 計算出來，用 varying 傳送給 fragment shader，fragment shader 收到的表面到相機方向就會是內插後的結果，我們可以把這個向量叫做 `v_surfaceToViewer`。讓 vertex shader 接收上面多傳的 uniform `u_worldMatrix` 以及 `u_worldViewerPosition`，計算出 `v_surfaceToViewer`：

```glsl
#version 300 es
in vec4 a_position;
in vec2 a_texcoord;
in vec3 a_normal;

uniform mat4 u_matrix;
uniform mat4 u_worldMatrix;
uniform mat4 u_normalMatrix;
uniform vec3 u_worldViewerPosition;

out vec2 v_texcoord;
out vec3 v_normal;
out vec3 v_surfaceToViewer;

void main() {
  gl_Position = u_matrix * a_position;
  v_texcoord = vec2(a_texcoord.x, 1.0 - a_texcoord.y);
  v_normal = mat3(u_normalMatrix) * a_normal;

  // 算出 頂點、表面 座標位置
  vec3 worldPosition = (u_worldMatrix * a_position).xyz;

  // 算出 表面到相機 方向向量
  v_surfaceToViewer = u_worldViewerPosition - worldPosition;
}
```

4 Lighting

⌘ Fragment Shader：實做反射光計算

到 fragment shader 實際計算顏色的時間了，照著上方『反射光的計算方法』所說來實做，先是從 vertex shader 傳來的 varying `v_surfaceToViewer` 轉成單位向量，而 `surfaceToLightDir` 在先前實做散射光時已經有算出來為單位向量可以直接使用，兩者長度皆為 1，加在一起除以二可以得到『中間向量』，但是之後也得轉換成為單位向量，除以二的步驟就可以省略，因此中間向量 `halfVector` 可以把兩個向量相加後取單位向量來算出，最後與法向量做內積，並讓結果 `specularBrightness` 直接加在顏色的所有 channel 上：

```glsl
#version 300 es
precision highp float;

in vec2 v_texcoord;
in vec3 v_normal;
in vec3 v_surfaceToViewer;

uniform vec3 u_color;
uniform sampler2D u_texture;
uniform vec3 u_lightDir;
uniform vec3 u_specular;
uniform float u_specularExponent;

out vec4 outColor;

void main() {
  vec3 color = u_color + texture(u_texture, v_texcoord).rgb;

  vec3 normal = normalize(v_normal);
  vec3 surfaceToLightDir = normalize(-u_lightDir);

  float colorLight = clamp(dot(surfaceToLightDir, normal), 0.0, 1.0);

  // 確保表面位置到相機方向為單位向量
  vec3 surfaceToViewerDirection = normalize(v_surfaceToViewer);

  // 『中間向量』
```

4-48

```
vec3 halfVector = normalize(surfaceToLightDir + surfaceToViewerDirection);

// 反射光亮度
float specularBrightness = dot(halfVector, normal);

outColor = vec4(
  color * colorLight + specularBrightness,
  1
);
}
```

⌘ 『反射範圍過大』問題

如果存檔去看渲染結果，看起來像是圖 4-21 這樣：

▲ 圖 4-21　反射光的效果範圍過大

顯然跟目標擷圖不一樣，為什麼呢？想想看，如果 halfVector 與法向量相差 60 度，那麼我們做完內積之後，可以獲得 0.5 的 specularBrightness，這樣的反射範圍顯然太大，我們希望內積之後的值非常接近 1 才能讓 specularBrightness 有值，有個簡單的方法可以解決這個問題：把內積結果的數值乘上 n 次方，這麼一來 0 還是 0、1 還是 1，只是改變 0 到 1 之間直線變成曲線，假設 n 為 40，圖 4-22 為畫出來的線圖，橫軸為內積完的數值，縱軸為乘上 n 次方後的數值，橫軸接近 0.9 時縱軸數值才開始明顯大於 0：

4 Lighting

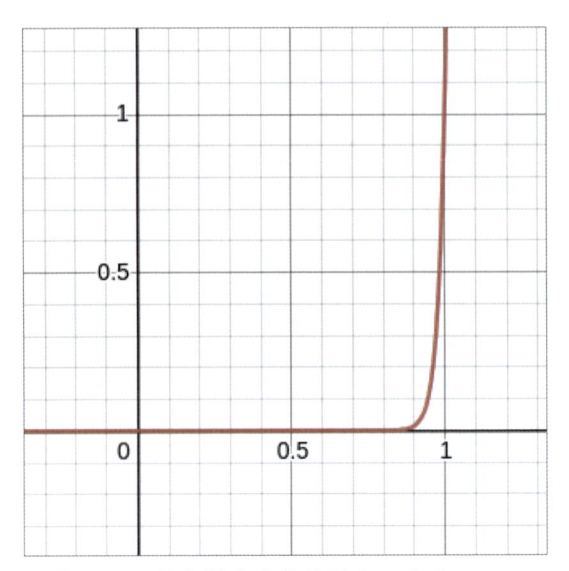

▲ 圖 4-22 將內積完的數值乘上 n 次方，n = 40

圖 4-22 乘上 n = 40 次方的曲線圖擷取自此數學式繪製網站：

https://www.desmos.com/calculator/yfa2jzzejm

讀者可以來這邊拉左方的 n 值感受一下曲線的變化。

其實這個 n 的值可以根據不同物件材質而有所不同，因此加上 uniform `u_specularExponent` 來控制，同時也加入控制反射光顏色的 uniform，稱為 `u_specular`，先在 `app.state` 中加入對球體、地板不同的 `specularExponent` 以及反射光顏色，一開始為白色：

```
async function setup() {
  // ...

  return {
    // ...
    state: {
      fieldOfView: degToRad(45),
      cameraPosition: [0, 0, 8],
```

4-50

4-4　Specular 反射光

```
      cameraVelocity: [0, 0, 0],
      sphereScaleX: 1,
      lightDir: [0, -1, 0],
      specular: [1, 1, 1],
      sphereSpecularExponent: 40,
      groundSpecularExponent: 100,
    },
    time: 0,
  };
}
```

並且把 `u_specularExponent` 以及反射光顏色部份 `u_specular uniform` 傳送進去：

```
function render(app) {
  // ...

  twgl.setUniforms(programInfo, {
    u_worldViewerPosition: state.cameraPosition,
    u_lightDir: state.lightDir,
    u_specular: state.specular,
  });

  // ...

  { // sphere
    // ...

    twgl.setUniforms(programInfo, {
      u_matrix: matrix4.multiply(viewMatrix, worldMatrix),
      u_worldMatrix: worldMatrix,
      u_normalMatrix: matrix4.transpose(matrix4.inverse(worldMatrix)),
      u_color: [0, 0, 0],
      u_texture: textures.steel,
      u_specularExponent: state.sphereSpecularExponent,
    });

    // ...
  }
```

4-51

4 Lighting

```
{ // ground
  // ...

  twgl.setUniforms(programInfo, {
    u_matrix: matrix4.multiply(viewMatrix, worldMatrix),
    u_worldMatrix: worldMatrix,
    u_normalMatrix: matrix4.transpose(matrix4.inverse(worldMatrix)),
    u_color: [0, 0, 0],
    u_texture: textures.wood,
    u_specularExponent: state.groundSpecularExponent,
  });

  // ...
 }
}
```

最後把『乘上 n 次方』實做到 fragment shader 內：

```
#version 300 es
precision highp float;

in vec2 v_texcoord;
in vec3 v_normal;
in vec3 v_surfaceToViewer;

uniform vec3 u_color;
uniform sampler2D u_texture;
uniform vec3 u_lightDir;
uniform vec3 u_specular;
uniform float u_specularExponent;

out vec4 outColor;

void main() {
  vec3 color = u_color + texture(u_texture, v_texcoord).rgb;

  vec3 normal = normalize(v_normal);
  vec3 surfaceToLightDir = normalize(-u_lightDir);
```

```glsl
float colorLight = clamp(dot(surfaceToLightDir, normal), 0.0, 1.0);

// 確保表面位置到相機方向為單位向量
vec3 surfaceToViewerDirection = normalize(v_surfaceToViewer);

// 『中間向量』
vec3 halfVector = normalize(surfaceToLightDir + surfaceToViewerDirection);

// 反射光亮度
float specularBrightness = dot(halfVector, normal);
float specularBrightness = pow(
  clamp(dot(halfVector, normal), 0.0, 1.0), u_specularExponent
);

outColor = vec4(
  color * colorLight +
  u_specular * specularBrightness,
  1
);
}
```

從 uniform 接收 `u_specular`、`u_specularExponent` 後，將原本的內積計算用 `clamp()` 包起來避免數值跑到負的，接著套上 GLSL 內建的 `pow()`[17]，乘上 `u_specularExponent` 次方成為反射光亮度，最後在輸出顏色的地方加上 `u_specular` 顏色乘上反射光亮度作為反射光部份顏色。

最後對 `app.state` 內新的狀態加入對應的 HTML：

```html
<body>
  <canvas id="canvas"></canvas>
  <form id="controls">
    <!-- ... -->
    <div class="py-1">
      <label for="sphere-specular-exponent">sphereSpecularExponent</label>
```

---

17 https://www.khronos.org/registry/OpenGL-Refpages/es3.0/html/pow.xhtml

4　Lighting

```html
  <input
    type="range" id="sphere-specular-exponent"
    name="sphere-specular-exponent"
    min="1" max="300" value="40"
  >
</div>
<div class="py-1">
  <label for="ground-specular-exponent">groundSpecularExponent</label>
  <input
    type="range" id="ground-specular-exponent"
    name="ground-specular-exponent"
    min="1" max="300" value="100"
  >
</div>
<div class="py-1">
  <label for="specular">specular</label>
  <input
    type="color" id="specular" name="specular"
    value="#ffffff"
  >
</div>
</form>
<script type="module" src="main.js"></script>
</body>
```

並且把 HTML input 資料串接到 app 狀態中：

```javascript
async function main() {
  // ...
  const controlsForm = document.getElementById('controls');
  controlsForm.addEventListener('input', () => {
    // formData ...

    // 將 input 的 specular 顏色色票字串轉換成 RGB 三個元素的陣列
    app.state.specular = formData.get('specular')
      .match(/#(\w{2})(\w{2})(\w{2})/)
      .slice(1,4).map(c => parseInt(c, 16) / 256);

    app.state.sphereSpecularExponent = parseFloat(
```

4-54

```
      formData.get('sphere-specular-exponent')
    );
    app.state.groundSpecularExponent = parseFloat(
      formData.get('ground-specular-exponent')
    );
  });
}
```

其中反射光顏色的部份 `formData.get('specular')` 從 input 取得的顏色資料為像是 `'#abcdef'` 這樣的顏色色票字串，但是 `app.state.specular` 需要為看起來像是 `[0.67, 0.80, 0.93]` 的 RGB 顏色資料，因此寫了 `/#(\w{2})(\w{2})(\w{2})/` 正規表達式[18]來把 RGB 取出、以 16 進位轉成數字除以 256。

完成之後成品如圖 4-23 所示，反射光效果正確的呈現了：

▲ 圖 4-23　將內積完的數值乘上 n 次方後，反射光效果可以正常呈現

18　正規表達式是被用來匹配字串中字元的一種語法：
　　https://developer.mozilla.org/zh-TW/docs/Web/JavaScript/Guide/Regular_Expressions

4 Lighting

到此進度的完整程式碼可以在這邊找到：

</>	ch4/05/index.html
</>	ch4/05/main.js

也可以參考上線版本，透過瀏覽器打開此頁面：

https://webgl-book.pastleo.me/ch4/05/index.html

讀者可以試著調整 `sphereSpecularExponent`、`groundSpecularExponent` 控制要乘上幾次方，進而控制反射光範圍，同時玩玩反射光的顏色，例如把 `sphereSpecularExponent` 調小並且把反射光顏色調成綠色，如圖 4-24：

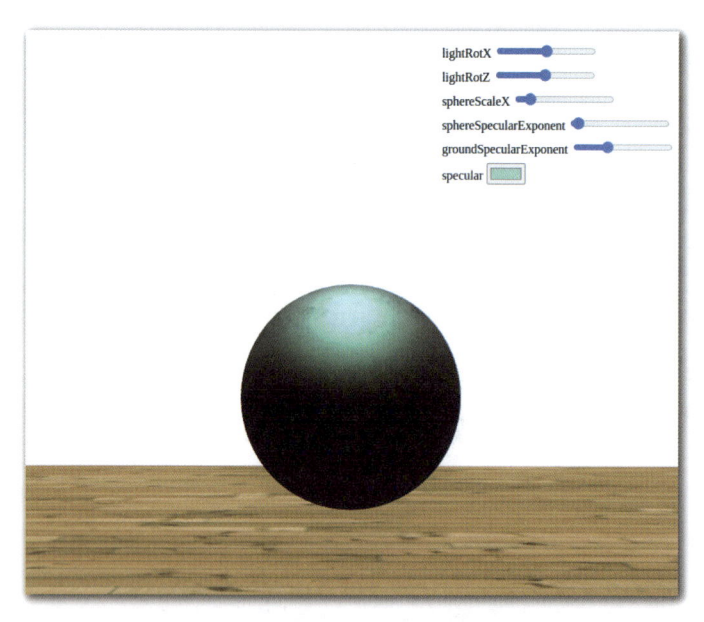

▲ 圖 4-24　玩轉球體反射光範圍以及顏色

4-5 點光源與自發光

從無限遠照射場景的平行光適合用來模擬太陽這類型的光源，本書到目前為止都是平行光，那如果是室內的燈泡光源呢？本篇將在場景中加入一個黃色自發光燈泡，並把平行光改成以這顆燈泡作為點光源。

⌘ 輸入光源位置並計算光線方向

在平行光的環境下，所有位置的光線方向都一樣，因此只需要一個 uniform `u_lightDir` 便可以，但是在點光源的情況下會因為頂點與表面位置不同而有不同的光線方向，這個光線方向可以透過 vertex shader 計算，並利用 varying 的內插功能使得 fragment shader 得到對應表面所接收到的光線方向，因此在 vertex shader 中使用 uniform 接收光源位置 `u_worldLightPosition`，並且計算出光線方向使用 varying `v_surfaceToLight` 傳給 fragment shader 使用：

```
#version 300 es
in vec4 a_position;
in vec2 a_texcoord;
in vec3 a_normal;

uniform mat4 u_matrix;
uniform mat4 u_worldMatrix;
uniform mat4 u_normalMatrix;
uniform vec3 u_worldViewerPosition;
uniform vec3 u_worldLightPosition;

out vec2 v_texcoord;
out vec3 v_normal;
out vec3 v_surfaceToViewer;
out vec3 v_surfaceToLight;

void main() {
  gl_Position = u_matrix * a_position;
  v_texcoord = vec2(a_texcoord.x, 1.0 - a_texcoord.y);
  v_normal = mat3(u_normalMatrix) * a_normal;
```

4 Lighting

```
  vec3 worldPosition = (u_worldMatrix * a_position).xyz;
  v_surfaceToViewer = u_worldViewerPosition - worldPosition;

  // 從頂點到光源，算出光線方向的反向
  v_surfaceToLight = u_worldLightPosition - worldPosition;
}
```

接著在 fragment shader 這邊，從 `u_lightDir` 改用 `v_surfaceToLight`，而且原本有一個負號 - 的反向也不需要了，`v_surfaceToLight` 已經是『表面到光源』向量：

```glsl
#version 300 es
precision highp float;

in vec2 v_texcoord;
in vec3 v_normal;
in vec3 v_surfaceToViewer;
in vec3 v_surfaceToLight;

uniform vec3 u_color;
uniform sampler2D u_texture;
uniform vec3 u_lightDir;
uniform vec3 u_specular;
uniform float u_specularExponent;

out vec4 outColor;

void main() {
  vec3 color = u_color + texture(u_texture, v_texcoord).rgb;

  vec3 normal = normalize(v_normal);
  vec3 surfaceToLightDir = normalize(v_surfaceToLight);

  float colorLight = clamp(dot(surfaceToLightDir, normal), 0.0, 1.0);

  // surfaceToViewerDirection, halfVector, specularBrightness ...

  outColor = vec4(
    color * colorLight +
    u_specular * specularBrightness,
    1
```

4-5 點光源與自發光

```
  );
}
```

接著加上點光源位置狀態，筆者設定光源的初始位置在 `[0, 2, 0]`：

```javascript
async function setup() {
  // ...

  return {
    // ...
    state: {
      // ...
      lightDir: [0, -1, 0],
      lightPosition: [0, 2, 0],
      // ...
    },
    time: 0,
  };
}
```

並且把狀態串到該設定的 uniform 上：

```javascript
function render(app) {
  // app, viewMatrix ...

  twgl.setUniforms(programInfo, {
    u_worldViewerPosition: state.cameraPosition,
    u_lightDir: state.lightDir,
    u_worldLightPosition: state.lightPosition,
    u_specular: state.specular,
  });

  // sphere, ground ...
}
```

然後老樣子加上對應的使用者控制，把控制光線方向的 HTML input 改成位置的控制：

4-59

4 Lighting

```html
<body>
  <canvas id="canvas"></canvas>
  <form id="controls">
    <div class="py-1">
      <label for="light-pos-x">lightPosX</label>
      <input
        type="range" id="light-pos-x" name="light-pos-x"
        min="-10" max="10" value="0" step="0.1"
      >
    </div>
    <div class="py-1">
      <label for="light-pos-y">lightPosY</label>
      <input
        type="range" id="light-pos-y" name="light-pos-y"
        min="-10" max="10" value="2" step="0.1"
      >
    </div>
    <div class="py-1">
      <label for="light-pos-z">lightPosZ</label>
      <input
        type="range" id="light-pos-z" name="light-pos-z"
        min="-10" max="10" value="0" step="0.1"
      >
    </div>
    <!-- ... -->
  </form>
  <script type="module" src="main.js"></script>
</body>
```

並且把 input 事件串到狀態中：

```javascript
async function main() {
  // ...

  const controlsForm = document.getElementById('controls');
  controlsForm.addEventListener('input', () => {
    const formData = new FormData(controlsForm);

    const lightRotXRad = degToRad(parseFloat(formData.get('light-rot-x')));
    const lightRotZRad = degToRad(parseFloat(formData.get('light-rot-z')));

    app.state.lightDir[0] = (
```

4-60

```
    -1 * Math.cos(lightRotXRad) * Math.sin(lightRotZRad)
  );
  app.state.lightDir[1] = (
    -1 * Math.cos(lightRotXRad) * Math.cos(lightRotZRad)
  );
  app.state.lightDir[2] = -1 * Math.sin(lightRotXRad);
  app.state.lightPosition[0] = parseFloat(formData.get('light-pos-x'));
  app.state.lightPosition[1] = parseFloat(formData.get('light-pos-y'));
  app.state.lightPosition[2] = parseFloat(formData.get('light-pos-z'));

  // ...
  });
}
```

調整完之後點光源就完成囉，讀者可以試著調整點光源的位置，觀察對於地板以及球體的反光效果，例如圖 4-24 把光源移動到球體右方，接近地板的位置：

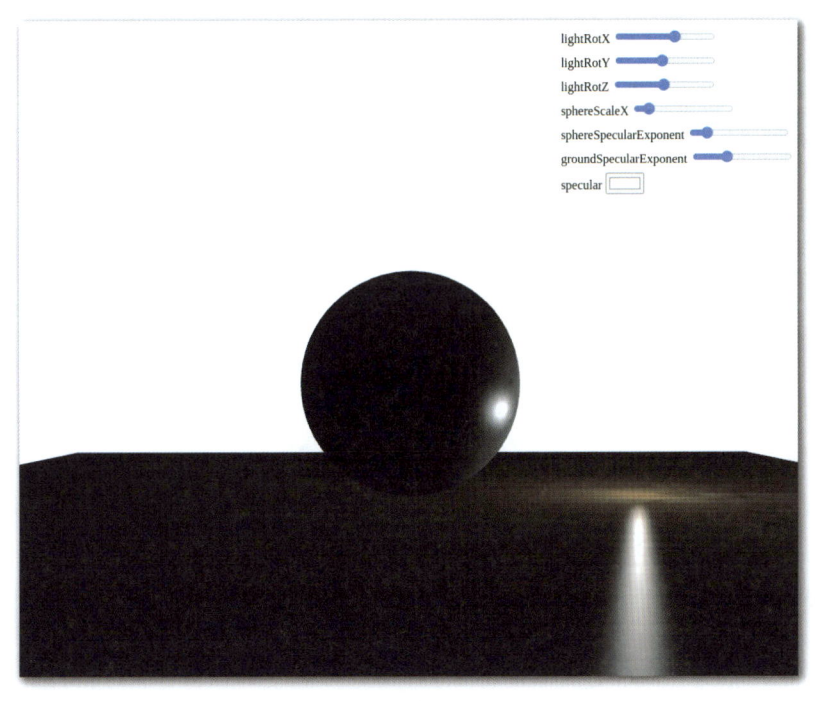

▲ 圖 4-24　點光源位於球體右方接近地板的位置

4　Lighting

　　散色光、反射光在地板形成的渲染結果是不是有真實世界中光滑地板的感覺呢？只可惜光源像是在空中隱形的燈泡，為了避免這樣的靈異現象，接著加入一個自發光的黃色純色燈泡。

⌘ 加入燈泡表示點光源位置

　　這個燈泡實質上就是一個球體，甚至可以重複使用現有的 `objects.sphere` 物件，因此不需要修改 `setup()`，直接在 `render()` 多渲染一次 `objects.sphere` 即可，透過輸入不一樣的 uniform 以及 `worldMatrix` 來 transform 使得燈泡看起來與原本 sphere 球體是不同物件，這邊把球體縮小到 0.1 倍大小並平移到狀態中的點光源位置：

```javascript
function render(app) {
  // ...

  { // light bulb
    gl.bindVertexArray(objects.sphere.vao);

    const worldMatrix = matrix4.multiply(
      matrix4.translate(...state.lightPosition),
      matrix4.scale(0.1, 0.1, 0.1),
    );

    twgl.setUniforms(programInfo, {
      u_matrix: matrix4.multiply(viewMatrix, worldMatrix),
      u_worldMatrix: worldMatrix,
      u_normalMatrix: matrix4.transpose(matrix4.inverse(worldMatrix)),
      u_color: [0, 0, 0],
      // u_texture: ??,
      u_specularExponent: 1000,
    });

    twgl.drawBufferInfo(gl, objects.sphere.bufferInfo);
  }
}
```

不過 `u_texture` 該輸入什麼呢？我們希望這個燈泡為自發光的黃色純色，不需要 texture，而且因為光源在球體內部，無論是散射光（為 `u_color` 加 `u_texture`）還是反射光（`u_specular`）亮度都會是零，設定 `u_color` 或是 `u_texture` 都不會有作用，球體會是全黑的，但是為了明確表示這個物體散色光部份為全黑，`u_color` 給上黑色 `[0, 0, 0]`，並且製作一個只有一個黑色 pixel 的 texture，稱為 `textures.null`，如果讀者不太記得手動輸入 texture 資料的使用方法，可以參考第二章的賽車格紋部份：

```javascript
async function setup() {
  // textures ...

  textures.null = gl.createTexture();
  gl.bindTexture(gl.TEXTURE_2D, textures.null);
  gl.texImage2D(
    gl.TEXTURE_2D,
    0, // level
    gl.RGB, // internalFormat
    1, 1, 0, // width, height, border
    gl.RGB, // format
    gl.UNSIGNED_BYTE, // type
    new Uint8Array([0, 0, 0, 255]), // data
  );

  // objects ...
}
```

有了這個 `textures.null`，未來如果真的有需要加上純色物件就可以設定 uniform 使用此 texture 作為 `u_texture`，因為 fragment shader 內散射光顏色 `color = u_color + texture(u_texture, v_texcoord).rgb` 的 texture 部份只有一個黑色 pixel 會回傳 `[0, 0, 0]`，該物體的散射光顏色就等於由 `u_color` 決定。

對於燈泡物件，`u_color`、`u_texture` uniform 就先都使用黑色：

```javascript
function render(app) {
  // ...
```

4 Lighting

```
{ // light bulb
  // ...
  twgl.setUniforms(programInfo, {
    // ...
    u_color: [0, 0, 0],
    u_texture: textures.null,
    // ...
  });
  // ...
}
```

讀者有興趣可以試著設定 `u_color`，甚至改 `textures.null` 那一個 pixel 的顏色，但是因為點光源在燈泡球體內部，無論如何此球體燈泡都會是全黑的，如圖 4-25 這樣：

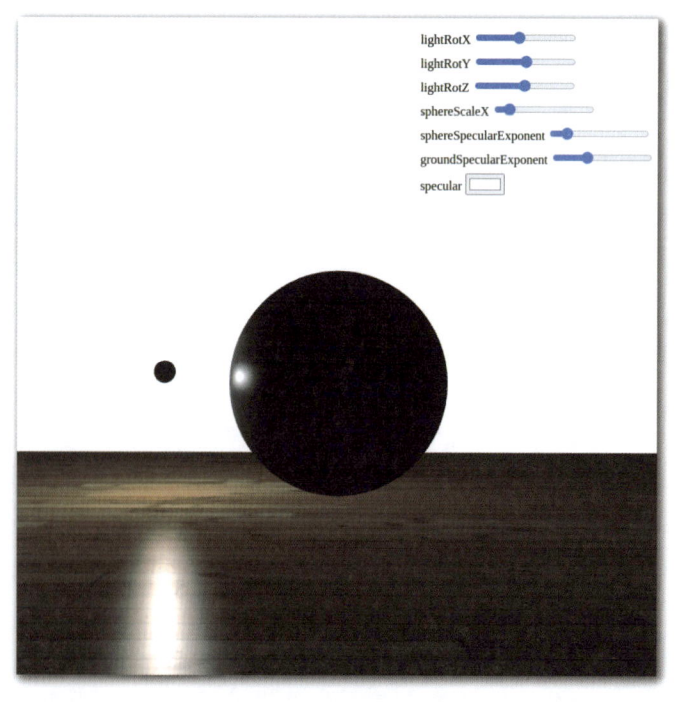

▲ 圖 4-25　因為點光源位於燈泡球體內部，燈泡球體為黑色

4-64

⌘ Emissive 自發光

為了讓燈泡球體有顏色，我們可以加上一個 uniform，在 fragment shader 中計算 `gl_FragColor` 時直接加上這個顏色，這個顏色即為自發光，uniform 變數名稱命名為 `u_emissive`：

```
// ...
uniform sampler2D u_texture;
uniform vec3 u_specular;
uniform float u_specularExponent;
uniform vec3 u_emissive;

out vec4 outColor;

void main() {
  // ...

  outColor = vec4(
    color * colorLight +
    u_specular * specularBrightness +
    u_emissive,
    1
  );
}
```

接下來對各個物件指定自發光顏色，筆者讓原本的球體也有一點點的亮度 `[0.15, 0.15, 0.15]`，而燈泡的部份就給黃色 `[1, 1, 0]`：

```
function render(app) {
  // ...

  { // sphere
    twgl.setUniforms(programInfo, {
      // ...
      u_emissive: [0.15, 0.15, 0.15],
    });
  }
  { // ground
    twgl.setUniforms(programInfo, {
      // ...
```

4-65

4 Lighting

```
      u_emissive: [0, 0, 0],
    });
  }
  { // light bulb
    twgl.setUniforms(programInfo, {
      // ...
      u_emissive: [1, 1, 0],
    });
  }
}
```

本篇的目標就完成啦，燈泡球體有了黃色的自發光，如圖 4-26：

▲ 圖 4-26　加上黃色自發光的燈泡球體

這麼一來，我們就探討了三種光：散射光、反射光以及自發光，使得物體表面可以有比較接近真實的光照反應；而散射光在 fragment shader 內計算時我們使用的變數名稱是 color，但是散射光應該叫做 diffuse 才對；事實上，在電腦 3D 運算物件『材質』對於光的反應、產生的顏色時不只本書提到的三種光，在其他的實做中還有會有環境光（ambient）、金屬質感、平滑度等等參數。

4-66

到此進度的完整程式碼可以在這邊找到：

</>	ch4/06/index.html
</>	ch4/06/main.js

也可以參考上線版本，透過瀏覽器打開此頁面：

https://webgl-book.pastleo.me/ch4/06/index.html

4-6　Normal Map

在尋找 3D 材質素材時，找到的素材包含的檔案常常不只有材質本身的顏色（diffuse），例如本章節 CH4 一開始的範例，其中在球體、地板使用之 steel、木紋材質的 texture，如果讀者自行下載素材包來用，會在壓縮檔中看到藍紫色的圖，如下圖 4-27 及圖 4-28：

▲ 圖 4-28　木質地板材質素材包中藍紫色的圖　　▲ 圖 4-27　球體 steel 材質資料夾中藍紫色的圖

4-67

4 Lighting

- 木質地板：https://opengameart.org/content/2048%C2%B2-wooden-texture
  - 選用壓縮檔中的 `wooden-floor/woodfloor_n.png`
- 球體：https://opengameart.org/content/commission-medieval
  - 選用 steel，於壓縮檔中的 `Medieval/steel/steel_normal.tga`

這些圖其實是一種稱為 normal map 的 texture，先前物件表面的法向量由頂點決定，因為 varying 的『內插』，使得光線照射物體時看起來很平順，而這些 normal map 則是可以讓物件表面有更豐富的光影細節。

圖 4-27 及圖 4-28 整體顏色為藍紫色，其 RGB 數值相當接近 `[127, 127, 255]`，減掉 127 再除以 128 會得到介於 -1 ~ +1 之間的數，這時與其說是顏色，一個 RGB 表示的其實是 `[x, y, z]` 單位向量來表示該表面位置的法向量，因此絕大部分的區域都是 `[0, 0, 1]` 指向 +z，但是可以在 normal map 這張 texture 圖片中看到有符合材質的紋路，這些紋路描述了不同於 +z 的表面法向量，可以使得物體表面法向量更有變化，讓光線反應在最後成像時使得物體表面可以呈現 normal map 的『凹凸感』。

⌘ 讀取、建立 Normal 法向量之 Texture

在 `setup()` 中讀取這兩張圖，筆者已經在範例包放置了經過轉換、最佳化的 normal map texture：

</>	assets/woodfloor_normal.webp
</>	assets/steel_normal.webp

```
async function setup() {
  // ...
  const textures = Object.fromEntries(
    await Promise.all(Object.entries({
      wood: '/assets/woodfloor.webp',
      steel: '/assets/steel.webp',
```

4-68

```
      woodNormal: '/assets/woodfloor_normal.webp',
      steelNormal: '/assets/steel_normal.webp',
    }).map(async ([name, url]) => {
      // image, texture ...
    }))
  );
  // ...
}
```

除此之外，也加入一個 null normal map，如果有物件不使用 normal map 時使用，使表面法向量一律指向 +z，RGB 值輸入 `[127, 127, 255]`：

```
async function setup() {
  // textures.null ...
  textures.nullNormal = gl.createTexture();
  gl.bindTexture(gl.TEXTURE_2D, textures.nullNormal);
  gl.texImage2D(
    gl.TEXTURE_2D,
    0, // level
    gl.RGBA, // internalFormat
    1, // width
    1, // height
    0, // border
    gl.RGBA, // format
    gl.UNSIGNED_BYTE, // type
    new Uint8Array([
      127, 127, 255, 255
    ])
  );
  // objects ...
}
```

⌘ Normal Map Transform

在本章節的一開始，我們有處理了 vertex attribute 中法向量的旋轉，但是現在得在原本 vertex 法向量的基礎上，再加上一層 normal map，也就是說 normal map 的 +z 要旋轉成 vertex 的法向量；舉一個例子，如果有一個 vertex

4 Lighting

資料形成之三角形的 normal 為 `[1, 0, 0]`，且在 fragment shader 從 normal map 取到的法向量為 `[0, 0, 1]`，那麼這個點的法向量應該 transform 為 `[1, 0, 0]`。

這樣的 transform 旋轉矩陣可以利用於 `matrix4.lookAt()` 內 `kHat`、`iHat`、`jHat`『基本向量』的產生方法，使用 vertex 的 normal 作為 `kHat`，將 up `[0, 1, 0]` 與 `kHat` 外積得到 `iHat`、`kHat` 與 `iHat` 外積得到 `jHat`，接著把 `iHat`、`jHat`、`kHat` 在矩陣中擺放好，此矩陣即為我們要的旋轉矩陣；同時就是傳入 `cameraPosition` 為 `[0, 0, 0]`、`target` 為 vertex 之 normal 的反向、`up` 為 `[0, 1, 0]` 到 `matrix4.lookAt()` 所產生的矩陣。

我們將在 vertex shader 產生這個矩陣，並且透過 varying 傳送到 fragment shader 來使用，這個 varying 筆者稱為 `v_normalMatrix`，這麼一來 vertex shader 的實做如下：

```
// ...
out vec2 v_texcoord;
out vec3 v_normal;
out vec3 v_surfaceToViewer;
out vec3 v_surfaceToLight;
out mat3 v_normalMatrix;

void main() {
  gl_Position = u_matrix * a_position;
  v_texcoord = vec2(a_texcoord.x, 1.0 - a_texcoord.y);
  v_normal = mat3(u_normalMatrix) * a_normal;

  vec3 normal = mat3(u_normalMatrix) * a_normal;
  vec3 normalMatrixI = normal.y >= 1.0 ?
    vec3(1, 0, 0) :
    normalize(cross(vec3(0, 1, 0), normal));
  vec3 normalMatrixJ = normalize(cross(normal, normalMatrixI));

  v_normalMatrix = mat3(
```

4-6　Normal Map

```
    normalMatrixI,
    normalMatrixJ,
    normal
  );

  // worldPosition, v_surfaceToViewer, v_surfaceToLight ...
}
```

- 首先 `vec3 normal = normalize((u_normalMatrix * a_normal).xyz);` 計算原本 vertex normal 要進行的旋轉。
- 原則上 `vec3 normalMatrixI` 為 `vec3(0, 1, 0)`（up）與 normal 的外積，但是為了避免 normal 為 `vec3(0, 1, 0)` 導致外積不出結果，遇到這樣的狀況時直接使得 `normalMatrixI` 為 `vec3(1, 0, 0)`。
- `vec3 normalMatrixJ` 為 normal 與 normalMatrixI 的外積。
- 這麼一來，`normalMatrixI`、`normalMatrixJ`、`normal` 作為變換矩陣的『基本向量』，擺放成矩陣 `v_normalMatrix`，可以把 normal map 法向量 transform 成以 vertex 法向量為基礎之向量。

完整、正確地轉換 normal map 法向量應該使用 TBN matrix，本書的範例由於 vertex 資料沒有 tangent attribute，因此稍微偷懶使用 lookAt 的方式來產生轉換矩陣，其中 up 寫死為 +z 方向，這意味著此轉換對於垂直於表面的旋轉都是同一個方向。

TBN 指的是 tangent、bitangent 以及 normal，如果有 tangent attribute，tangent 應與 normal 垂直，並且可以描述『垂直於表面的旋轉』，這時就可以透過 normal 與 tangent 做外積來得到 bitangent，把 tangent、bitangent 以及 normal 當成這邊的 normalMatrixI、normalMatrixJ、normal，作為『基本向量』放置在矩陣中來對 normal map 法向量進行轉換。

有興趣的讀者可以參考一下資料：

https://webgl2fundamentals.org/webgl/lessons/webgl-load-obj-w-mtl.html
https://learnopengl.com/Advanced-Lighting/Normal-Mapping

4 Lighting

⌘ 在 fragment shader 對 normal map 進行 transform

在 fragment shader 內，將使用 uniform u_normalMap 來傳送 normal map texture，取樣得到 normal map 法向量之後，使用 `v_normalMatrix` 進行矩陣運算得到該點的最後法向量：

```glsl
// ...
in vec2 v_texcoord;
in vec3 v_normal;
in vec3 v_surfaceToViewer;
in vec3 v_surfaceToLight;
in mat3 v_normalMatrix;

// ...
uniform float u_specularExponent;
uniform vec3 u_emissive;
uniform sampler2D u_normalMap;

out vec4 outColor;

void main() {
  vec3 color = u_color + texture(u_texture, v_texcoord).rgb;

  vec3 normal = normalize(v_normal);
  vec3 normal = texture(u_normalMap, v_texcoord).xyz * 2.0 - 1.0;
  normal = normalize(v_normalMatrix * normal);

  vec3 surfaceToLightDir = normalize(v_surfaceToLight);

  float colorLight = clamp(dot(surfaceToLightDir, normal), 0.0, 1.0);

  // surfaceToViewerDirection, specularBrightness, outColor ...
}
```

值得注意的是，在 normal map 如果一個點的 RGB 資料為 `[127, 127, 255]`，那麼其法向量為 `[0, 0, 1]`，但在 `texture(u_normalMap, v_texcoord).xyz` 取出資料時會得到 `[0.5, 0.5, 1]`，因此乘以 2 減 1 轉回 `[0, 0, 1]`。

最後當然得在對應的物件渲染時設定好 u_normalMap 要使用的 normal map：

```javascript
function render(app) {
  // ...
  { // sphere
    // ...
    twgl.setUniforms(programInfo, {
      // ...
      u_normalMap: textures.steelNormal,
    });
    // ...
  }

  { // ground
    // ...
    twgl.setUniforms(programInfo, {
      // ...
      u_normalMap: textures.woodNormal,
    });
    // ...
  }

  { // light bulb
    // ...
    twgl.setUniforms(programInfo, {
      // ...
      u_normalMap: textures.nullNormal,
    });
    // ...
  }
}
```

存檔重整之後，可以看到因為木質地板光澤有更多細節，使得這個『平面』立體了起來，如圖 4-27：

4 Lighting

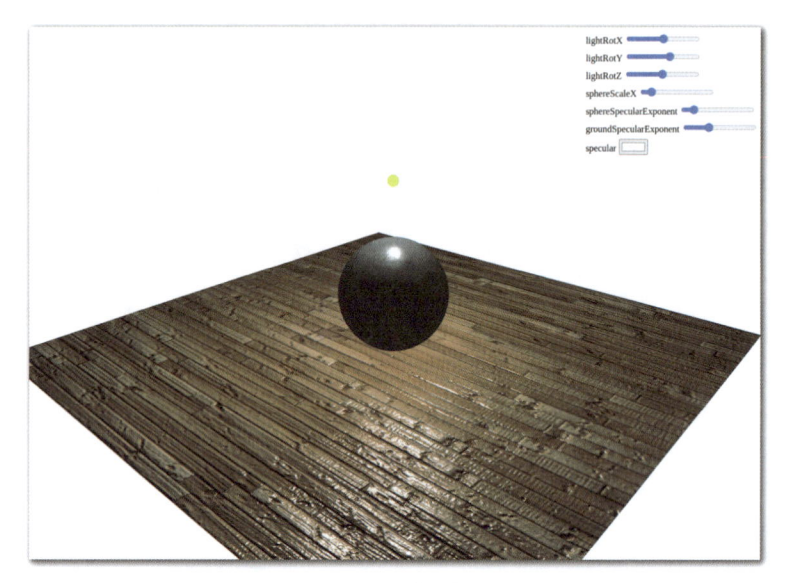

▲ 圖 4-27　使用 normal map 讓物體表面對光線產生凹凸感

我們也有對球體使用 `textures.steelNormal` 作為 normal map，調整一下光源位置可以觀察到其『感覺』凹下去的地方，如圖 4-28：

▲ 圖 4-28　normal map 在球體表面產生凹凸不平的『感覺』

到此進度的完整程式碼可以在這邊找到：

</>	ch4/07/index.html
</>	ch4/07/main.js

也可以參考上線版本，透過瀏覽器打開此頁面：

https://webgl-book.pastleo.me/ch4/07/index.html

在場景中加入光以及物體表面上的散射、反射光，甚至還有 normal map 呈現凹凸感之後，物體是否看起來更加真實了呢？針對光的討論差不多就到這邊，既然有了光，那麼影子呢？在實做陰影之前，要先學會如何讓 WebGL 渲染到 texture 上，使我們可以請 GPU 先進行一些運算，並在實際渲染畫面時取用先運算好的資料。

4 Lighting

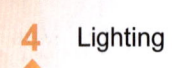

5

Framebuffer & 陰影

在 3D 場景中,『光影』效果的『光』的部份已經好好説明一輪了,那麼陰『影』的部份呢?這部份會比較複雜,會需要去設定 WebGL 的 framebuffer,讓 GPU 可以預先做一些背景運算。在本章節將介紹並使用這個技巧,製作出鏡面以及陰影效果!

5 Framebuffer & 陰影

在上個章節的最後我們不僅有散射光、反射光，還有使得物體表面有更多凹凸細節的 normal map，筆者從這個實做成果進行修改成為本章節的起始點，渲染與操作部份如下：

- 改回平行光源，並且會隨著時間擺動角度，也因此移除了標示點光源位置的燈泡物件。
- 上個章節的 `u_color` 與 `u_texture` 事實上是描述 diffuse 散射光的顏色，因此將 `u_color` 改名為 `u_diffuse`、`u_texture` 改名為 `u_diffuseMap`。
- 加入環境光 `u_ambient`，其顏色與 diffuse 顏色相乘後為物體的最低亮度。
- 讓使用者可以用比較直覺的方式控制視角，而非各種拉桿，現在開始相機有一個看著的位置為圓心，以一個距離對著該位置並旋轉，此旋轉軌道可以形成部份的球面：
  - 滑鼠左鍵、觸控拖曳移動相機看著的位置；滑鼠右鍵、雙指拖曳進行旋轉；滾輪、縮放手勢調整距離。
  - 建立 lib/input.ts 把各種事件的監聽、在動畫迴圈 `startLoop()` 內的相關程式抽出，控制 `app.state` 中的這幾個數值：
    - `cameraViewing`：相機看著的位置、圓心。
    - `cameraDistance`：相機與 `cameraViewing` 的距離、旋轉半徑。
    - `cameraRotationXY`：視角，以 x 與 y 軸進行的旋轉弧度。

本章節起始點完整程式碼可以在這邊找到，不知道讀者是否能找到以上修改項目所對應的程式碼呢？

</>	ch5/00/index.html
</>	ch5/00/main.js
</>	lib/input.ts

也可以參考上線版本，透過瀏覽器打開此頁面：

https://webgl-book.pastleo.me/ch5/00/index.html

物件部份，地板顏色選用全黑、並使用空的 diffuse、normal map；球體則改用 CH4 時 CC0 Medieval 材質包中的 scale，再附上一次材質包頁面網址：

https://opengameart.org/content/commission-medieval

- `scale` diffuse texture 使用 `Medieval/scale/scale_diffuse.tga`
- `scaleNormal` texture 使用 `Medieval/scale/scale_normal.tga`

筆者將這兩個檔案稍做轉換以及壓縮，放置在範例資料夾中的這兩個位置：

</>	assets/scale_diffuse.webp
</>	assets/scale_normal.webp

這個起始點運作起來如圖 5-1 所示：

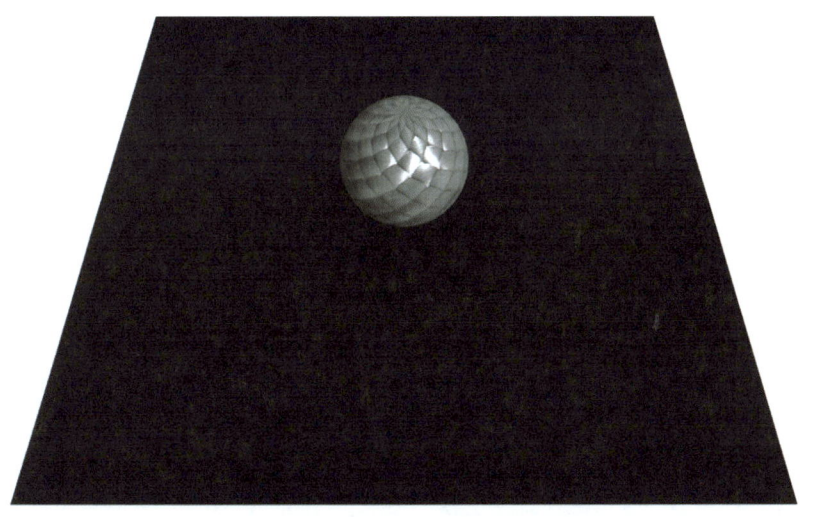

▲ 圖 5-1　CH5 起始點之渲染結果

5　Framebuffer & 陰影

為什麼地板要用全黑的呢？因為這是本章接下來要實做的，我們將利用 framebuffer 的功能，陸續在地板上加入鏡面、陰影效果！

5-1　Framebuffer 是什麼？

簡單來說，framebuffer 讓我們可以渲染到畫面以外的目標，讓 GPU 做一些背景運算。本書到目前為止渲染的目標皆為畫面上給使用者看的 `<canvas />` 元件，而 framebuffer 可以改變這件事，其中一個選項是使 WebGL 渲染到 texture 上。

為什麼要渲染到 texture 上呢？假設今天有一面鏡子，所看到最終的成像，等同於從鏡子中的相機看回原本場景，因此可以先從鏡子內繪製一次場景到一個 texture 上，接著繪製鏡子時就可以拿此 texture 來繪製；甚至感覺上比較沒有關聯的陰影效果，也需要透過 framebuffer 的功能，事先請 GPU 做一些運算，在正式『畫』到 `<canvas />` 的時候使用。

⌘ 初嘗 Framebuffer

在實做鏡面或是陰影之前，先來專注在 framebuffer 這個功能上，畢竟想想也知道鏡子、陰影需要的不會只是 framebuffer，還需要一些能夠讓物件位置成像能對得起來的方法，因此先把目標設定的簡單一點：渲染到 texture 上，接著渲染地板時使用該 texture，效果上來說像是把畫面上的球體變到黑色地板這福『畫』中。

首先在 `setup()` 中建立 framebuffer，呼叫 `gl.createFramebuffer()`[1] 即可建立 framebuffer，再呼叫 `gl.bindFramebuffer()`[2] 把目標對準（bind）新建立的

---

[1] https://developer.mozilla.org/en-US/docs/Web/API/WebGLRenderingContext/createFramebuffer

[2] https://developer.mozilla.org/en-US/docs/Web/API/WebGLRenderingContext/bindFramebuffer

5-1 Framebuffer 是什麼？

framebuffer：

```javascript
async function setup() {
  // canvas, gl, textures ...

  const framebuffer = gl.createFramebuffer();
  gl.bindFramebuffer(gl.FRAMEBUFFER, framebuffer);

  // objects ...
}
```

同時也建立好 texture 作為 framebuffer 渲染的目標，筆者命名為 textures.fb，fb 為 framebuffer 的縮寫：

```javascript
async function setup() {
  // canvas, gl, textures ...

  textures.fb = gl.createTexture();
  gl.bindTexture(gl.TEXTURE_2D, textures.fb);

  const width = 2048;
  const height = 2048;

  gl.texImage2D(
    gl.TEXTURE_2D,
    0, // level
    gl.RGBA, // internalFormat
    width,
    height,
    0, // border
    gl.RGBA, // format
    gl.UNSIGNED_BYTE, // type
    null, // data
  );

  gl.texParameteri(gl.TEXTURE_2D, gl.TEXTURE_MIN_FILTER, gl.NEAREST);
  gl.texParameteri(gl.TEXTURE_2D, gl.TEXTURE_MAG_FILTER, gl.NEAREST);
  gl.texParameteri(
    gl.TEXTURE_2D, gl.TEXTURE_WRAP_S, gl.CLAMP_TO_EDGE,
```

5 Framebuffer & 陰影

```
);
gl.texParameteri(
  gl.TEXTURE_2D, gl.TEXTURE_WRAP_T, gl.CLAMP_TO_EDGE,
);

// framebuffer, objects ...
}
```

可以看到建立了一個 2048x2048 大小的 texture，並且傳 `null` 讓資料留白，同時也得關閉 mipmap 功能，畢竟渲染到 texture 上之後，如果還要呼叫 `gl.generateMipmap()` 計算縮圖就太浪費資源了，有需要的話可以回去參考 CH2 對於 texture 的講解。

`texture.fb` 與 `framebuffer` 都建立好之後，建立 framebuffer 與 texture 的關聯：

```
async function setup() {
  // canvas, gl, textures.fb, framebuffer ...

  gl.framebufferTexture2D(
    gl.FRAMEBUFFER,
    gl.COLOR_ATTACHMENT0, // attachment
    gl.TEXTURE_2D,
    textures.fb,
    0, // level
  );

  // objects ...
}
```

呼叫 `gl.framebufferTexture2D()`[3] 使得當下對準（bind）之 framebuffer 的一個 attachment 對準指定的 texture，因為 WebGL 繪製輸出除了畫面顏色之外還有深度資訊，attachment 這個參數就是用來指定顏色或是深度，現在關心的

---

3 https://developer.mozilla.org/en-US/docs/Web/API/WebGLRenderingContext/framebufferTexture2D

是顏色，attachment 參數輸入 `gl.COLOR_ATTACHMENT0` 使得渲染到 framebuffer 時『顏色』部份會寫入 texture，最後 `level` 表示要寫入 mipmap 的哪一層，輸入 0 寫入原圖。這麼一來整組 framebuffer-texture 其實就建立出來了，只是說很快我們就會有多個 framebuffer，因此也建立 `framebuffers` 物件來統整：

```javascript
async function setup() {
  // canvas, gl, textures.null, textures.nullNormal ...

  const framebuffers = {};

  {
    textures.fb = gl.createTexture();
    gl.bindTexture(gl.TEXTURE_2D, textures.fb);

    // width, height
    // gl.texImage2D( ... ); gl.texParameteri( ... ); ...

    const framebuffer = gl.createFramebuffer();
    gl.bindFramebuffer(gl.FRAMEBUFFER, framebuffer);

    gl.framebufferTexture2D(
      // target, attachment, textarget, texture, level ...
    );

    framebuffers.fb = {
      framebuffer, width, height,
    };
  }

  // objects ...
}
```

這邊可以看到筆者讓 `framebuffers.fb` 成為一個物件將 `framebuffer`、width 與 `height` 包起來，同時把上面寫好建立 `texture.fb`、`framebuffer` 以及兩者的關聯的程式碼以 `{}` 圈起來形成程式碼區塊。

這樣就都建立完成，把 `framebuffers` 物件放置到 `app` 中：

5 Framebuffer & 陰影

```
async function setup() {
  // canvas, gl, textures, framebuffers, objects ...

  return {
    gl,
    programInfo,
    textures, framebuffers, objects,
    state: {
      // ...
    },
    time: 0,
  };
}
```

⌘ 渲染到 Framebuffer

如果接下來會需要先渲染到 framebuffer，再渲染到畫面，那麼可以想見某些物體會需要繪製兩次，為了避免重複程式碼，筆者把標注 sphere、ground 的花括弧 {} 區域抽出來獨立成兩個 function：

```
function renderSphere(app, viewMatrix) {
  const { /* ... */ } = app;

  gl.bindVertexArray(objects.sphere.vao);

  const worldMatrix = matrix4.multiply( /* ... */ );

  twgl.setUniforms(programInfo, {
    // ...
  });

  twgl.drawBufferInfo(gl, objects.sphere.bufferInfo);
}

function renderGround(app, viewMatrix) {
  const { /* ... */ } = app;
```

5-1　Framebuffer 是什麼？

```
  gl.bindVertexArray(objects.ground.vao);

  const worldMatrix = matrix4.multiply( /* ... */ );

  twgl.setUniforms(programInfo, { /* ... */ });

  twgl.drawBufferInfo(gl, objects.ground.bufferInfo);
}
```

這麼一來在 `render()` 內調整 `renderSphere()`、`renderGround()` 要執行多次或是執行的先後順序就變得容易許多了。

回到 `render()` 來渲染寫入至 texture 吧！設定好全域 uniform 之後，呼叫 `gl.bindFramebuffer()` 切換到 `framebuffers.fb`，在接下來繪製時就會寫入至 `textures.fb`：

```
function render(app) {
  const {
    gl, programInfo, framebuffers, state,
  } = app;

  // useProgram, cameraMatrix, viewMatrix, twgl.setUniforms() ...

  gl.bindFramebuffer(gl.FRAMEBUFFER, framebuffers.fb.framebuffer);
  gl.viewport(0, 0, framebuffers.fb.width, framebuffers.fb.height);

  renderSphere(app, viewMatrix);
}
```

要記得使用 `gl.viewport()` 設定渲染長寬跟 texture 一樣大，這時當初包的 `width`、`height` 就能派上用場了，接著就跟原本渲染到畫面上一樣，因此呼叫 `renderSphere()` 渲染球體。

渲染到 `textures.fb` 完成之後，要怎麼讓渲染目標切換回 `<canvas />` 呢？呼叫 `gl.bindFramebuffer()` 並傳入 `null` 即可，不過一樣要記得把渲染長寬設定好：

5　Framebuffer & 陰影

```javascript
function render(app) {
  // useProgram, cameraMatrix, viewMatrix, twgl.setUniforms() ...
  // bindFramebuffer to framebuffers.fb and renderSphere ...

  gl.bindFramebuffer(gl.FRAMEBUFFER, null);

  gl.canvas.width = gl.canvas.clientWidth;
  gl.canvas.height = gl.canvas.clientHeight;
  gl.viewport(0, 0, canvas.width, canvas.height);

  renderGround(app, viewMatrix);
}
```

呼叫 `renderGround()` 渲染地板的同時，設定其 uniform 時在 `u_diffuseMap` 填上 `textures.fb` 使地板顯示在 framebuffer 時渲染的樣子：

```javascript
function renderGround(app, viewMatrix) {
  // ...
  twgl.setUniforms(programInfo, {
    // ...
    u_normalMap: textures.nullNormal,
    u_diffuse: [0, 0, 0],
    u_diffuseMap: textures.fb,
    u_specularExponent: 200,
    u_emissive: [0, 0, 0],
  });

  twgl.drawBufferInfo(gl, objects.ground.bufferInfo);
}
```

存檔重整，轉一下的確看得出來球體渲染在一幅畫上的感覺，但是平移會看到如圖 5-2 的殘影現象：

5-1 Framebuffer 是什麼？

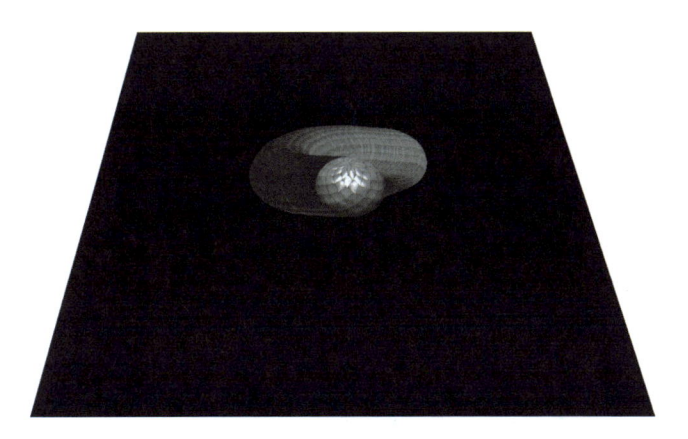

▲ 圖 5-2 使用 framebuffer 把球體變到地板中，但是有平移後會有殘影

有殘影是因為上一次渲染到 texture 的東西不會被自動清除，因此透過 CH1 一開始的油漆工具清除 `textures.fb` 中上個 frame 渲染的結果：

```javascript
async function setup() {
  // ...

  gl.enable(gl.CULL_FACE);
  gl.enable(gl.DEPTH_TEST);
  gl.clearColor(1, 1, 1, 1);

  return {
    gl,
    programInfo,
    textures, framebuffers, objects,
    state: {
      // ...
    },
    time: 0,
  };
}

// ...

function render(app) {
```

5-11

5 Framebuffer & 陰影

```
// useProgram, cameraMatrix, viewMatrix, twgl.setUniforms() ...

gl.bindFramebuffer(gl.FRAMEBUFFER, framebuffers.fb.framebuffer);
gl.viewport(0, 0, framebuffers.fb.width, framebuffers.fb.height);

gl.clear(gl.COLOR_BUFFER_BIT);

renderSphere(app, viewMatrix);

gl.bindFramebuffer(gl.FRAMEBUFFER, null);

gl.canvas.width = gl.canvas.clientWidth;
gl.canvas.height = gl.canvas.clientHeight;
gl.viewport(0, 0, canvas.width, canvas.height);

renderGround(app, viewMatrix);
}
```

　　這邊可以注意到清除的顏色是白色使得地板的底色不再是黑色，但是因為光線方向會移動導致地板的散射（`textures.fb` 設定在 `u_diffuseMap` 上）亮度下降，大多數時候地板底色看起來是灰色，並且在平移之後地板不再有殘影，如圖 5-3 所示：

▲ 圖 5-3　使用 framebuffer 把球體變到地板中，並且平移後不會有殘影

5-12

總結一下目前在 `render()` 內的渲染流程：

1. 計算 `cameraMatrix`、`viewMatrix` 這類繪製各個物件之間共用的 transform 矩陣。

2. 設定各個物件之間共用的 uniform 如相機位置、平行光方向等。

3. 切換 framebuffer 渲染至 `textures.fb` 並設定渲染長寬與 `textures.fb` 相同。

4. 使用油漆工具清除 framebuffer 目標上的內容為白色純色，也就是把 `textures.fb` 清除為白色。

5. 繪製球體，繪製結果會寫入到當前 framebuffer 指定的 `textures.fb`。

6. 切換 framebuffer 渲染至給使用者觀看的 `<canvas />`，並設定渲染長寬與畫布大小相同。

7. 繪製地板，此時地板的 `u_diffuseMap texture` 會使用第三、四步有繪製球體的 `textures.fb` 使得球體呈現在地板中。

到此進度的完整程式碼可以在這邊找到：

</>	ch5/01/index.html
</>	ch5/01/main.js

也可以參考上線版本，透過瀏覽器打開此頁面：
https://webgl-book.pastleo.me/ch5/01/index.html

因為渲染球體與地板使用同一組 shader program，因此地板同樣具有反射光效果，只是需要轉到適當角度，如圖 5-4 反射光蓋過球體的區域：

5 Framebuffer & 陰影

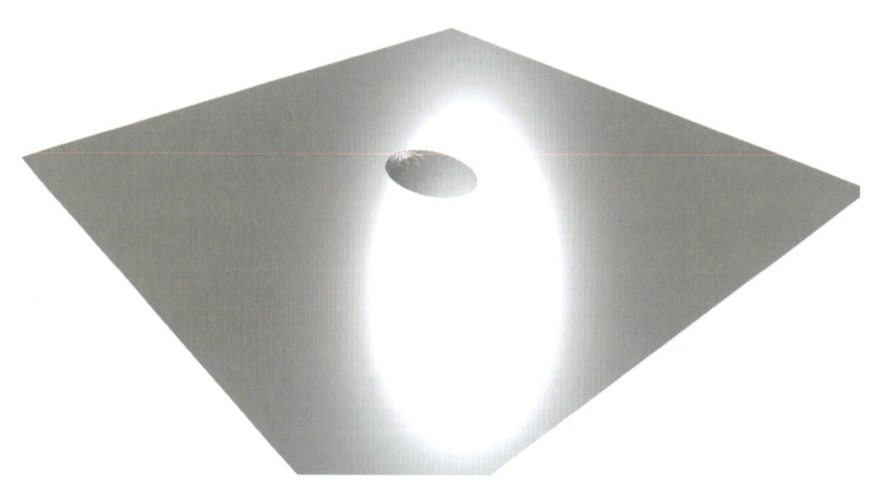

▲ 圖 5-4 地板同樣具有反射光效果，這時球體在地板中就變得非常明顯

5-2 鏡面效果

取得了 framebuffer 這個工具，把球體畫在在一幅畫中，已經完成鏡面效果所需要的基石，接著來探討如何把鏡面效果實做出來。

⌘ 透過 TWGL 簡化建立 framebuffer 的程式碼

上篇建立 framebuffer 時直接使用 WebGL API 來建立 framebuffer，其實 TWGL 已經有實做好一定程度的包裝，我們可以呼叫 twgl.createFramebufferInfo()[4]，它會建立好 framebuffer、textures 並且為他們建立關聯，並且回傳叫做 framebufferInfo 的資料結構：

```
async function setup() {
  // canvas, gl, textures ...

  const framebuffers = {};
```

---

[4] https://twgljs.org/docs/module-twgl_framebuffers.html#.createFramebufferInfo

```
framebuffers.mirror = twgl.createFramebufferInfo(
  gl,
  null, // attachments
  2048, // width
  2048, // height
);

// objects ...
}
```

這邊的 `attachments` 參數可以傳入一個陣列指定顏色、深度所要寫入之 texture 的細部設定，不過 TWGL 也幫我們定義好一組『預設』的組合，傳 `null` 即可讓 twgl 使用這個預設值，此預設值的顏色輸出到 texture、深度輸出到可以存放深度資訊的物件，因為接下來要實做的功能為鏡面，把此 framebuffer 命名為 `framebuffers.mirror`。

那麼要怎麼取得預設建立會被寫入顏色的 texture 呢？嘗試在 Console 中輸入 `app.framebuffers.mirror` 查看建立的物件 framebufferInfo，看起來如圖 5-5 所示：

```
> app.framebuffers.mirror
< ▼ {framebuffer: WebGLFramebuffer, attachments: Array(2), width: 2048, height: 2048}
    ⓘ
    ▼ attachments: Array(2)
      ▶ 0: WebGLTexture {}
      ▶ 1: WebGLRenderbuffer {}
        length: 2
      ▶ [[Prototype]]: Array(0)
    ▼ framebuffer: WebGLFramebuffer
      ▶ [[Prototype]]: WebGLFramebuffer
        height: 2048
        width: 2048
      ▶ [[Prototype]]: Object
```

▲ 圖 5-5　app.framebuffers.mirror 中的 framebufferInfo

看起來就放在 attachments 下呢，那麼把 texture 指定到 `textures.mirror` 以便之後取用：

5 Framebuffer & 陰影

```
async function setup() {
  // canvas, gl, textures, framebuffers ...

  framebuffers.mirror = twgl.createFramebufferInfo(
    gl,
    null, // attachments
    2048, // width
    2048, // height
  );
  textures.mirror = framebuffers.mirror.attachments[0];

  // objects ...
}
```

值得注意的是，framebufferInfo 同時包含了長寬資訊，如果使用 twgl.bindFramebufferInfo()[5] 來做 framebuffer 的切換，它同時會幫我們呼叫 gl.viewport() 調整渲染區域，因此在繪製階段也改用 twgl 所提供的工具來做 framebuffer 的切換：

```
function render(app) {
  // gl.useProgram(), cameraMatrix, viewMatrix, twgl.setUniforms() ...

  gl.bindFramebuffer(gl.FRAMEBUFFER, framebuffers.fb.framebuffer);
  gl.viewport(0, 0, framebuffers.fb.width, framebuffers.fb.height);
  twgl.bindFramebufferInfo(gl, framebuffers.mirror);

  gl.clear(gl.COLOR_BUFFER_BIT | gl.DEPTH_BUFFER_BIT);

  // render objects ...
}
```

可以發現在 `gl.clear()` 時除了清除 `gl.COLOR_BUFFER_BIT`，筆者也加入 `gl.DEPTH_BUFFER_BIT` 清除『深度』，這是因為 `twgl.createFramebufferInfo()` 所建立的 `attachments` 組合預設包含了一張深度資訊，這個資訊也得清除以避免第二次渲染到 framebuffer 時產生問題。

---

[5] https://twgljs.org/docs/module-twgl_framebuffers.html#.bindFramebufferInfo

渲染鏡面中看到的世界

目前在 framebuffer 中繪製的球體就是正常狀態下看到的球體,那麼要怎麼繪製『鏡像』的樣子呢?想像一個觀察著看著鏡面中的一顆球如圖 5-6 所示:

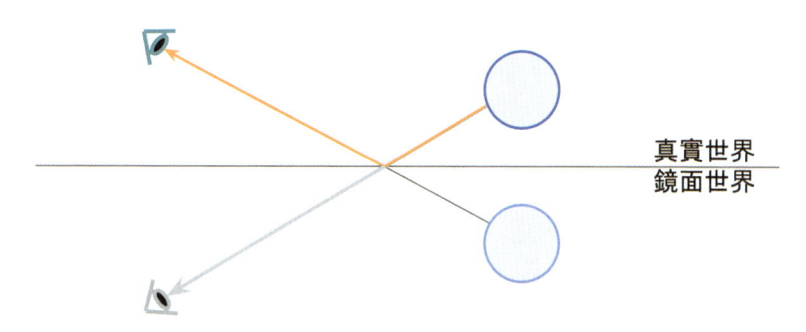

▲ 圖 5-6　觀察著看著鏡面中的一顆球,鏡面世界的想像位置圖

橘色箭頭為實際光的路線,把光線打直可以獲得一個鏡面中的觀察者看著真實世界(灰色箭頭與眼睛),因此繪製鏡像中的世界時,把相機移動到鏡面中拍一次,我們就獲得了鏡面世界看回真實世界的成像,準備好在繪製地板時使用。

筆者為本章節範例起始點實做的相機控制方式使用了不同於 `matrix4.lookAt()` 的 `cameraMatrix` 產生方式:

```
const cameraMatrix = matrix4.multiply(
  matrix4.translate(...state.cameraViewing),
  matrix4.yRotate(state.cameraRotationXY[1]),
  matrix4.xRotate(state.cameraRotationXY[0]),
  matrix4.translate(0, 0, state.cameraDistance),
);
```

用白話文來說:

1. 相機一開始在 `[0, 0, 0]` 看著 -z 方向。

5 Framebuffer & 陰影

2. 往 +z 方向移動 `state.cameraDistance`。
3. 轉動 x 軸 `state.cameraRotationXY[0]`。
4. 轉動 y 軸 `state.cameraRotationXY[1]`，這時相機會在半徑為 `state.cameraDistance` 的球體表面上看著原點。
5. 平移 `state.cameraViewing` 移動相機到所看的目標。

如果使用 y = 0 形成的平面作為鏡面，只要讓轉動 x 軸時反向，就變成對應在鏡面中的相機；為了鏡面中的視角建立另一組鏡面使用的 `cameraMatrix` 稱為 `mirrorCameraMatrix`，並且進而算出鏡面使用的 viewMatrix，命名為 `mirrorViewMatrix`：

```
function render(app) {
  // gl.useProgram(), cameraMatrix, viewMatrix ...

  const mirrorCameraMatrix = matrix4.multiply(
    matrix4.translate(...state.cameraViewing),
    matrix4.yRotate(state.cameraRotationXY[1]),
    matrix4.xRotate(-state.cameraRotationXY[0]), // 反向旋轉 x 軸
    matrix4.translate(0, 0, state.cameraDistance),
  );

  const mirrorViewMatrix = matrix4.multiply(
    matrix4.perspective(
      state.fieldOfView, gl.canvas.width / gl.canvas.height, 0.1, 2000,
    ),
    matrix4.inverse(mirrorCameraMatrix),
  );

  // twgl.setUniforms(), render objects ...
}
```

接著我們要讓地板為 y = 0 形成的平面，因為原本有一個往 -y 方向上平移 1 單位的 transform，我們得將之去除，同時也得把球體向上、往 +y 方向平移 1 單位，否則有一半的球體會陷在地板中；也就是地板由 `[0, -1, 0]` 改放在 `[0, 0, 0]`、球體由 `[0, 0, 0]` 改放在 `[0, 1, 0]`：

```
function renderSphere(app, viewMatrix) {
  // gl.bindVertexArray() ...

  const worldMatrix = matrix4.multiply(
    matrix4.translate(0, 1, 0),
    matrix4.scale(1, 1, 1),
  );

  // twgl.setUniforms(); twgl.drawBufferInfo(); ...
}

function renderGround(app, viewMatrix) {
  // gl.bindVertexArray() ...

  const worldMatrix = matrix4.multiply(
    matrix4.translate(0, 0, 0),
    matrix4.scale(10, 1, 10),
  );

  // twgl.setUniforms(); twgl.drawBufferInfo(); ...
}
```

在繪製從鏡像世界中看的出來世界時使用 mirrorViewMatrix，而球體是這邊唯一需要繪製的物件：

```
function render(app) {
  // useProgram, cameraMatrix, viewMatrix ...
  // mirrorCameraMatrix, mirrorViewMatrix ...
  // twgl.setUniforms() ...

  twgl.bindFramebufferInfo(gl, framebuffers.mirror);

  gl.clear(gl.COLOR_BUFFER_BIT | gl.DEPTH_BUFFER_BIT);

  renderSphere(app, mirrorViewMatrix);

  // switch framebuffer, render to <canvas /> ...
}
```

5 Framebuffer & 陰影

　　到這邊，從鏡像世界中看出來的成像就已經拍下來存放在 `textures.mirror` 中，準備好讓 `<canvas />` 渲染時使用；這時可以先把 `renderGround()` 所使用的 `u_diffuseMap` 改用 `textures.mirror`：

```javascript
function renderGround(app, viewMatrix) {
  // ..
  twgl.setUniforms(programInfo, {
    // ...
    u_diffuseMap: textures.mirror,
    // ...
  });
  twgl.drawBufferInfo(gl, objects.ground.bufferInfo);
}
```

　　看看目前這樣渲染出來的樣子，如圖 5-7 所示，地板所呈現的樣子為鏡面世界的相機看回真實世界所拍下來的樣子：

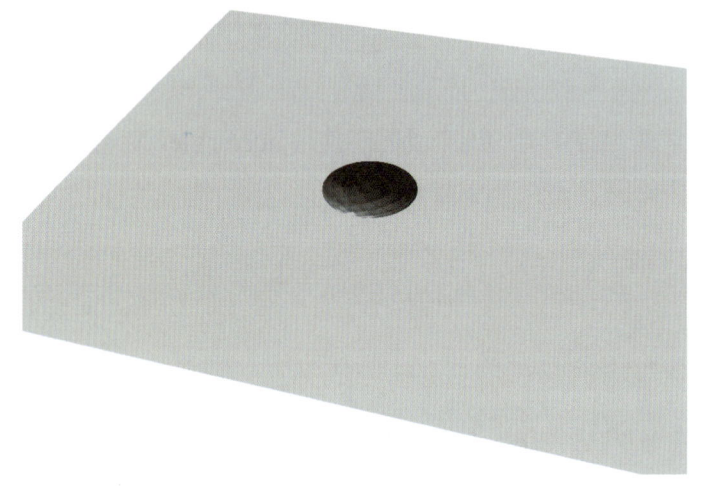

▲ 圖 5-7　鏡面世界的相機看回真實世界所拍下來的樣子，可以看到球體的底下

5-20

⌘ 繪製鏡面地板時取用 texture 中對應的位置

有了鏡像世界中看回來的樣子，但是要怎麼在繪製鏡面地板到 `<canvas />` 的時候找到鏡面世界的照片對應的位置呢？請看下圖 5-8：

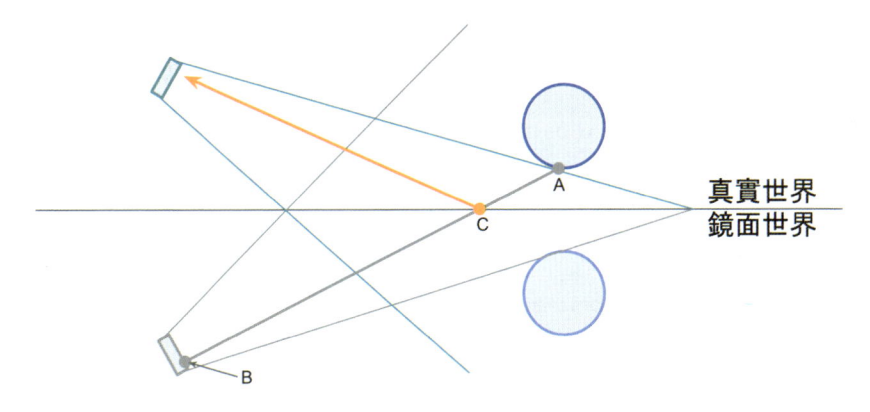

▲ 圖 5-8　繪製鏡面（地板）到 `<canvas />` 投影位置示意圖

物件上的一個點 A，是經過物件自身 `worldMatrix` transform 的位置，再經過 `mirrorViewMatrix` transform 到鏡面世界的相機底片上（點 B）；繪製 `<canvas />` 具有鏡面效果的地板物件時，我們知道的是 C 點的位置（`worldPosition`），而這個點座落在 A 與 B 點之間，因此拿著 C 點做 `mirrorViewMatrix` transform 便可以獲得對應的 B 點，這時 B 點準確來說是 clip space 中的位置，不過只要只取 .xy 二維的位置就是底片上的 B 點囉，這時再將此位置向量加一除以二就能得到 texture 上的位置。

也就是說，在繪製 `<canvas />` 的時候也會需要 `mirrorViewMatrix`，uniform 命名為 `u_mirrorMatrix`，並且在 vertex shader 中計算出 B 點，透過 varying `v_mirrorTexcoord` 傳送給 fragment shader：

```
// ...
uniform mat4 u_matrix;
uniform mat4 u_worldMatrix;
uniform mat4 u_normalMatrix;
```

5 Framebuffer & 陰影

```glsl
uniform vec3 u_worldViewerPosition;
uniform mat4 u_mirrorMatrix;

out vec2 v_texcoord;
out vec3 v_surfaceToViewer;
out mat3 v_normalMatrix;
out vec4 v_mirrorTexcoord;

void main() {
  gl_Position = u_matrix * a_position;

  // v_texcoord ...
  // normal, v_normalMatrix ...

  vec4 worldPosition = u_worldMatrix * a_position;
  v_surfaceToViewer = u_worldViewerPosition - worldPosition.xyz;

  v_mirrorTexcoord = u_mirrorMatrix * worldPosition;
}
```

這邊算出來的 worldPosition 在繪製地板時就是圖 5-8 的 C 點，拿這個點再做 u_mirrorMatrix transform 就可以投影到 B 點（clip space 位置），只不過原本 worldPosition 是 vec3，我們必須使之為 vec4 才能與 mat4 的 u_mirrorMatrix 做矩陣運算。

到 fragment shader，鏡面世界拍攝到的照片放在 `u_diffuseMap`，不過鏡面物體取用 texture 的方式將會與其他物件不同，因此加入一個 uniform `u_useMirrorTexcoord` 來控制是否要使用 v_mirrorTexcoord：

```glsl
// ...
in vec2 v_texcoord;
in vec3 v_surfaceToViewer;
in vec3 v_surfaceToLight;
in mat3 v_normalMatrix;
in vec4 v_mirrorTexcoord;

uniform sampler2D u_normalMap;
```

```glsl
uniform vec3 u_diffuse;
uniform sampler2D u_diffuseMap;
uniform vec3 u_lightDir;
uniform vec3 u_specular;
uniform float u_specularExponent;
uniform vec3 u_emissive;
uniform vec3 u_ambient;
uniform bool u_useMirrorTexcoord;

out vec4 outColor;

void main() {
  vec2 texcoord = u_useMirrorTexcoord ? (
    (v_mirrorTexcoord.xy / v_mirrorTexcoord.w) * 0.5 + 0.5
  ) : v_texcoord;
  vec3 normal = texture(u_normalMap, v_texcoord).xyz * 2.0 - 1.0;
  normal = normalize(v_normalMatrix * normal);

  vec3 diffuse = u_diffuse + texture(u_diffuseMap, texcoord).rgb;

  // surfaceToLightDir ... outColor ...
}
```

可以注意到 u_useMirrorTexcoord 為 true 時使用 v_mirrorTexcoord.xy，這邊 .xy 取二維的位置成為圖 5-8 底片上的 B 點，不過還有個 / v_mirrorTexcoord.w，為什麼要除以 .w 呢？還記得 CH3 做 perspective 投影時，頂點位置在進入 clip space 之前，WebGL 內建的行為會把 gl_Position.x、gl_Position.y、gl_Position.z 都除以 gl_Position.w，而 v_mirrorTexcoord 是 varying 就沒有這樣的行為了，我們得自己實做，最後 * 0.5 + 0.5 就是把 clip space 位置（-1 ~ +1）轉換成 texture 上的 texcoord（0 ~ +1）。

完成 shader 的修改，繪製地板時把需要餵進去的 uniform 餵進去：

```javascript
function renderGround(app, viewMatrix, mirrorViewMatrix) {
  // ...
  twgl.setUniforms(programInfo, {
    // ...
```

5 Framebuffer & 陰影

```
    u_diffuseMap: textures.mirror,
    u_specularExponent: 200,
    u_emissive: [0, 0, 0],
    u_useMirrorTexcoord: true,
    u_mirrorMatrix: mirrorViewMatrix,
  });

  twgl.drawBufferInfo(gl, objects.ground.bufferInfo);

  twgl.setUniforms(programInfo, {
    u_useMirrorTexcoord: false,
  });
}
```

在呼叫 `twgl.drawBufferInfo()` 進行『畫』這個動作之後，需要特地把 `u_useMirrorTexcoord` 關閉，因為只有地板物件會需要這個特殊的模式，而 uniform 是跟著 program 的，畫完此物件立刻關閉可以避免影響到其他物件的渲染。

並且在切回 framebuffer、繪製到 `<canvas />` 的時候也畫出球體，並為 `renderGround()` 傳入 `mirrorViewMatrix`：

```
function render(app) {
  // viewMatrix, mirrorViewMatrix, twgl.setUniforms ...
  // render to framebuffers.mirror, textures.mirror ...

  gl.bindFramebuffer(gl.FRAMEBUFFER, null);

  gl.canvas.width = gl.canvas.clientWidth;
  gl.canvas.height = gl.canvas.clientHeight;
  gl.viewport(0, 0, canvas.width, canvas.height);

  renderSphere(app, viewMatrix);
  renderGround(app, viewMatrix, mirrorViewMatrix);
}
```

鏡面效果就完成囉！如圖 5-9、圖 5-10：

▲ 圖 5-9　球體與鏡面效果中的球體

▲ 圖 5-10　把視角調低一點可以看到更明顯的鏡面效果

到此進度的完整程式碼可以在這邊找到：

</>	ch5/02/index.html
</>	ch5/02/main.js

也可以參考上線版本，透過瀏覽器打開此頁面：
https://webgl-book.pastleo.me/ch5/02/index.html

5-25

5 Framebuffer & 陰影

5-3 陰影―拍攝深度資訊

有了 framebuffer 的幫助，我們可以動用 GPU 的力量事先運算，在正式繪製到 `<canvas />` 給使用者觀看時使用。繼鏡面完成之後，根據本章一開始所說，另一個 framebuffer 的應用是陰影，接下來就來介紹如何製作出陰影效果。

⌘ 拍攝深度照片

陰影的產生是因為物體表面到光源之間有其他物體而被遮住，為了得知有沒有被遮住，我們可以從光源出發『拍攝』一次場景，從上篇改用 TWGL 時有提到 framebuffer 也可以包含深度資訊，實際繪製畫面時就可以利用深度資訊來得知是否在陰影下，我們以下面圖 5-11 示意來舉例一個平行光下球體產生陰影的狀況：

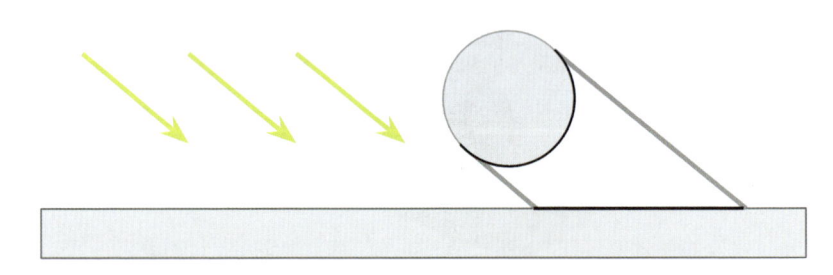

▲ 圖 5-11　平行光下在地板上產生陰影示意圖

但是這個舉例的光源是平行光，要怎麼拍攝？首先，利用 CH3 的 Orthogonal 3D 投影，如果光線是直直往 +y 的方向與地面垂直倒是蠻容易想像的，不過如果不是的時候，那麼拍攝的範圍可能就沒辦法很大，如下圖 5-12 淡藍色區域為投影區域，藍色面為成像面：

5-26

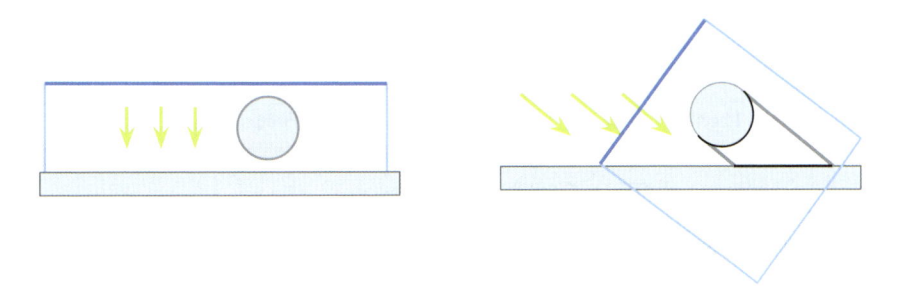

▲ 圖 5-12　平行光使用 Orthogonal 投影來拍攝『深度』

在矩陣運算中還有一個叫做 shear[6]，可以把一個空間中的矩形轉換成平行四邊形，透過這個工具，可以使得投影區域為平行四邊形，如下圖 5-13：

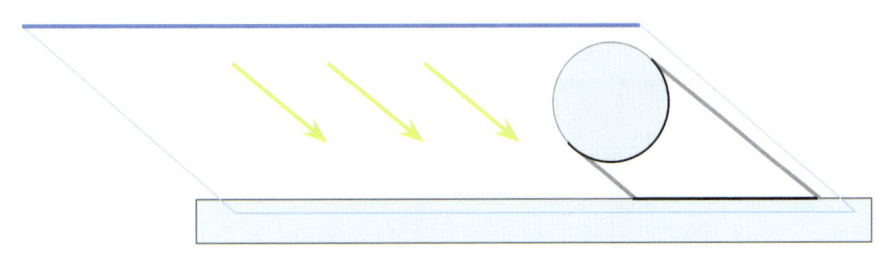

▲ 圖 5-13　使用 shear 將 orthogonal 投影區域轉成平行四邊形

⌘ 建立存放深度資訊的 texture

在前面有看過 `twgl.createFramebufferInfo()` attachments 給 `null` 時，預設建立的 framebuffer 與 textures 組合（圖 5-5），會建立一個存放顏色的 texture 以及一個存放深度的 `WebGLRenderbuffer`，但是 `WebGLRenderbuffer` 所保存的深度資訊無法給 shader 當成 texture 使用，為了建立可以存放深度資訊又可以給 shader 用的 texture，需要指定 attachments 詳細參數給 `twgl.createFramebufferInfo()`：

6　https://zh.wikipedia.org/wiki/錯切

5 Framebuffer & 陰影

```
async function setup() {
  // const framebuffers = {} ...

  framebuffers.lightProjection = twgl.createFramebufferInfo(gl, [{
    attachmentPoint: gl.DEPTH_ATTACHMENT,
    internalFormat: gl.DEPTH_COMPONENT32F,
    minMag: gl.NEAREST,
  }], 2048, 2048);
  textures.lightProjection = (
    framebuffers.lightProjection.attachments[0]
  );

  // objects ...
}
```

這邊 attachments 只有一個，其設定 `attachmentPoint: gl.DEPTH_ATTACHMENT` 表示給 framebuffer 寫入深度用的，並由於此 texture 是存放深度，與一般存放顏色的 texture 不同，設定格式為 `internalFormat: gl.DEPTH_COMPONENT32F`，格式名稱中的 32F 表示長度 32bit 的浮點數，為什麼要使用這樣的設定呢？想想如果我們只是使用顏色 texture 來放置深度，那就勢必要選擇一個顏色 channel 來放置深度，除了剩下的 channel 被浪費之外，只有 0-255 整數的精度，設定為浮點數可以使得深度比較更精準，其他 texture 的參數（`gl.texImage2D()` 之 `format`、`type` 等）則由 TWGL 從我們設定的 `internalFormat: gl.DEPTH_COMPONENT32F` 來自動推得適用的設定；最後則是傳入 `minMag: gl.NEAREST` 來關閉 texture 縮圖功能。

⌘ 建立一組拍攝深度用的 shader

在拍攝深度時，顏色計算就變成多餘的，因此建立了一個新的、簡單的 fragment shader 稱為 `depthFragmentShaderSource`，待會會與現有的 `vertexShaderSource` 連結：

```
// fragmentShaderSource ...
```

```
const depthFragmentShaderSource = `#version 300 es
precision highp float;

in float v_depth;

out vec4 outColor;

void main() {
  outColor = vec4(v_depth, v_depth, v_depth, 1);
}
`;

// setup, render, main ...
```

這個 `depthFragmentShaderSource` 其實可以就只有什麼都不做的 `main()`，不過為了預覽深度照片的成像，我們加入 varying `v_depth`，將表示 clip space 中的遠近，將此深度直接當成灰階顏色輸出，接下來讓 vertex shader 計算此 `v_depth`：

```
// ...
out vec2 v_texcoord;
out vec3 v_surfaceToViewer;
out mat3 v_normalMatrix;
out vec4 v_mirrorTexcoord;
out float v_depth;

void main() {
  gl_Position = u_matrix * a_position;

  // ...

  v_depth = gl_Position.z / gl_Position.w * 0.5 + 0.5;
}
```

在 vertex shader 中寫的 `gl_Position.z / gl_Position.w` 可以得到在 clip space 中的 z 值，範圍是 -1 ~ +1，需要乘以 0.5 並加上 0.5 使之介於 0 ~ +1 用於灰階顏色輸出。

5 Framebuffer & 陰影

使用 `twgl.createProgramInfo()` 建立 `depthProgramInfo`：

```js
async function setup() {
  const canvas = document.getElementById('canvas');
  const gl = canvas.getContext('webgl2');

  twgl.setAttributePrefix('a_');

  const programInfo = twgl.createProgramInfo(gl, [
    vertexShaderSource,
    fragmentShaderSource,
  ]);
  const depthProgramInfo = twgl.createProgramInfo(gl, [
    vertexShaderSource,
    depthFragmentShaderSource,
  ]);

  // textures, framebuffers, objects ...

  return {
    gl,
    programInfo, depthProgramInfo,
    textures, framebuffers, objects,
    state: { /* ... */ },
    time: 0,
  };
}
```

從現在開始，我們有第二組 shader、programInfo 了！也就是說，我們接下來要開始注意繪製前最後一次呼叫 `gl.useProgram()` 是使用哪組 shader 以及 uniform 要設定到哪個 shader。

⌘ 產生平行光線投影用的 view matrix

本章節的平行光光線方向向量是由 state.lightRotationXY 所控制，光線一開始向著 -y 方向，依序旋轉 x 軸 state.lightRotationXY[0]、旋轉 y 軸 state.lightRotationXY[1]，使用三角函數將向量 [0, -1, 0] 進行這兩個旋轉之後可以

5-30

得到一組以 state.lightRotationXY 為參數的算式，在一開始的範例就實做好了：

```
twgl.setUniforms(programInfo, {
  u_lightDir: [
    -1 *
      Math.sin(state.lightRotationXY[0]) *
      Math.sin(state.lightRotationXY[1]),
    -1 * Math.cos(state.lightRotationXY[0]),
    -1 *
      Math.sin(state.lightRotationXY[0]) *
      Math.cos(state.lightRotationXY[1]),
  ],
});
```

有興趣的讀者可以自己動筆或是想想看是怎麼得到這樣的算式；回到拍攝深度用的 view matrix — `lightProjectionViewMatrix`，想像一台掃描器，向著 -y 方向進行 orthogonal 投影，並加入以平行光方向偏移的錯切（shear），錯切偏移的角度為旋轉 x 軸的 `state.lightRotationXY[0]`，旋轉 y 軸的 `state.lightRotationXY[1]` 則是旋轉掃描器（也就是視角）即可，整個 light projection 用的 view matrix 將經過以下步驟：

1. 移動視角，因為接下來 orthogonal projection 的 `matrix4.projection()` 捕捉的正面看著 +z，需要先旋轉使之看著 -y，接著旋轉 y 軸 `state.lightRotationXY[1]`，這兩個轉換就是 CH3 的 cameraMatrix、視角 transform，需要包起來做反矩陣。

2. 進行錯切，同樣因為 `matrix4.projection()` 捕捉的正面看著 +z，依據角度從 z 偏移到 y：。

3. 使用 `matrix4.projection()` 進行投影，捕捉場景中 xz 介於 0～20，y（深度）介於 0～10 的物件。

4. `matrix4.projection()` 會把原點偏移到左上，透過 `matrix4.translate (1, -1, 0)` 轉換回來，最後捕捉場景中 xz 介於 -10～+10，y 介於 -5～+5 的物件。

5 Framebuffer & 陰影

把這些 transform 通通融合進 `lightProjectionViewMatrix`，寫成程式如下：

```javascript
function render(app) {
  // cameraMatrix, viewMatrix
  // mirrorCameraMatrix, mirrorViewMatrix ...

  const lightProjectionViewMatrix = matrix4.multiply(
    matrix4.translate(1, -1, 0),
    matrix4.projection(20, 20, 10),
    [ // shearing
      1, 0, 0, 0,
      0, 1, 0, 0,
      0, Math.tan(state.lightRotationXY[0]), 1, 0,
      0, 0, 0, 1,
    ],
    matrix4.inverse(
      matrix4.multiply(
        matrix4.yRotate(state.lightRotationXY[1]),
        matrix4.xRotate(degToRad(90)),
      )
    ),
  );

  // twgl.setUniforms() ...
}
```

⌘ 把深度視覺化，畫到畫面上

因為畫到 framebuffer、texture 的深度到時候繪製陰影時在畫面上是看不到的，為了先驗證『深度』拍攝結果正確，我們先把此平行光照射下去的深度畫到畫面上，暫時不進行 perspective 場景的繪製。

因為現在有多個 program，得在 renderSphere() 以及 renderGround() 時指定使用的 program 以利 uniform 的設定，因此加入 programInfo 參數到這兩個 function，取代原本在 function 內固定使用的 programInfo：

5-3 陰影─拍攝深度資訊

```javascript
function renderSphere(app, viewMatrix, programInfo) {
  const {
    gl,
    programInfo,
    textures, objects,
  } = app;

  // ...

  twgl.setUniforms(programInfo, {
    // ...
  });

  twgl.drawBufferInfo(gl, objects.sphere.bufferInfo);
}

function renderGround(app, viewMatrix, mirrorViewMatrix, programInfo) {
  const {
    gl,
    programInfo,
    textures, objects,
  } = app;

  // ...

  twgl.setUniforms(programInfo, {
    // ...
  });

  twgl.drawBufferInfo(gl, objects.ground.bufferInfo);
  // ...
}
```

這樣就都準備完成了，來到 render() 把 depthProgramInfo 引入，並把原本 perspective 場景的繪製改用深度 shader 以及平行光線投影之 view matrix：

```javascript
function render(app) {
  const {
    gl,
```

5-33

5 Framebuffer & 陰影

```
    programInfo,
    depthProgramInfo,
    framebuffers,
    state,
} = app;

// cameraMatrix, viewMatrix ...
// twgl.setUniforms() ...

twgl.bindFramebufferInfo(gl, framebuffers.mirror);

gl.clear(gl.COLOR_BUFFER_BIT | gl.DEPTH_BUFFER_BIT);

renderSphere(app, mirrorViewMatrix, programInfo);

gl.bindFramebuffer(gl.FRAMEBUFFER, null);

// gl.canvas, gl.viewport() ...

gl.useProgram(depthProgramInfo.program);

renderGround(
  app, lightProjectionViewMatrix, mirrorViewMatrix, depthProgramInfo,
);
renderSphere(app, lightProjectionViewMatrix, depthProgramInfo);
}
```

存檔回到瀏覽器，現在看到的是平行光照射、投影下去所得到的深度，顏色越深越接近掃描器，如圖 5-14：

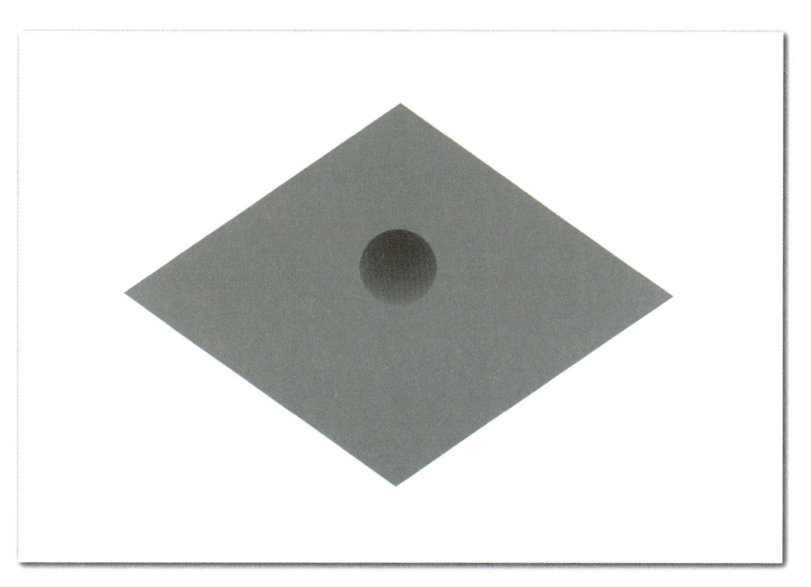

▲ 圖 5-14　平行光照射、投影下去所得到的深度圖

到此進度的完整程式碼可以在這邊找到：

</>	ch5/03/index.html
</>	ch5/03/main.js

也可以參考上線版本，透過瀏覽器打開此頁面：

https://webgl-book.pastleo.me/ch5/03/index.html

⌘ 在 WebGL1 使用深度 texture 功能

在 WebGL1 的 spec 中並沒有支援深度的 texture，也就是說 `gl.texImage2D()` 的 `internalFormat`、`format` 並不能使用深度相關的格式，不過幸好也有一個 WebGL extension 稱為 `WEBGL_depth_texture`，在主流瀏覽器中都有支援[7]，只需要寫點程式來啟用：

7　https://caniuse.com/?search=WEBGL_depth_texture

5 Framebuffer & 陰影

```
async function setup() {
  const gl = canvas.getContext('webgl');
  // ...

  const webglDepthTexExt = gl.getExtension('WEBGL_depth_texture');
  if (!webglDepthTexExt) {
    throw new Error(
      'Your browser does not support WebGL ext: WEBGL_depth_texture'
    );
  }

  // ...
}
```

同時，在 WebGL1 的環境下 `twgl.createFramebufferInfo()` 沒辦法自動偵測正確的 `format` 設定，需要明確告知使用 `gl.DEPTH_COMPONENT`：

```
async function setup() {
  // const framebuffers = {} ...

  framebuffers.lightProjection = twgl.createFramebufferInfo(gl, [{
    attachmentPoint: gl.DEPTH_ATTACHMENT,
    internalFormat: gl.DEPTH_COMPONENT32F,
    format: gl.DEPTH_COMPONENT,
    minMag: gl.NEAREST,
  }], 2048, 2048);
  textures.lightProjection = (
    framebuffers.lightProjection.attachments[0]
  );

  // objects ...
}
```

5-4　陰影—深度 Framebuffer 與 Texture

渲染好深度並繪製到畫面上，可以看到中間一顆球的輪廓，並且在其頂部的地方顏色深度更深，表示更接近深度投影的投影面，接下來讓這個拍攝深度的目標移動到 framebuffer、texture 去，並且在陰影效果時使用。

⌘ 把拍攝深度的目標改成 framebuffer

因為除了前面鏡面的 framebuffer 渲染之外又要多了光源投影，為了讓渲染到不同 framebuffer 之程式能夠在程式碼中比較好分辨，使用 `{}` 建立一個程式碼區塊來區分，首先切換到 `framebuffers.lightProjection`，並呼叫 `renderGround()`、`renderSphere()` 繪製物件進行平行光投影，也就是把深度視覺化的 program 設定、繪製動作拿過來：

```
function render(app) {
  // useProgram, cameraMatrix, viewMatrix ...
  // mirrorCameraMatrix, mirrorViewMatrix ...

  { // lightProjection
    twgl.bindFramebufferInfo(gl, framebuffers.lightProjection);

    gl.clear(gl.COLOR_BUFFER_BIT | gl.DEPTH_BUFFER_BIT);

    gl.useProgram(depthProgramInfo.program);

    renderGround(
      app, lightProjectionViewMatrix,
      mirrorViewMatrix, depthProgramInfo,
    );
    renderSphere(app, lightProjectionViewMatrix, depthProgramInfo);
  }

  // switch framebuffer, render to framebuffers.mirror ...
}
```

5 Framebuffer & 陰影

平行光深度投影這個動作在渲染鏡面世界之前執行，畢竟在鏡面世界也可以看到陰影，這樣的執行順序才能使得在渲染鏡面世界時有深度可用。

因為鏡面世界與 `<canvas />` 的渲染都使用主要的 shader（`programInfo.program`），把 `gl.useProgram(programInfo.program)` 與設定全域 uniform 的程式碼移動下來到渲染鏡面世界之前，記得要先呼叫 `gl.useProgram()` 切換 shader 再設定 uniforms，否則會設定到錯誤的 shader 上；另外也把鏡面世界的渲染用程式碼區塊 `{}` 包起來，與平行光深度投影區塊一樣：

```javascript
function render(app) {
  // ... render to framebuffers.lightProjection

  gl.useProgram(programInfo.program);

  twgl.setUniforms(programInfo, {
    /* u_worldViewerPosition, u_lightDir, u_specular, u_ambient */
  });

  { // mirror
    twgl.bindFramebufferInfo(gl, framebuffers.mirror);

    gl.clear(gl.COLOR_BUFFER_BIT | gl.DEPTH_BUFFER_BIT);

    renderSphere(app, mirrorViewMatrix, programInfo);
  }

  gl.bindFramebuffer(gl.FRAMEBUFFER, null);
  // ...
}
```

並且讓繪製到 `<canvas />` 的程式回去使用 viewMatrix、programInfo：

```javascript
function render(app) {
  // ...
  // render to lightProjection, mirror

  gl.bindFramebuffer(gl.FRAMEBUFFER, null);
```

5-38

```
gl.canvas.width = gl.canvas.clientWidth;
gl.canvas.height = gl.canvas.clientHeight;
gl.viewport(0, 0, canvas.width, canvas.height);

renderGround(app, viewMatrix, mirrorViewMatrix, programInfo);
renderSphere(app, viewMatrix, programInfo);
}
```

為了確認渲染到 `framebuffers.lightProjection`、`textures.lightProjection` 能夠正常使用，我們先把地板渲染時 `u_diffuseMap` 指定的顏色 texture 改成 `textures.lightProjection`，並且取消使用鏡面對應的 texcoord：

```
function renderGround(app, viewMatrix, mirrorViewMatrix, programInfo) {
  // ...
  twgl.setUniforms(programInfo, {
    u_matrix: matrix4.multiply(viewMatrix, worldMatrix),
    u_worldMatrix: worldMatrix,
    u_normalMatrix: matrix4.transpose(matrix4.inverse(worldMatrix)),
    u_normalMap: textures.nullNormal,
    u_diffuse: [0, 0, 0],
    u_diffuseMap: textures.lightProjection,
    u_specularExponent: 200,
    u_emissive: [0, 0, 0],
    u_useMirrorTexcoord: false,
    u_mirrorMatrix: mirrorViewMatrix,
  });
  // ...
}
```

存檔之後，可以看到渲染了原本的場景，只是地板從鏡面變成由不同亮度的紅色來表示的平行光投影深度圖，如圖 5-15 所示：

5 Framebuffer & 陰影

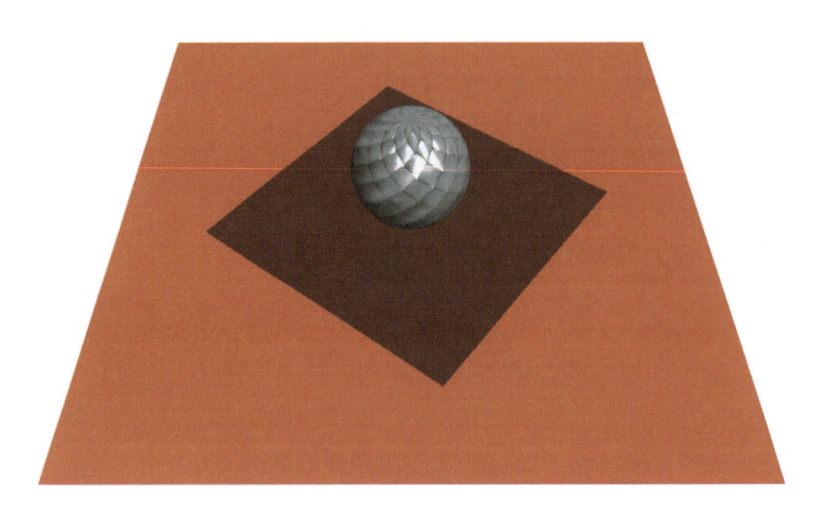

▲ 圖 5-15　平行光投影深度圖繪製在地板上

稍微把視角拉低，可以看到深度圖中的球，如圖 5-16 所示：

▲ 圖 5-16　深度圖中的球

因為我們把地板的 `u_useMirrorTexcoord` 關閉，也就是取消使用鏡面對應的 `texcoord`，這時 `u_diffuseMap` 使用的 `textures.lightProjection` 就只是一般的 texture 貼圖，也就是把前面預覽深度圖時所看到的樣子貼在地板上，至於為何是紅色？因為這個 texture 使用的格式是 `gl.DEPTH_COMPONENT32F`，並沒有 RGB 三個 channel，所以每個 pixel 唯一的深度數值就跑到紅色 channel 去，剩下綠色、藍色 channel 為 0，因此為不同亮度的紅色，到此進度的完整程式碼可以在這邊找到：

</>	ch5/04/index.html
</>	ch5/04/main.js

也可以參考上線版本，透過瀏覽器打開此頁面：

https://webgl-book.pastleo.me/ch5/04/index.html

5-5 陰影—計算是否產生陰影

有了深度圖，但是渲染到 `<canvas />` 的時候要怎麼運用這個深度圖來判斷物體表面是否要產生陰影？請參考圖 5-17，B 點所在的灰色面為平行光投影掃描器的底面、A 點為平行光可以到達位於球體的亮面一處、C 點為地板上平行光無法到達、應產生陰影的一處，B 到 A 再到 C 所產生的直線與平行光方向平行：

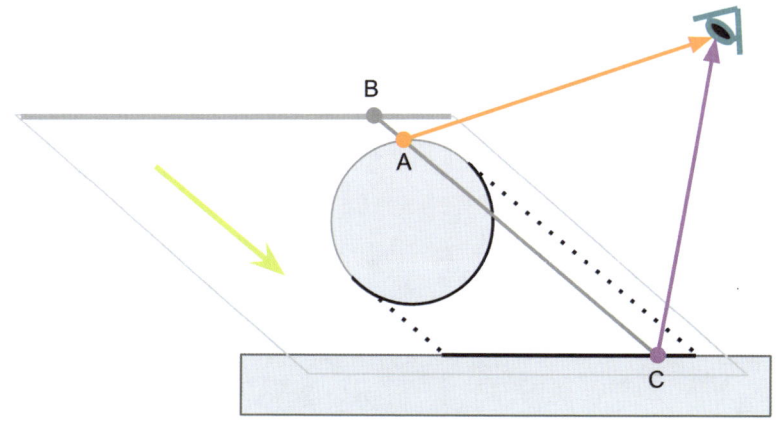

▲ 圖 5-17 陰影產生示意圖

5　Framebuffer & 陰影

在 `textures.lightProjection` 的深度資料中，B 點上的深度來自 A 點，可以說是將 A 點做 `lightProjectionViewMatrix` 的轉換後得到 B 點，這時有趣的來了，在渲染 `<canvas />` 中地板的 C 點時如果對其位置進行 `lightProjectionViewMatrix` 的轉換，同樣可以到達 B 點，與 A 點的差別就是 C 點做完『投影』的深度會比 A 點來的深，我們可以利用這一點來檢查是否在陰影下。

⌘ 在 shader 中進行投影並比較深度產生陰影

對於將圖 5-17 A 點、C 點投影到 B 點這個技巧，陰影與鏡面非常相似，在 shader 中對表面位置進行 framebuffer 的 view matrix 轉換，也就是說我們又需要給 shader 再傳一個 matrix 了，筆者稱之為 `u_lightProjectionMatrix`。

先從 vertex shader 開始，接收這個平行光投影用的 `u_lightProjectionMatrix` uniform，在 vertex shader 中傳換成 `v_lightProjection` varying 表示投影後的位置（圖 5-17 的 B 點）給 fragment shader 使用：

```
// ...

uniform mat4 u_matrix;
uniform mat4 u_worldMatrix;
uniform mat4 u_normalMatrix;
uniform vec3 u_worldViewerPosition;
uniform mat4 u_mirrorMatrix;
uniform mat4 u_lightProjectionMatrix;

out vec2 v_texcoord;
out vec3 v_surfaceToViewer;
out mat3 v_normalMatrix;
out vec4 v_mirrorTexcoord;
out float v_depth;
out vec4 v_lightProjection;

void main() {
  // gl_Position ...
```

5-42

```
vec4 worldPosition = u_worldMatrix * a_position;
v_surfaceToViewer = u_worldViewerPosition - worldPosition.xyz;

v_mirrorTexcoord = u_mirrorMatrix * worldPosition;

v_depth = gl_Position.z / gl_Position.w * 0.5 + 0.5;
v_lightProjection = u_lightProjectionMatrix * worldPosition;
}
```

在 fragment shader 方面，接收深度 texture `u_lightProjectionMap` uniform 以及平行光投影後的位置 `v_lightProjection`，並且跟 `v_mirrorTexcoord` 一樣要除以 `.w` 使之與 WebGL 計算成 clip space 中的位置相同，這邊需要取得兩個深度來進行比較，藉此知道是否產生陰影：

- 由 `v_lightProjection.z / v_lightProjection.w` 計算而來的 `lightToSurfaceDepth`：表示該點（可能為 A 或是 C 點）投影下去的深度。
- 從深度 texture 查詢到的 `lightProjectedDepth`：平行光投影時該點的深度，也就是 B 點上的值。
  - 為了從深度 texture 查詢，還需要其 texcoord，而這就是該點投影下去的 `.xy` 位置，即為 `v_lightProjection.xy / v_lightProjection.w`。

由於 `v_lightProjection.z / v_lightProjection.w` 與 `v_lightProjection.xy / v_lightProjection.w` 都是 clip space 中的位置，範圍為 -1 ~ +1，為了符合深度 texture 取出的『紅色』顏色亮度與 texcoord 的範圍 0 ~ +1，因此需要對這兩者乘以 0.5 加上 0.5，這麼一來整體 `lightToSurfaceDepth` 與 `lightProjectedDepth` 的計算過程寫到 fragment shader 如下：

```
// ...
in mat3 v_normalMatrix;
in vec4 v_mirrorTexcoord;
in vec4 v_lightProjection;

// ...
```

5 Framebuffer & 陰影

```glsl
uniform vec3 u_ambient;
uniform bool u_useMirrorTexcoord;
uniform sampler2D u_lightProjectionMap;

out vec4 outColor;

void main() {
  // ...

  vec2 lightProjectionCoord = (
    v_lightProjection.xy / v_lightProjection.w * 0.5 + 0.5
  );
  float lightToSurfaceDepth = (
    v_lightProjection.z / v_lightProjection.w * 0.5 + 0.5
  );
  float lightProjectedDepth = texture(
    u_lightProjectionMap,
    lightProjectionCoord
  ).r;

  // ...
}
```

好了，兩個深度資料準備就緒，進行深度比較：

```glsl
void main() {
  // ... lightToSurfaceDepth, lightProjectedDepth
  // ... diffuseLight, specularBrightness ...

  float occlusion = (
    lightToSurfaceDepth > lightProjectedDepth ? 0.9 : 0.0
  );

  diffuseLight *= 1.0 - occlusion;
  specularBrightness *= 1.0 - clamp(occlusion * 2.0, 0.0, 1.0);

  // ... outColor = ...
}
```

5-5 陰影—計算是否產生陰影

筆者在這邊建立 `occlusion` 變數表示『有多少成的光源被遮住』，使用三元運算子比較，當該點（可能為 A 或是 C 點）投影下去的深度比平行光投影時該點的深度來的深（值較大、顏色較亮），我們就可以知道這個表面在陰影下，並在陰影下時設定 90% 的 `occlusion`，讓散射光減少 90%、反射光減少 100%，反射光的部份把 `occlusion` 乘以 2 並限制在 0 ~ +1 之間，就可以避免減少超過 100%；這麼一來 shader 的部份就改好囉。

剩下得把 shader 所需要的 `u_lightProjectionMatrix`、`u_lightProjectionMap` 傳入：

```javascript
function render(app) {
  const {
    gl,
    programInfo,
    depthProgramInfo,
    framebuffers, textures,
    state,
  } = app;

  // ... lightProjectionViewMatrix ...

  twgl.setUniforms(programInfo, {
    // ...
    u_specular: [1, 1, 1],
    u_ambient: [0.4, 0.4, 0.4],
    u_lightProjectionMatrix: lightProjectionViewMatrix,
    u_lightProjectionMap: textures.lightProjection,
  });
}
```

前面為了在地板預覽紅色深度有修改 uniform，在這邊把設定還原回鏡面：

```javascript
function renderGround(
  app, viewMatrix, mirrorViewMatrix, programInfo,
) {
```

5 Framebuffer & 陰影

```
// ...
twgl.setUniforms(programInfo, {
  // ...
  u_diffuse: [0, 0, 0],
  u_diffuseMap: textures.mirror,
  u_specularExponent: 200,
  u_emissive: [0, 0, 0],
  u_useMirrorTexcoord: true,
  u_mirrorMatrix: mirrorViewMatrix,
});
// ...
}
```

好的，存檔回瀏覽器看看結果，如圖 5-18 所示[8]：

▲ 圖 5-18 直接比較 lightToSurfaceDepth、lightProjectedDepth 產生的陰影

顯然怪怪的，不過也是可以看到有部份陰影正確地出現，如果把視角調低，在特定的平行光方向下可以看到如圖 5-19 紅色框框中正確的陰影：

---

8 對於浮點數誤差所導致的渲染結果在不同瀏覽器、顯示卡之間可能會不同

5-46

5-5　陰影—計算是否產生陰影

▲ 圖 5-19　直接比較 lightToSurfaceDepth、lightProjectedDepth 產生部份正確的陰影

⌘ 深度比較的浮點數誤差值與平滑化

除了整個地板有一半都被陰影所壟罩著，球體上光照的區域也有一點一點的陰影，為什麼會這樣呢？因為我們比較的是兩個深度的『浮點數』，儘管像是上方示意圖 5-17 中的 A 點，光源投影下來的深度與後來重算的深度可能因為 GPU 計算過程中浮點數的微小差異而導致 `lightToSurfaceDepth > lightProjectedDepth` 成立，這個現象的會隨著不同的 GPU、作業系統等環境而有所不同，為了避免這個問題我們讓 `lightToSurfaceDepth` 必須比 `lightProjectedDepth` 還要大出一定的數值才判定為有陰影，筆者讓這個值為 0.01，寫成程式：

```
void main() {
  // ... lightToSurfaceDepth, lightProjectedDepth ...
  float occlusion = (
    lightToSurfaceDepth > lightProjectedDepth + 0.01 ? 0.75 : 0.0
  );
  // ...
}
```

5-47

5　Framebuffer & 陰影

這麼一來陰影就正常多囉，如圖 5-20 所示：

▲ 圖 5-20　深度差距必須大於 0.01 才產生陰影

在大部份的狀況下這個陰影已經很夠用，不過把視角調到很近的時候，在球面上可以看到一些不平整的陰影，如圖 5-21：

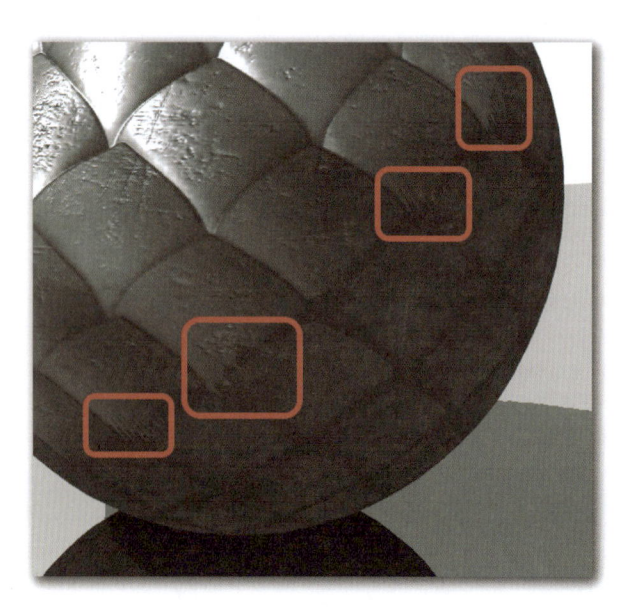

▲ 圖 5-21　不平整的陰影

這樣的原因是因為我們使用了三元運算子，陰影只有『有』跟『無』兩種狀況，能不能讓深度差距大於 0.01 的時候開始漸漸使得 occlusion 有數值，在差距接近一個最大值的時候平滑地接近 1.0？在 GLSL 中有一個內建的 function 叫做 smoothstep[9]，就可以做到這件事，給它兩個邊緣 edge0、edge1 與要加工的數值 x（在這邊也就是深度差異），輸入的 x 在小於 edge0 的時候回傳 0、大於 edge1 的時候回傳 1，介於 edge0 與 edge1 之間會平滑地接近兩個邊緣，畫成線圖[10] 如圖 5-22 所示：

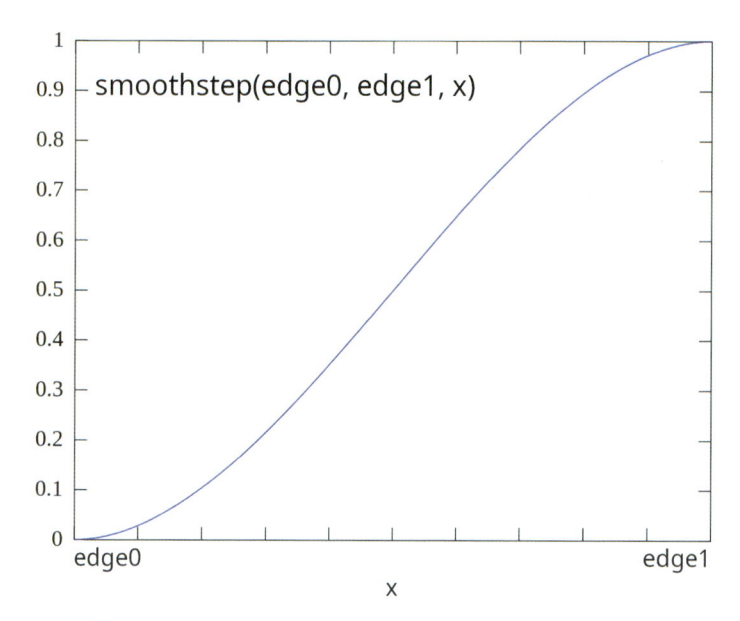

▲ 圖 5-22　smoothstep(edge0, edge1, x) 的 x 輸入與回傳值

將 smoothstep 套用在 occlusion 的計算，寫成程式如下：

```
void main() {
  // ... lightToSurfaceDepth, lightProjectedDepth ...
  float occlusion = smoothstep(
```

9　https://registry.khronos.org/OpenGL-Refpages/es3.0/html/smoothstep.xhtml
10　圖 5-22 修改參考修改自 smoothstep 維基百科：
　　https://en.wikipedia.org/wiki/Smoothstep

5　Framebuffer & 陰影

```
    0.01, 0.1, lightToSurfaceDepth - lightProjectedDepth
);
// ...
}
```

這時陰影就滑順囉，如圖 5-23 所示、完整的鏡面及陰影效果如圖 5-24 所示：

▲ 圖 5-23　套用 smoothstep 作為陰影『程度』

▲ 圖 5-24　鏡面及陰影效果成品

5-50

本章節的完整程式碼可以在這邊找到：

</>	ch5/05/index.html
</>	ch5/05/main.js

也可以參考上線版本，透過瀏覽器打開此頁面：

https://webgl-book.pastleo.me/ch5/05/index.html

如果有讀者覺得圖 5-23 在球上的陰影範圍不夠明顯，想要確認陰影範圍，可以把陰影顯現方式改成限制顏色的最大亮度，像是這樣使得有陰影的地方會只剩下 RGB 的 B 藍色：

```
void main() {
  // ...
  outColor = vec4(
    clamp(
      diffuse * diffuseLight +
      u_specular * specularBrightness +
      u_emissive,
      ambient, vec3(1.0 - occlusion, 1.0 - occlusion, 1)
    ),
    1
  );
}
```

這麼一來，就可以好好比較使用三元運算子判斷有或是無陰影，以及使用 smoothstep 所形成的 occlusion 範圍，分別如圖 5-25、圖 5-26 所示：

5 Framebuffer & 陰影

▲ 圖 5-25　使用三元運算子的 occlusion 範圍

▲ 圖 5-26　使用 smoothstep 的 occlusion 範圍

　　CH5 主要的內容就到此，使用 framebuffer 延伸了更多光線效果，這麼一來就從散射光、反射光到鏡面與陰影都有了，這些效果加在一起可以製作出頗生動的畫面，不覺得圖 5-24 的畫面蠻漂亮的嗎？接下來我們再讓鏡面地板可以使用 normal map，為下個篇章『帆船與海』的水面做好準備。

5-6　毛玻璃效果—使用 Normal Map 的鏡面

　　在 CH4 的結尾我們為地板加上 normal map 使得散射光、反射光可以在表面形成更多細節，看起來就如同真的有凹凸的細節，在這邊也來幫鏡面加上 normal map，對鏡面反射、陰影產生干擾，形成一個常見的使用情境所需的效果——波光粼粼的水面，不過由於目前要使用的 normal map 是靜態的，不會像水面那樣有波動，反而比較像是毛玻璃，看起來如圖 5-27：

5-6 毛玻璃效果—使用 Normal Map 的鏡面

▲ 圖 5-27 鏡面表面使用 normal map，產生毛玻璃效果

本篇接下來就以圖 5-27 為目標開始實做吧，首先這個毛玻璃的 normal map 是筆者使用往後會實做的水面波紋產生的，如圖 5-28：

▲ 圖 5-28 毛玻璃 normal map

在 `setup()` 中讀取這張圖，筆者已經在範例包放置了經過轉換、最佳化的 normal map texture：

</> assets/water_normal.webp

```javascript
async function setup() {
  // ...
  const textures = Object.fromEntries(
    await Promise.all(Object.entries({
      scale: '/assets/scale_diffuse.webp',
      scaleNormal: '/assets/scale_normal.webp',
      groundNormal: '/assets/water_normal.webp',
    }).map(async ([name, url]) => {
      // image, texture ...
    }))
  );
  // ...
}
```

這樣就可以讀取毛玻璃 normal map 進來了，不過稍微一看會發現在 `setup()` 中讀取、設定 `textures` 的程式碼已經累積不少，TWGL 剛好有提供 `twgl.createTextures()`[11]，注意這個 function 名稱最後的 `Textures` 有一個 s，表示可以一次讀取複數張 texture，一次把所有需要的讀取、設定好：

```javascript
async function setup() {
  // canvas, gl, programInfo ...
  const textures = twgl.createTextures(gl, {
    scale: {
      src: '/assets/scale_diffuse.webp',
      min: gl.LINEAR_MIPMAP_LINEAR, mag: gl.LINEAR,
    },
    scaleNormal: {
      src: '/assets/scale_normal.webp',
      min: gl.LINEAR_MIPMAP_LINEAR, mag: gl.LINEAR,
```

---

11 https://twgljs.org/docs/module-twgl.html#.createTextures

```
  },
  groundNormal: {
    src: '/assets/water_normal.webp',
    min: gl.LINEAR_MIPMAP_LINEAR, mag: gl.LINEAR,
    wrap: gl.REPEAT,
  },
  null: { src: [0, 0, 0, 255] },
  nullNormal: { src: [127, 127, 255, 255] },
});
// framebuffers ...
}
```

twgl.createTextures() 除了一次讀取多張圖之外，還可以個別設定每個 texture 的縮放參數、渲染超過邊緣的行為等，甚至手動指定的 null、nullNormal texture 資料都可以透過簡單的陣列一行搞定，透過給定的參數自動偵測或是使用預設值把 gl.texImage2D() 包裝好，可以少寫許多手動指定的參數，最後回傳的 textures 也是 Javascript object 物件，對應傳入的 key-value 結構，因此與我們目前其他地方的使用方式吻合，可以把原本手動 loadImage()、gl.texImage2D() 冗長的程式碼刪除了：

```
async function setup() {
  // canvas, gl, programInfo …

  // const textures = twgl.createTextures({ ... })
  const textures = Object.fromEntries(
    await Promise.all(Object.entries({ /* ... */ }).map(() => {
      // loadImage, gl.createTexture, gl.bindTexture, gl.texImage2D ...
    }))
  );

  textures.null = gl.createTexture();
  gl.bindTexture(/* ... */);
  gl.texImage2D(/* ... */);

  textures.nullNormal = gl.createTexture();
  gl.bindTexture(/* ... */);
  gl.texImage2D(/* ... */);

  // framebuffers ...
```

5 Framebuffer & 陰影

⌘ 建立毛玻璃地板專用的 shader

毛玻璃 normal map 讀取好，並且使用 TWGL 做了重構，接下來要對陰影、鏡面反射進行干擾，這個效果只需要對毛玻璃地板作用，在先前我們使用 `u_useMirrorTexcoord` 來控制是否使用鏡面投影的 `v_mirrorTexcoord`，如果想要繼續讓地板與場景物件共用 shader（精確來說是 WebGL `program`、`programInfo`），那就勢必要加上更多判斷式，如果今天是一個在 CPU 上運算的程式，使用 if 或是三元運算子產生程式執行的岔路是司空見慣，不過 shader 在 GPU 中執行時的硬體等級並不存在真正的岔路，shader 中寫的 if 或是三元運算子的兩種可能路徑 GPU 都會執行，並且根據岔路的條件留下想要的結果，因為 GPU 的硬體設計是一大堆運算核心同時做『一樣』的事情，沒辦法因為某一筆資料符合特定條件而只走岔路的其中一條，所有核心都必須兩條都走過，過多複雜的岔路很可能會對性能造成影響；因此這邊或許是一個好時機把地板所使用的 shader 獨立出來，要對這個特別的物件做更細部的調整也更容易，不需要顧忌其他物件。

先整個複製 `fragmentShaderSource` 字串，因為要給地板使用，筆者命名為 `groundFragmentShaderSource`：

```
// depthFragmentShaderSource ...
const groundFragmentShaderSource = `#version 300 es
precision highp float;

// ...

out vec4 outColor;

void main() {
  vec2 texcoord = u_useMirrorTexcoord ? (
    (v_mirrorTexcoord.xy / v_mirrorTexcoord.w) * 0.5 + 0.5
  ) : v_texcoord;
  vec3 normal = texture(u_normalMap, v_texcoord).xyz * 2.0 - 1.0;
  normal = normalize(v_normalMatrix * normal);
```

```
// ...
outColor = vec4(
  clamp(
    diffuse * diffuseLight +
    u_specular * specularBrightness +
    u_emissive,
    ambient, vec3(1, 1, 1)
  ),
  1
);
}
`;
```

接著 groundFragmentShaderSource 中 texcoord 一律使用 v_mirrorTexcoord 反射對應的鏡面世界相片之 texture 位置：

```
// groundFragmentShaderSource ...
uniform vec3 u_ambient;
uniform bool u_useMirrorTexcoord;
uniform sampler2D u_lightProjectionMap;

out vec4 outColor;

void main() {
  vec2 texcoord = (v_mirrorTexcoord.xy / v_mirrorTexcoord.w) * 0.5 + 0.5;
  vec3 normal = texture(u_normalMap, v_texcoord).xyz * 2.0 - 1.0;

  // lightProjectionCoord, outColor ...
}
```

這邊要注意 normal map 的取樣使用原始的 v_texcoord 而非 v_mirrorTexcoord，畢竟 normal map 的位置是物件本體的表面 texcoord，不是鏡面投影的位置；原本的 fragmentShaderSource 中，v_mirrorTexcoord、u_useMirrorTexcoord 以及 texcoord 都不需要了，可以移除：

```
// fragmentShaderSource ...
```

5 Framebuffer & 陰影

```glsl
in mat3 v_normalMatrix;
in vec4 v_mirrorTexcoord;
in vec4 v_lightProjection;

// ...

uniform vec3 u_ambient;
uniform bool u_useMirrorTexcoord;
uniform sampler2D u_lightProjectionMap;

out vec4 outColor;

void main() {
  vec2 texcoord = u_useMirrorTexcoord ? (
    (v_mirrorTexcoord.xy / v_mirrorTexcoord.w) * 0.5 + 0.5
  ) : v_texcoord;
  vec3 normal = texture(u_normalMap, v_texcoord).xyz * 2.0 - 1.0;
  normal = normalize(v_normalMatrix * normal);

  // lightProjectionCoord, lightToSurfaceDepth ...

  vec3 diffuse = u_diffuse + texture(u_diffuseMap, v_texcoord).rgb;

  // ... outColor
}
```

　　vertex shader 的部份倒是可以共用，編譯、連結毛玻璃地板專用的 groundProgramInfo 並好好放置在 app 中：

```js
async function setup() {
  // canvas, gl ...

  const programInfo = twgl.createProgramInfo(gl, [
    vertexShaderSource, fragmentShaderSource,
  ]);
  const depthProgramInfo = twgl.createProgramInfo(gl, [
    vertexShaderSource, depthFragmentShaderSource,
  ]);
  const groundProgramInfo = twgl.createProgramInfo(gl, [
```

```
    vertexShaderSource, groundFragmentShaderSource,
  ]);

  // textures, framebuffers, objects ...

  return {
    gl,
    programInfo, depthProgramInfo,
    groundProgramInfo,
    textures, framebuffers, objects,
    state: { /* ... */ },
    time: 0,
  };
}
```

再到 `render()`，這邊有一個小問題，因為 uniform 是跟著 program 的，意味著對於 `groundProgramInfo` 我們也必須把原先共用的 uniform 設定上去，為此把單獨設定給 `programInfo` 的 uniform 內容拉出來成為變數 `globalUniforms` 給兩個 program 共用：

```
function render(app) {
  // cameraMatrix, viewMatrix ...
  // mirrorCameraMatrix, mirrorViewMatrix ...
  // lightProjectionViewMatrix ...
  const globalUniforms = {
    u_worldViewerPosition: cameraMatrix.slice(12, 15),
    u_lightDir: [
      -1 * Math.sin(state.lightRotationXY[0]) *
        Math.sin(state.lightRotationXY[1]),
      -1 * Math.cos(state.lightRotationXY[0]),
      -1 * Math.sin(state.lightRotationXY[0]) *
        Math.cos(state.lightRotationXY[1]),
    ],
    u_specular: [1, 1, 1],
    u_ambient: [0.4, 0.4, 0.4],
    u_lightProjectionMatrix: lightProjectionViewMatrix,
    u_lightProjectionMap: textures.lightProjection,
  }
```

5　Framebuffer & 陰影

```
{ // lightProjection
  // ...
}

gl.useProgram(programInfo.program);
twgl.setUniforms(programInfo, globalUniforms);

// ...
}
```

　　修改繪製流程，將地板繪製獨立出來，先設定切換好 program、設定 uniform 再呼叫 `renderGround()` 使用新建立的 `groundProgramInfo` 繪製地板：

```
function render(app) {
  const {
    gl,
    programInfo,
    depthProgramInfo,
    groundProgramInfo,
    framebuffers, textures,
    state,
  } = app;

  // ...

  gl.bindFramebuffer(gl.FRAMEBUFFER, null);

  gl.canvas.width = gl.canvas.clientWidth;
  gl.canvas.height = gl.canvas.clientHeight;
  gl.viewport(0, 0, canvas.width, canvas.height);

  renderGround(app, viewMatrix, mirrorViewMatrix, programInfo);
  renderSphere(app, viewMatrix, programInfo);

  gl.useProgram(groundProgramInfo.program);
  twgl.setUniforms(groundProgramInfo, globalUniforms);
  renderGround(app, viewMatrix, mirrorViewMatrix, groundProgramInfo);
}
```

5-6 毛玻璃效果—使用 Normal Map 的鏡面

順便把 `renderGround()` 內對 `u_useMirrorTexcoord uniform` 的設定移除：

```
function renderGround(app, viewMatrix, mirrorViewMatrix, programInfo) {
  // ...

  twgl.setUniforms(programInfo, {
    // ...
    u_specularExponent: 200,
    u_emissive: [0, 0, 0],
    u_useMirrorTexcoord: true,
    u_mirrorMatrix: mirrorViewMatrix,
  });

  twgl.drawBufferInfo(gl, objects.ground.bufferInfo);

  twgl.setUniforms(programInfo, {
    u_useMirrorTexcoord: false,
  });
}
```

到這邊增加地板專用 shader、`groundProgramInfo` 的調整就完成了，存檔重整結果與先前圖 5-24 相同。

⌘ 根據 Normal Map 對鏡面反射、陰影產生干擾

如果要讓鏡面實際依據 normal map 模擬光線反射方向去找到成像，現有『使用 framebuffer 先拍攝鏡面世界作為正式渲染地板反射成像』的做法就基本上不能使用了；陰影部份也沒辦法單純地使用法向量做表面位置的改變，進而對陰影深度比較產生影響。

但是我們還是可以利用現有的 normal map 來製造『干擾』，方法也很簡單，就是直接把 normal map 的 xy 方向當成從鏡面世界相片、深度照片取樣時的偏移值，搭配得當的數值大小調整，還是可以製造足以以假亂真的效果，實做到地板 shader 內：

5 Framebuffer & 陰影

```
// groundFragmentShaderSource
// ...

void main() {
  vec2 texcoord = (v_mirrorTexcoord.xy / v_mirrorTexcoord.w) * 0.5 + 0.5;
  vec3 normal = texture(u_normalMap, v_texcoord).xyz * 2.0 - 1.0;
  vec2 distortion = normalize(normal).xy;
  normal = normalize(v_normalMatrix * normal);

  vec2 lightProjectionCoord =
    v_lightProjection.xy / v_lightProjection.w * 0.5 + 0.5;
  float lightToSurfaceDepth =
    v_lightProjection.z / v_lightProjection.w * 0.5 + 0.5;
  float lightProjectedDepth = texture(
    u_lightProjectionMap,
    lightProjectionCoord + distortion * 0.01
  ).r;

  vec3 diffuse = u_diffuse + texture(
    u_diffuseMap,
    texcoord + distortion * 0.1
  ).rgb;

  // surfaceToLightDir ...
  // outColor ...
}
```

在 normal 做 `v_normalMatrix` 轉換、旋轉之前取得其單位向量化的 xy 值作為干擾值 `distortion`，並且把此干擾值加到對鏡面世界相片 `u_diffuseMap`、深度照片 `u_lightProjectionMap` 取樣時的取樣位置，筆者這邊分別乘上 0.01、0.1 來縮小，要不然 normal 如果是 `[1, 0, 0]` 那不就偏到很遠的地方去了。

最後在繪製地板之前把 normal map 指定到 uniform 上：

```
function renderGround(app, viewMatrix, mirrorViewMatrix, programInfo) {
  // ...
```

5-62

5-6 毛玻璃效果—使用 Normal Map 的鏡面

```
twgl.setUniforms(programInfo, {
  // ...
  u_normalMap: textures.groundNormal,
  // ...
});

twgl.drawBufferInfo(gl, objects.ground.bufferInfo);
}
```

毛玻璃的效果就完成囉，

如圖 5-29：

▲ 圖 5-29　使用 normal map 『干擾』鏡面、陰影深度取樣產生的毛玻璃效果

　　讀者可以自行調整上面 shader 中 `distortion` 干擾取樣位置的程度，大約感受一下適用的倍數，筆者在這邊也是經過數次調整之後得到這兩個感覺效果比較好的數值。

5-63

5 Framebuffer & 陰影

⌘ 鏡面顏色

目前這個地板的本質還是一個鏡面，在下個章節會需要作為水面，希望可以透過指定 u_diffuse 顏色來決定其基本顏色，為此需要把 gl.clearColor() 清除 framebuffer 的顏色設定成全黑，鏡面相片 u_diffuseMap 在沒有物件的地方呈現黑色，才不會導致其與 diffuse 顏色相加之後大部分區域皆為全白：

```javascript
async function setup() {
  // canvas, gl ...
  // textures, framebuffers, objects ...

  gl.enable(gl.CULL_FACE);
  gl.enable(gl.DEPTH_TEST);
  gl.clearColor(0, 0, 0, 1);

  return { /* ... */ };
}
```

測試一下，把地板物件設定一個灰色：

```javascript
function renderGround(app, viewMatrix, mirrorViewMatrix, programInfo) {
  // ...
  twgl.setUniforms(programInfo, {
    // ...
    u_diffuse: [0.75, 0.75, 0.75],
    // ...
  });

  twgl.drawBufferInfo(gl, objects.ground.bufferInfo);
}
```

先前最亮的時候整片是白色的（圖 5-30），現在（圖 5-31）空白區域最亮就頂多是 u_diffuse 指定的顏色：

5-6　毛玻璃效果—使用 Normal Map 的鏡面

▲ 圖 5-30　先前以白色作為清除顏色、鏡面照片底色

▲ 圖 5-31　改用黑色為底色之後可指定鏡面基本顏色

⌘ 解析度與 HiDPI

　　在下個章節，我們會在 shader 中加入較為複雜的運算，因為在 Web 平台上本身就會受到系統、瀏覽器比較嚴苛的效能限制，可能會有運算效能比較不足的裝置開始出現卡頓、每秒畫面張數達不到 60 FPS 的狀況；在光譜的另一端是則是具備 HiDPI 螢幕的裝置，HiDPI 為 High Dots Per Inch 的簡稱，蘋果的是視網膜（Retina）螢幕就是 HiDPI，表示實際解析度高於作業系統、CSS 的 pixel 單位，這些螢幕的畫面會看起來更細緻，在 Web 平台上可以透過 `window.devicePixelRatio`[12] 取得這個比例。

　　為了滿足這兩種狀況，我們可以在畫面上加入一個簡單的解析度控制，就像是許多遊戲中的 Options 選項，提供『低，0.75 倍』、『普通，1 倍』解析度以及『Rentina』，選擇 Rentina 時則根據裝置系統提供的 `window.devicePixelRatio` 來設定，這些選項的 UI 直接使用 HTML 放在畫面右上角：

12　https://developer.mozilla.org/en-US/docs/Web/API/Window/devicePixelRatio

5 Framebuffer & 陰影

```html
<style>
  html, body { /* ... */}
  #canvas { /* ... */ }
  #options {
    position: fixed;
    top: 0;
    right: 0;
    margin: 1rem;
  }
</style>
<body>
  <canvas id="canvas"></canvas>
  <div id="options">
    <label for="resolution-ratio">解析度</label>
    <select id="resolution-ratio" name="resolution-ratio">
      <option value="0.75">低</option>
      <option value="1" selected>普通</option>
      <option
        id="resolution-ratio-retina"
        value="1" disabled
      >Retina</option>
    </select>
  </div>
  <script type="module" src="main.js"></script>
</body>
```

　　Rentina 選項預設是關閉的，透過 Javascript 檢查發現是 HiDPI 螢幕時啟用這個選項並設定此選項的值為 `window.devicePixelRatio`：

```javascript
async function main() {
  // setup, listenToInputs ...

  if (window.devicePixelRatio > 1) {
    const retinaOption = document.getElementById('resolution-ratio-retina');
    retinaOption.value = window.devicePixelRatio;
    retinaOption.disabled = false;
  }

  startLoop(app);
}
```

5-66

5-6 毛玻璃效果—使用 Normal Map 的鏡面

接著回到 `setup()`，在 `state` 中加入 `resolutionRatio`，預設值為 1，也就是普通解析度：

```
async function setup() {
  // ...
  return {
    // ...
    state: {
      // ...
      cameraViewing: [0, 0, 0],
      lightRotationXY: [0, 0],
      resolutionRatio: 1,
    },
    time: 0,
  };
}
```

把 `<select />` 元素的 `change` 事件與 `app.state.resolutionRatio` 串起來：

```
async function main() {
  // setup, listenToInputs ...

  const resolutionSelect = document.getElementById('resolution-ratio');
  resolutionSelect.addEventListener('change', () => {
    app.state.resolutionRatio = parseFloat(resolutionSelect.value);
  });
  if (window.devicePixelRatio > 1) { /* ... */ }

  startLoop(app);
}
```

最後則是實際把 `app.state.resolutionRatio` 反應到 `<canvas />`，在 CH1 我們要使得畫布充滿整個畫面時將 `clientWidth`、`clientHeight` 指定到畫布的實際寬高 `width`、`height`，這件事 TWGL 也有 function[13] 可以幫忙，並且接收一個倍數參數讓我們設定要是 CSS pixel 解析度的幾倍：

13　https://twgljs.org/docs/module-twgl.html#.resizeCanvasToDisplaySize

5 Framebuffer & 陰影

```
function render(app) {
  // ...

  gl.bindFramebuffer(gl.FRAMEBUFFER, null);

  twgl.resizeCanvasToDisplaySize(gl.canvas, state.resolutionRatio);
  gl.canvas.width = gl.canvas.clientWidth;
  gl.canvas.height = gl.canvas.clientHeight;
  gl.viewport(0, 0, canvas.width, canvas.height);

  renderSphere(app, viewMatrix, programInfo);

  // renderSphere, renderGround ...
}
```

使用 iPad Pro 分別對三種選項進行測試如下圖 5-32、圖 5-33、圖 5-34，可以比較出細緻度的差異：

▲ 圖 5-32　低解析度　　　▲ 圖 5-26　普通解析度　　　▲ 圖 5-26　Retina 解析度

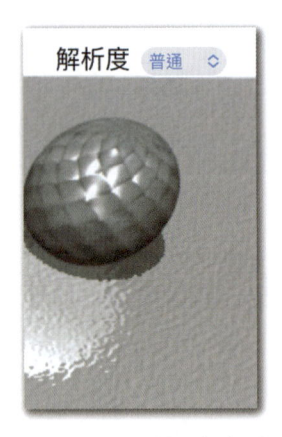

好的，這麼一來我們透過 normal map 法向量干擾了鏡面反射以及陰影，形成往後將用於水面的毛玻璃效果，同時也做了解析度的選項控制以增進對應裝置的使用者體驗，到此進度的完整程式碼可以在這邊找到：

5-6　毛玻璃效果─使用 Normal Map 的鏡面

</>	ch5/06/index.html
</>	ch5/06/main.js

也可以參考上線版本，透過瀏覽器打開此頁面：

https://webgl-book.pastleo.me/ch5/06/index.html

在本章提到的鏡面實做方式，是使用 framebuffer 拍攝鏡面世界中的相機所看到的，當成地面物件之 texture 並搭配鏡面相機的 transform 使用，雖然成果非常精準，但是這個做法很不通用，如果鏡面不是 y = 0 的平面，或是複數個鏡子，都會讓此方法變得非常複雜，無法重複使用拍攝的結果。

實務製作複雜 3D 場景需要鏡面反射時有一個更通用的方法，稱為反射貼圖[14]，與其拍攝一張特定角度的照片，不如使用『環境』的 360 度環景相片，反射時使用反射角度從環景照片中取樣，就可以得到大致的反射結果，雖然不會非常精確，但是這個方式很通用，可以提供各種狀況使用，甚至毛玻璃這邊想要實際使用法向量計算反射對應角度也可以使用。

對於這張反射貼圖，我們可以在場景搭建時拍攝好提供遊戲運作時使用，大部分的遊戲引擎或是建模工具相信都有提供對應的工具，也是大多數遊戲採用的方式，不過也可以利用本章介紹的 framebuffer 在每個 frame 拍照以提供即時的反射效果。

14　https://zh.wikipedia.org/wiki/反射貼圖

5 Framebuffer & 陰影

6

帆船與海

基於製作好的光影效果與 shader，本章節將製作一個完整的場景，作為本書所要製作的小遊戲的場景建構：主角是一艘帆船，在一片看不到邊的海面上，天氣晴朗。

6 帆船與海

在 CH5 結尾有稍微為這邊的開頭做好準備，因此這邊的起始點與 CH5 結尾相差不大，跑起來如圖 6-1 所示：

▲ 圖 6-1　CH6 起始點，海上的一顆球

可以看到畫面上有一顆從上上個章節就一直存在的球體，第一個目標就是將之換成新主角—帆船，值得注意的有：

- 地板（ground）改成海面（ocean），物件更名為 `objects.plane` 物件，使用 scale 放大 xz 方向為 4000x4000 的平面，並使用藍色 `#2d8da9` 為 diffuse。
  - normal map 以 `u_normalMapSize` 設定 16x16 為單位重複，為此加上 varying `v_worldPosition` 使 fragment shader 能取得場景位置。
- mirror 改稱 reflection。

起始點完整程式碼可以在這邊找到：

</>	ch6/00/index.html
</>	ch6/00/main.js

也可以參考上線版本，透過瀏覽器打開此頁面：
https://webgl-book.pastleo.me/ch6/00/index.html

6-1 主角『帆船』— obj 3D 模型檔案的讀取與繪製

建構 3D 場景的過程中,往往不會只需要球體、長方體、平面這種平淡無奇的物件,更多時候想放的是複雜的 3D 物件,而這些物件通常會用專用的建模軟體進行製作,完成之後輸出成某種檔案格式,讀取這些資料放置到 GPU buffer 之後,在渲染場景時繪製,就可以在場景中放置各式各樣豐富的物件。

⌘ `.obj` & `.mtl` 檔案

`.obj` 就是一種 3D 模型的檔案格式,精確的說,經過讀取後可以成為 vertex attribute 的資料來源,包含了各個頂點的位置、texcoord、法向量,成為 3D 場景中的一個物件,而 `.mtl` 則是存放材質資料,像是散射光、反射光的顏色等。

筆者先前在練習 `.obj` 的讀取時,順便小小學習了 Blender[1] 這套開源且強大的 3D 建模軟體,並且製作了一艘帆船,渲染出來如圖 6-2:

▲ 圖 6-2 主角『帆船』

---

1 https://www.blender.org/

6 帆船與海

筆者也將此 3D 模型上傳到 Sketchfab，一個分享 3D 模型的網站：

https://sketchfab.com/3d-models/my-first-boat-f505dd73384245e08765ea6824b12644

網站上提供了各式各樣的格式，不過在撰文的當下需要登入才能下載就是了。

這個模型就成了接下來練習時需要模型時使用的素材，同時也是我們要放入場景中的帆船，匯出成 `.obj`、`.mtl` 之後，放置在範例包中的這個位置：

</>	assets/sailboat.obj
</>	assets/sailboat.mtl

用文字編輯器打開 `sailboat.obj` 可以看到：

```
# Blender v3.2.2 OBJ File: 'sailboat.blend'
# www.blender.org
mtllib sailboat.mtl
o Cube_Cube.001
v -0.245498 -0.021790 2.757482
v -0.551836 0.552017 2.746644
v -0.371110 -0.118091 0.326329
...
vt 0.559949 0.000000
vt 0.625000 0.000000
vt 0.625000 0.250000
...
vn -0.7759 -0.6250 0.0861
vn 0.0072 -0.0494 -0.9988
vn 0.7941 -0.6020 0.0836
...
usemtl body
...
f 17/1/1 2/2/1 4/3/1 18/4/1
```

```
f 18/4/2 4/3/2 8/5/2 19/6/2
f 19/6/3 8/5/3 6/7/3 20/8/3
...
o Cylinder.004_Cylinder.009
v 0.000000 0.308823 0.895517
v 0.000000 0.640209 0.895517
...
```

　　`.obj` 要紀錄 3D 物件的每個頂點資料，想當然爾檔案通常不小，這個模型有 20.6k 個三角形，檔案大小約 1.3MB，這邊不會看全部的細節，只擷取了一些小片段來觀察其內容：

- `mtllib sailboat.mtl` 的 `mtllib` 開頭表示使用了 `sailboat.mtl` 這個檔案來描述材質。

- `o Cube_Cube.001` 的 `o` 開頭表示一個子物件的開始，`Cube_Cube.001` 這個名字來自於 Blender 中的物件名稱，可見筆者當時建構子物件時是從一個立方體開始，不過沒有使用良好的習慣好好命名就是了。

- `v -0.245498 -0.021790 2.757482`、`vt 0.559949 0.000000`、`vn -0.7759 -0.6250 0.0861` 的 `v`、`vt`、`vn` 開頭分別為位置（position）、texcoord、法向量資料，實際去打開檔案可以看到 `.obj` 絕大部分的內容都是這些三角形頂點資料。

- `usemtl body` 表示這個子物件要使用的材質的名字，理論上可以在 `.mtl` 中找到對應的名字。

- `f 17/1/1 2/2/1 4/3/1 18/4/1` 表示一個『面』，這邊是一個四邊形，有四個頂點，每個頂點分別用一個 index 數字表示使用的哪一筆位置（position）、texcoord、法向量，類似於 indexed element 的指標。

- 接下來看到另一個 `o` 開頭 `o Cylinder.004_Cylinder.009` 表示另一個子物件的開始，看來下個子物件筆者從圓柱體開始，一樣沒有好好命名。

　　同樣地，看一下 `.mtl` 的片段：

6 帆船與海

```
# Blender MTL File: 'sailboat.blend'
# Material Count: 11

newmtl mast
Ns 225.000000
Ka 1.000000 1.000000 1.000000
Kd 0.352941 0.196078 0.047058
Ks 0.500000 0.500000 0.500000
Ke 0.000000 0.000000 0.000000
Ni 1.450000
d 1.000000
illum 2
...
newmtl flag-pastleo
Ns 225.000000
Ka 1.000000 1.000000 1.000000
Kd 0.000000 0.000000 0.000000
Ks 0.500000 0.500000 0.500000
Ke 0.000000 0.000000 0.000000
Ni 1.450000
d 1.000000
illum 2
map_Kd pastleo.jpg
...
```

　　這個檔案就相對小很多，`newmtl mast` 對應 `.obj` 中的材質名稱，接下來則是對於不同光線的顏色或參數，這邊提幾個本書有關注過的材質參數[2]，Ka 表示環境光顏色、Kd 散射光顏色、Ks 反射光顏色、Ns 反射光『範圍參數』（u_specularExponent），根據不同 3D 模型、渲染環境會有更多不同的參數；其中一個命名為 `flag-pastleo` 的材質有一個參數 `map_Kd pastleo.jpg` 表示要使用 `pastleo.jpg` 這個圖檔作為散射光 texture，稍微觀察圖 6-2 帆船模型中間船桅上有一面旗子，旗子中的圖案就是筆者的大頭貼，對應範例包中的 `assets/pastleo.jpg`，可見那面旗子就是使用了 `flag-pastleo` 這個材質。

2　關於 `.mtl` 各個開頭所表示材質參數的定義：
　https://people.sc.fsu.edu/~jburkardt/data/mtl/mtl.html

6-1 主角『帆船』— obj 3D 模型檔案的讀取與繪製

範例包中的 .mtl 檔案與上傳到 sketchfab.com 的有點不同,因為本系列文實做的 shader 會導致一些材質顏色不明顯,因此筆者有手動調整 .mtl 部份材質的顏色。

⌘ 讀取 .obj & .mtl

好的,綜觀來看,自己寫讀取程式的話,除了 .obj、.mtl parser 之外,得從 f 開頭的『面』資料展開成一個個三角形,接著取得要使用的位置、texcoord、法向量,轉換成 buffer 作為 vertex attribute 使用,除此之外還要處理 .mtl 的對應、建立子物件等,顯然是個不小的工程;既然 .obj 是一種公用格式,那麼應該可以找到現成的讀取工具,筆者找到的是這款:

https://github.com/frenchtoast747/webgl-obj-loader

原作者以 MIT 許可[3]開源,可惜沒有提供 ES module 的方式引入,因此筆者 fork 此專案並且修改使之可以產出 ES module 的版本:

https://github.com/pastleo/webgl-obj-loader

建置好的 ES module `.js` 檔案放置在此專案中的 `dist/webgl-obj-loader.esm.js`,並且已經放置在範例程式碼包當中:

> lib/vendor/webgl-obj-loader.esm.js

因此使用 ES import 語法就可以直接引入:

```
import * as WebGLObjLoader from '../../lib/vendor/webgl-obj-loader.esm.js';
```

---

3 https://github.com/frenchtoast747/webgl-obj-loader/blob/master/LICENSE.md

6 帆船與海

接著建立一個 function 來串接 `WebGLObjLoader` 讀取 `.obj`、`.mtl` 並傳入 WebGL context gl 變數，稱為 `loadSailboatModel`，使用 webgl-obj-loader 的 `WebGLObjLoader.downloadModels()`，可以同時下載所有需要的檔案並解析好，包含 `.mtl` 甚至 texture 圖檔，先看一下經過 `WebGLObjLoader` 讀取好的資料看起來如何：

```
// setup ...

async function loadSailboatModel(gl) {
  const { boatModel } = await WebGLObjLoader.downloadModels([{
    name: 'boatModel',
    obj: '/assets/sailboat.obj',
    mtl: true,
  }]);

  console.log(boatModel);
}
```

在此傳入的 name 將對應回傳物件的 key 值，因此直接用 Javascript 解構語法把 `boatModel` 取出，同時可以注意到 `WebGLObjLoader.downloadModels()` 接受一個陣列，意味著這個 function 可以用來一次讀取許多 `.obj` 檔案。

在 setup() 中呼叫：

```
async function setup() {
  // canvas, gl ...
  // textures, framebuffers, objects ...

  await loadSailboatModel(gl);

  // ...
}
```

在瀏覽器的 Console 中可以看到 `console.log()` 印出 `boatModel` 的內容如圖 6-3：

6-8

```
main.js:299
n {name: 'boatModel', indicesPerMaterial: Array(1
▼ 0), materialsByIndex: {…}, tangents: Array(0), bit
  angents: Array(0), …} ⓘ
  ▶ bitangents: []
  ▶ indices: (34560) [14290, 14291, 14292, 14292, 142
  ▶ indicesPerMaterial: (10) [Array(168), Array(552),
  ▶ materialIndices: {body: 0, building: 1, building_
  ▶ materialNames: (10) ['body', 'building', 'buildin
  ▶ materialsByIndex: {0: l, 1: l, 2: l, 3: l, 4: l,
    name: "boatModel"
  ▶ tangents: []
    textureStride: 2
  ▶ textures: (74660) [0.559949, 0, 0.625, 0, 0.625,
  ▶ vertexMaterialIndices: (37330) [0, 0, 0, 0, 0, 0,
  ▶ vertexNormals: (111990) [-0.7759, -0.625, 0.0861,
  ▶ vertices: (111990) [-0.374793, 0.372056, 2.750663
  ▶ [[Prototype]]: Object
```

▲ 圖 6-3 使用 webgl-obj-loader 讀取好的 boatModel 內容

配合 webgl-obj-loader README 文件的說明[4]，我們需要關注的資料分別是 `.vertices` 對應 `a_position`、`.vertexNormals` 對應 `a_normal`、`.textures` 對應 `a_texcoord`，但是這邊的資料沒有子物件的概念，有一個 `.indices` 已經是 indexed element 的 `ELEMENT_ARRAY_BUFFER` 指標陣列，把整個 .obj 檔案的三角形頂點資料合併成一組，不像是 `.obj` 中一個個 `f` 開頭的頂點指向不同組子物件的 position、texcoord、normal。

那材質的部分呢？在當前的渲染實作中同一個物件一次渲染只能指定一組 `u_diffuse`、`u_specular` 等 uniform 讓物件為一個單色，直接使用 `.indices` 讓整艘船作為一個 3D 物件、VAO 的話便無法使不同子物件使用不同的材質，幸好 `WebGLObjLoader.downloadModels()` 所回傳的物件中有 `.indicesPerMaterial`，裡面包含了一個個的 indices 陣列，分別對應一組材質設定，我們可以利用這個建立以材質劃分的子物件陣列，有趣的事情是，這些 indices 所對應的實際 vertex attribute 是共用的，也就是說 position、texcoord、

4　https://github.com/frenchtoast747/webgl-obj-loader#meshobjstr

6 帆船與海

normal 的 buffer 只要建立一組，接下來每個子物件建立各自的 indices buffer 並與共用 position、texcoord、normal 的 buffer 組成一個個『物件』、VAO，最後渲染時各個物件設定好各自的 uniform 後進行繪製即可。

因此在 `WebGLObjLoader.downloadModels()` 之後建立共用的 bufferInfo：

```javascript
async function loadSailboatModel(gl) {
  const { boatModel } = await WebGLObjLoader.downloadModels([{
    // ...
  }]);

  const sharedBufferInfo = twgl.createBufferInfoFromArrays(gl, {
    position: { numComponents: 3, data: boatModel.vertices },
    texcoord: { numComponents: 2, data: boatModel.textures },
    normal: { numComponents: 3, data: boatModel.vertexNormals },
  });

  console.log(boatModel);
}
```

接下來讓 app.objects.sailboat 表示整艘帆船，但是要一個一個繪製子物件，因此使 `app.objects.sailboat` 將會是一個陣列，每一個元素包含子物件的 bufferInfo、VAO 以及 uniforms，從 `boatModel.indicesPerMaterial.map()` 出發：

```javascript
async function setup() {
  // canvas, gl ...
  // textures, framebuffers, objects ...
  await loadSailboatModel(gl, programInfo);
  // ...
}

async function loadSailboatModel(gl, programInfo) {
  // boatModel ...
  // sharedBufferInfo ...
```

6-1 主角『帆船』— obj 3D 模型檔案的讀取與繪製

```javascript
const parts = boatModel.indicesPerMaterial.map((indices, mtlIdx) => {
  const material = boatModel.materialsByIndex[mtlIdx];

  const bufferInfo = twgl.createBufferInfoFromArrays(gl, {
    indices,
  }, sharedBufferInfo);

  return {
    bufferInfo,
    vao: twgl.createVAOFromBufferInfo(gl, programInfo, bufferInfo),
  }
});
}
```

這邊東西有點多，我們一個一個來：

- Javascript 的 `.map()` 內 function 第一個參數 `indices` 為陣列元素，也就是各個材質 indices 指標陣列，第二個參數 `mtlIdx` 為該元素在陣列中的第幾個 index。

- 在 `.indicesPerMaterial` 陣列中，第幾個 indices 陣列就使用第幾個 material 材質，因此 `boatModel.materialsByIndex[mtlIdx]` 取得對應的材質設定。

- 使用 `twgl.createBufferInfoFromArrays()`[5] 的第三個參數 `srcBufferInfo` 來共用剛才建立的 `sharedBufferInfo`，這感覺其實有點像是 Map 的 merge 或是 `Object.assign()`。

- 為 `loadSailboatModel()` 傳入 `programInfo`，並以 `twgl.createVAOFromBufferInfo()` 建立 VAO。

- `.map()` 內 function 為每個子物件回傳 Javascript 物件來放置該子物件到時候繪製所需要的資料，這邊包含了 `bufferInfo` 與 vao，`.map()` 完回傳 parts 陣列表示帆船的部件子物件們。

5　https://twgljs.org/docs/module-twgl.html#.createBufferInfoFromArrays

6 帆船與海

這樣一來物件們的三角形頂點資料 vertex attribute buffer、indices buffer 以及 VAO 就準備好了，剩下的就是把材質資料轉換成我們可以拿來設定到 uniform 的格式，把這邊取得的 material 印出來看：

```javascript
async function loadSailboatModel(gl, programInfo) {
  // boatModel ...
  // sharedBufferInfo ...

  const parts = boatModel.indicesPerMaterial.map((indices, mtlIdx) => {
    const material = boatModel.materialsByIndex[mtlIdx];
    console.log(material);

    // ...
  });
}
```

　　console.log() 在這個位置會把所有的材質內容依序顯示在 Console 上，圖 6-4 為 Console 上展開 `flag-pastleo` 材質的內容，雖然這個物件有不少東西，不過要找到 `u_diffuse`、`u_specular` 對應的資料不會很困難，名字幾乎能夠直接對起來；如果是有 diffuse texture 的，可以在 `material.mapDiffuse.texture` 找到，如圖 6-5，而且已經是 `HTMLImageElement`[6] 物件。

6　https://developer.mozilla.org/en-US/docs/Web/API/HTMLImageElement

6-1　主角『帆船』— obj 3D 模型檔案的讀取與繪製

```
                                                                    main.js:307
▶ 1 {name: 'sails-front', ambient: Array(3), diffuse: Array(3), specula
  r: Array(3), emissive: Array(3), …}
                                                                    main.js:307
▼ 1 {name: 'flag-pastleo', ambient: Array(3), diffuse: Array(3), specula
  r: Array(3), emissive: Array(3), …} ℹ
   ▶ ambient: (3) [1, 1, 1]
     antiAliasing: false
   ▶ diffuse: (3) [0, 0, 0]
     dissolve: 1
   ▶ emissive: (3) [0, 0, 0]
     illumination: 2
   ▶ mapAmbient: {colorCorrection: false, horizontalBlending: true, verti
   ▶ mapBump: {colorCorrection: false, horizontalBlending: true, vertical
   ▶ mapDecal: {colorCorrection: false, horizontalBlending: true, vertica
   ▶ mapDiffuse: {colorCorrection: false, horizontalBlending: true, verti
   ▶ mapDisplacement: {colorCorrection: false, horizontalBlending: true,
   ▶ mapDissolve: {colorCorrection: false, horizontalBlending: true, vert
   ▶ mapEmissive: {colorCorrection: false, horizontalBlending: true, vert
   ▶ mapReflections: []
   ▶ mapSpecular: {colorCorrection: false, horizontalBlending: true, vert
   ▶ mapSpecularExponent: {colorCorrection: false, horizontalBlending: tr
     name: "flag-pastleo"
     refractionIndex: 1.45
     sharpness: 0
   ▶ specular: (3) [0.5, 0.5, 0.5]
     specularExponent: 225
   ▶ transmissionFilter: (3) [0, 0, 0]
     transparency: 0
   ▶ [[Prototype]]: Object
```

▲ 圖 6-4　flag-pastleo 材質的內容

```
▼ mapDiffuse:
     boostMipMapSharpness: 0
     bumpMultiplier: 1
     clamp: false
     colorCorrection: false
     filename: "pastleo.jpg"
     horizontalBlending: true
     imfChan: null
   ▶ modifyTextureMap: {brightness: 0, contrast: 1}
   ▶ offset: {u: 0, v: 0, w: 0}
   ▶ scale: {u: 1, v: 1, w: 1}
   ▶ texture: img
     textureResolution: null
   ▶ turbulence: {u: 0, v: 0, w: 0}
     verticalBlending: true
   ▶ [[Prototype]]: Object
```

▲ 圖 6-5　material.mapDiffuse.texture 為讀取好的 HTMLImageElement

6-13

6 帆船與海

到時候渲染帆船的時候，我們需要對各個子物件設定多項 uniform，也就是使用 `twgl.setUniforms()`，這個 function 設定的方式是傳入以 uniform 為 key 的 Javascript 物件，那麼我們就可以在 `loadSailboatModel()` 的階段就把這樣的 key-value uniform 物件準備好，把需要的材質資料從 webgl-obj-loader 的 material 取出：

```javascript
async function setup() {
  // canvas, gl ...
  // textures, framebuffers, objects ...
  await loadSailboatModel(gl, textures, programInfo);
  // ...
}

async function loadSailboatModel(gl, textures, programInfo) {
  // ...
  const parts = boatModel.indicesPerMaterial.map((indices, mtlIdx) => {
    // material, bufferInfo ...

    let u_diffuseMap = textures.null;
    if (material.mapDiffuse.texture) {
      u_diffuseMap = twgl.createTexture(gl, {
        wrapS: gl.CLAMP_TO_EDGE, wrapT: gl.CLAMP_TO_EDGE,
        min: gl.LINEAR_MIPMAP_LINEAR,
        src: material.mapDiffuse.texture,
      });
    }

    return {
      // bufferInfo, vao ...
      uniforms: {
        u_diffuse: material.diffuse,
        u_diffuseMap,
        u_specular: material.specular,
        u_specularExponent: material.specularExponent,
        u_emissive: material.emissive,
        u_ambient: [0.6, 0.6, 0.6],
      },
    }
  });
}
```

如果 `material.mapDiffuse.texture` 存在，也就是有 diffuse texture，因為已經是 `HTMLImageElement`，使用 `twgl.createTexture()` 對其建立 WebGL texture，對於沒有使用 texture 的子物件，就跟之前一樣要設定成 `textures.null` 避免影響到單色渲染；另一個比較特別的是 `u_ambient`，因為筆者為此系列文撰寫的 shader 運作方式與 Blender、Sketchfab 網站上看到的不同，會顯得特別暗，同時 `u_ambient` 這邊實做的功能是基於 diffuse 的最低亮度，因此筆者一律設定成 `[0.6, 0.6, 0.6]`。

最後讓 `loadSailboatModel()` 回傳 `parts`，並在 `setup()` 這邊以 `app.objects.sailboat` 放好：

```
async function setup() {
  // canvas, gl ...
  // textures, framebuffers, objects ...
  objects.sailboat = await loadSailboatModel(gl, textures, programInfo);
  console.log(objects.sailboat);
  // ...
}

async function loadSailboatModel(gl, textures, programInfo) {
  // ...
  const parts = boatModel.indicesPerMaterial.map((indices, mtlIdx) => {
    // ...
  });

  return parts;
}
```

這樣一來 `app.objects.sailboat` 就準備好帆船物件了，同時筆者也加上一行 `console.log()` 來檢視讀取、整理好的帆船部件子物件陣列，如圖 6-6：

6 帆船與海

```
▼ (10) [{…}, {…}, {…}, {…}, {…}, {…}, {…}, {…}, {…}, {…}] ⓘ    main.js:271
    ▼ 0:
      ▼ bufferInfo:
        ▶ attribs: {a_position: {…}, a_texcoord: {…}, a_normal: {…}}
          elementType: 5123
        ▶ indices: WebGLBuffer {}
          numElements: 168
        ▶ [[Prototype]]: Object
      ▼ uniforms:
        ▶ u_ambient: (3) [0.6, 0.6, 0.6]
        ▶ u_diffuse: (3) [0.303197, 0.118964, 0.079075]
        ▶ u_diffuseMap: WebGLTexture {}
        ▶ u_emissive: (3) [0, 0, 0]
        ▶ u_specular: (3) [0.977941, 0.977941, 0.977941]
          u_specularExponent: 66
        ▶ [[Prototype]]: Object
      ▶ vao: WebGLVertexArrayObject {}
      ▶ [[Prototype]]: Object
    ▼ 1:
      ▶ bufferInfo: {attribs: {…}, numElements: 552, indices: WebGLBuffer,
      ▶ uniforms: {u_diffuse: Array(3), u_diffuseMap: WebGLTexture, u_specu
      ▶ vao: WebGLVertexArrayObject {}
      ▶ [[Prototype]]: Object
    ▶ 2: {bufferInfo: {…}, vao: WebGLVertexArrayObject, uniforms: {…}}
    ▶ 3: {bufferInfo: {…}, vao: WebGLVertexArrayObject, uniforms: {…}}
    ▶ 4: {bufferInfo: {…}, vao: WebGLVertexArrayObject, uniforms: {…}}
    ▶ 5: {bufferInfo: {…}, vao: WebGLVertexArrayObject, uniforms: {…}}
    ▶ 6: {bufferInfo: {…}, vao: WebGLVertexArrayObject, uniforms: {…}}
    ▶ 7: {bufferInfo: {…}, vao: WebGLVertexArrayObject, uniforms: {…}}
    ▶ 8: {bufferInfo: {…}, vao: WebGLVertexArrayObject, uniforms: {…}}
    ▶ 9: {bufferInfo: {…}, vao: WebGLVertexArrayObject, uniforms: {…}}
      length: 10
```

▲ 圖 6-6　app.objects.sailboat 帆船部件子物件陣列

到此進度的完整程式碼可以在這邊找到：

</>	ch6/01/index.html
</>	ch6/01/main.js

也可以參考上線版本，透過瀏覽器打開此頁面：

https://webgl-book.pastleo.me/ch6/01/index.html

帆船 3D 模型之繪製

繪製的部份，建立一個 `renderSailboat()` function 負責，因為對於各個物件所需設定的 uniform、VAO 切換與繪製指令類似，可以把 `renderSphere()` 當成模板複製出來改：

```javascript
function renderSailboat(app, viewMatrix, programInfo) {
  const { gl, textures, objects } = app;

  gl.bindVertexArray(objects.sphere.vao);

  const worldMatrix = matrix4.multiply( /* ... */ );

  twgl.setUniforms(programInfo, {
    // ...
  });

  twgl.drawBufferInfo(gl, objects.sphere.bufferInfo);
}

// renderOcean, render ...
```

把原本繪製球體用的 VAO 切換以及 twgl.drawBufferInfo() 移除後，因為帆船 3D 模型 app.objects.sailboat 是子物件陣列，使用 .forEach() 依序繪製子物件：

```javascript
function renderSailboat(app, viewMatrix, programInfo) {
  const { gl, textures, objects } = app;

  // worldMatrix, twgl.setUniforms() ...

  objects.sailboat.forEach(({ bufferInfo, vao, uniforms }) => {

  });
}
```

6 帆船與海

.forEach() 的 function 在參數上直接進行 Javascript 物件解構把 loadSailboatModel() 的準備好的子物件內容 bufferInfo、vao、uniforms 取出，接著使用這些資料切換 VAO、設定 uniforms 並呼叫繪製指令：

```javascript
function renderSailboat(app, viewMatrix, programInfo) {
  const { gl, textures, objects } = app;

  // worldMatrix, twgl.setUniforms() ...

  objects.sailboat.forEach(({ bufferInfo, vao, uniforms }) => {
    gl.bindVertexArray(vao);
    twgl.setUniforms(programInfo, uniforms);
    twgl.drawBufferInfo(gl, bufferInfo);
  });
}
```

回到 render()，把渲染球體改成渲染帆船：

```javascript
function render(app) {
  // cameraMatrix, viewMatrix ... globalUniforms

  { // lightProjection
    // ...

    renderOcean(
      app, lightProjectionViewMatrix, reflectionViewMatrix, depthProgramInfo
    );
    renderSailboat(app, lightProjectionViewMatrix, depthProgramInfo);
  }

  gl.useProgram(programInfo.program);
  twgl.setUniforms(programInfo, globalUniforms);

  { // reflection
    // ...

    renderSailboat(app, reflectionViewMatrix, programInfo);
  }
```

6-18

6-1 主角『帆船』— obj 3D 模型檔案的讀取與繪製

```javascript
gl.bindFramebuffer(gl.FRAMEBUFFER, null);

twgl.resizeCanvasToDisplaySize(gl.canvas, state.resolutionRatio);
gl.viewport(0, 0, canvas.width, canvas.height);

renderSailboat(app, viewMatrix, programInfo);

// ... renderOcean
}
```

帆船就出現在場景中了！成功讀取、繪製 3D obj 檔案，如圖 6-7：

▲ 圖 6-7　3D obj 帆船繪製在畫面中

6-19

6 帆船與海

　　除此之外，也可以看到帆船的陰影與水面倒影，因為在陰影深度投影與鏡面世界都有渲染帆船，現在可說是坐收前面辛苦實做的成果，只是轉一下會發現帆船漂浮在空中，並且可以看到之前球體使用的 normal map 產生之光線反應，如圖 6-8：

▲ 圖 6-8　漂浮在空中、有套用 scale normal map 的船體

　　因為我們還沒去修改 uniform，球體的 `worldMatrix` 轉換被套用到帆船上了，當時透過 `matrix4.translate(0, 1, 0)` 往 y 方向平移 1 單位，在這邊就變成讓帆船漂浮在空中，先著手修改這個部份：

```
function renderSailboat(app, viewMatrix, programInfo) {
  const { gl, textures, objects } = app;

  const worldMatrix = matrix4.multiply(
    matrix4.yRotate(degToRad(45)),
    matrix4.translate(0, 0, 0),
    matrix4.scale(1, 1, 1),
  );

  twgl.setUniforms(programInfo, {
    u_matrix: matrix4.multiply(viewMatrix, worldMatrix),
```

```
  u_worldMatrix: worldMatrix,
  u_normalMatrix: matrix4.transpose(matrix4.inverse(worldMatrix)),
  // ...
});

// objects.sailboat.forEach() ...
}
```

取消平移,同時筆者加上 `matrix4.yRotate(degToRad(45))` 讓帆船一開始可以稍微轉一下不要背對著使用者。

對於帆船身上還保留著球體 scale normal map 的問題,顯然得把 `u_normalMap: textures.scaleNormal` 改用空白 normal map,除此之外,在 `renderSailboat()` 頂層 `twgl.setUniforms()` 現在的工作變成設定帆船所有子物件要共用的 uniform,可以看到有許多 uniform 已經與子物件的 uniform 重複,例如 `u_diffuse`、`u_diffuseMap` 等:

```javascript
function renderSailboat(app, viewMatrix, programInfo) {
  // ... worldMatrix
  twgl.setUniforms(programInfo, {
    u_matrix: matrix4.multiply(viewMatrix, worldMatrix),
    u_worldMatrix: worldMatrix,
    u_normalMatrix: matrix4.transpose(matrix4.inverse(worldMatrix)),
    u_normalMap: textures.nullNormal,
    u_diffuse: [0, 0, 0],
    u_diffuseMap: textures.scale,
    u_specularExponent: 40,
    u_emissive: [0.15, 0.15, 0.15],
    u_ambient: [0.4, 0.4, 0.4],
  });
  // objects.sailboat.forEach() ...
}
```

帆船的位置以及表面法向量就正常囉,如圖 6-9、圖 6-10:

6　帆船與海

▲ 圖 6-9　3D obj 帆船取消平移後正確放置在水面上

▲ 圖 6-10　網頁載入後帆船稍微以 y 軸旋轉展示給使用者觀看

最後把不再使用的 `renderSphere()` 以及 `console.log()` 移除，到此進度的完整程式碼可以在這邊找到：

</>	ch6/02/index.html
</>	ch6/02/main.js

6-22

也可以參考上線版本，透過瀏覽器打開此頁面：
https://webgl-book.pastleo.me/ch6/02/index.html

一般能幫助 obj 或是其他格式的 3D 檔案讀取的函式庫，或許是因為絕大部分都會直接與一個渲染框架（如 three.js）或是遊戲引擎（如 Unity）整合，基本上沒有一個能夠讓 WebGL 直接使用的，筆者想這是因為 WebGL 是一個底層的 API，函式庫不知道我們會怎麼設定 uniform、呼叫繪製指令做繪製流程，因此頂多就是幫忙把 obj、mtl 檔案解析好回傳頂點以及材質資料，整合的工作還是不少，不過透過這個流程相信讀者也了解到其中的細節，知道 obj 的頂點資料是如何最後渲染在畫面上的。

6-2 Skybox

現在有了各種光線、陰影、水面倒影效果，物體也從球體換成 3D 帆船模型，整個場景已經可以看得出來是一艘帆船在海上囉；但是如果把視角調低，就可以立刻看到我們缺少的東西：天空。

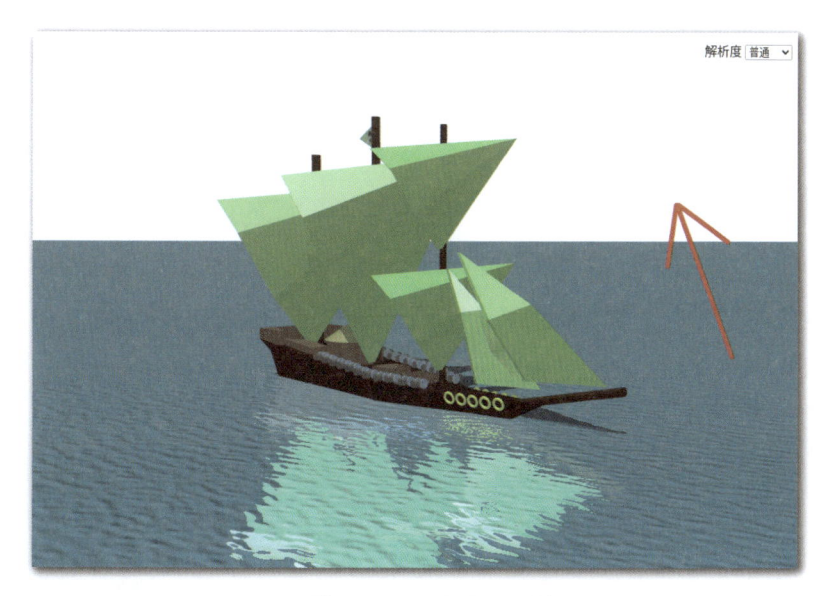

▲ 圖 6-11　天空是全白的

6 帆船與海

3D 場景中的天空事實上只是一個背景，但是要符合視角方向，因此這個背景就成了一張 360 度的照片，類似於 Google 街景那樣，在 WebGL 中要做出這樣效果可以透過 `gl.TEXTURE_CUBE_MAP` 的 texture 來做到；我們常用的 texture 形式是 `gl.TEXTURE_2D`，在 shader 中對著 `sampler2D` 使用 vec2 來取樣，使用 `gl.TEXTURE_CUBE_MAP` 的 texture 時，要給他 6 張圖，分別為 +x、-x、+y、-y、+z、-z，貼在下圖立方體的 6 個面，在 shader 中對著 `samplerCube` 使用 vec3 來取樣（理論上此 vec3 應為單位向量），這個向量稱為 normal 法向量，類似於一顆球體表面上某個位置的法向量，取樣的結果將是從立方體中心點往該向量方向出發，其延伸的線與面相交的點的顏色，請見下圖 6-12：

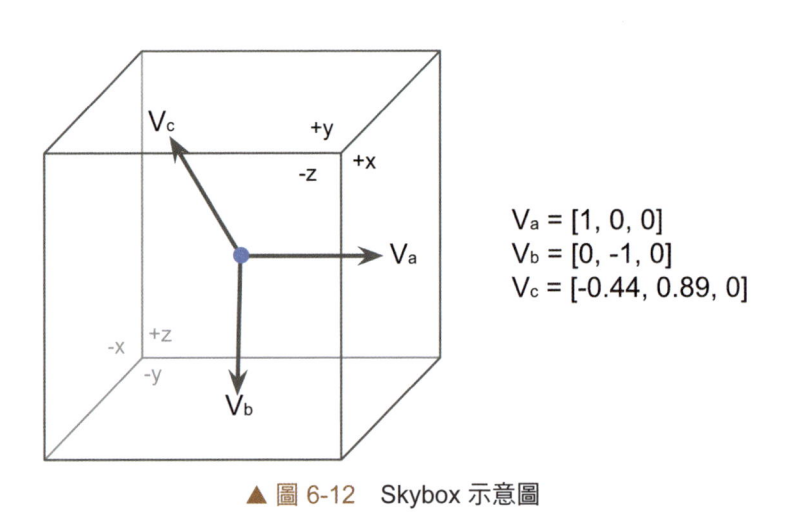

▲ 圖 6-12　Skybox 示意圖

若使用 V_a 取樣，會取樣到 +x 圖的正中央，V_b 的話會取樣到 -y 圖的正中央，V_c 的話會取樣到 +y 圖的中間偏左。這樣一來這個天空就像是一個盒子一樣，因此這樣的效果叫做 skybox。

6-2 Skybox

⌘ 讀取圖檔並建立 texture cube map

筆者在遊戲素材分享網站 https://opengameart.org/ 找到名為『clouds skybox 1』[7] 的 CC0 Skybox 作為接下來實做 skybox 之素材，經果筆者調色使之更鮮艷後，其提供的 +x、-x、+y、-y、+z、-z 六個面如圖 6-13 至圖 6-18：

▲ 圖 6-13　clouds skybox 1 +x

▲ 圖 6-14　clouds skybox 1 -x

▲ 圖 6-15　clouds skybox 1 +y

▲ 圖 6-16　clouds skybox 1 -y

▲ 圖 6-17　clouds skybox 1 +z

▲ 圖 6-18　clouds skybox 1 -z

7　https://opengameart.org/content/clouds-skybox-1

6-25

6 帆船與海

這六張圖放置在範例程式碼包中的這些位置：

</>	assets/skybox/east.webp
</>	assets/skybox/west.webp
</>	assets/skybox/up.webp
</>	assets/skybox/down.webp
</>	assets/skybox/normal.webp
</>	assets/skybox/south.webp

把 skybox 所需的圖片載入，使用 WebGL API 建立、載入 cube texture 圖片的方法與 2D texture 差在兩個地方：

- `gl.bindTexture()` 時目標為 `gl.TEXTURE_CUBE_MAP` 而非 `gl.TEXTURE_2D`。
- `gl.texImage2D()` 需要呼叫 6 次，把圖片分別輸入到 +x、-x、+y、-y、+z、-z

首先在 setup() 中建立一個程式碼區塊，先用 `loadImage()` 把六張圖片讀取進來：

```
import { degToRad, loadImage } from '../../lib/utils.js';
// ...
async function setup() {
  // canvas, gl ...
  // programInfo ...
  // textures ...
  {
    const images = await Promise.all([
      '/assets/skybox/east.webp',
      '/assets/skybox/west.webp',
      '/assets/skybox/up.webp',
      '/assets/skybox/down.webp',
```

6-26

```
    '/assets/skybox/north.webp',
    '/assets/skybox/south.webp',
  ].map(loadImage));
}
// ...
}
```

建立 skybox cube texture，texture 類別為 gl.TEXTURE_CUBE_MAP：

```
async function setup() {
  // ...
  { // skybox texture
    const images = await Promise.all([
      // ...
    ].map(loadImage));

    const texture = gl.createTexture();
    gl.bindTexture(gl.TEXTURE_CUBE_MAP, texture);
  }
  // ...
}
```

把六張圖片輸入到 cube texture 中，先指定第一張：

```
async function setup() {
  // ...
  { // skybox texture
    // images, texture
    gl.texImage2D(
      gl.TEXTURE_CUBE_MAP_POSITIVE_X,
      /* level: */ 0, /* internalFormat: */ gl.RGBA,
      /* format: */ gl.RGBA, /* type: */ gl.UNSIGNED_BYTE,
      images[0],
    );
  }
  // ...
}
```

gl.texImage2D() 在這邊第二個以後的參數與先前 2D texture 使用方式一樣，注意到第一個參數傳入 `gl.TEXTURE_CUBE_MAP_POSITIVE_X`，表示指定的 `images[0]` 是 skybox 的 +x 面，接著其他五個也要分別對應：`gl.TEXTURE_CUBE_MAP_NEGATIVE_X`、`gl.TEXTURE_CUBE_MAP_POSITIVE_Y`、`gl.TEXTURE_CUBE_MAP_NEGATIVE_Y`、`gl.TEXTURE_CUBE_MAP_POSITIVE_Z`、`gl.TEXTURE_CUBE_MAP_NEGATIVE_Z`：

```javascript
async function setup() {
  // ...
  { // skybox texture
    // images, texture
    // TEXTURE_CUBE_MAP_POSITIVE_X -> images[0]
    gl.texImage2D(
      gl.TEXTURE_CUBE_MAP_NEGATIVE_X,
      0, gl.RGBA, gl.RGBA, gl.UNSIGNED_BYTE,
      images[1],
    );
    gl.texImage2D(
      gl.TEXTURE_CUBE_MAP_POSITIVE_Y,
      0, gl.RGBA, gl.RGBA, gl.UNSIGNED_BYTE,
      images[2],
    );
    gl.texImage2D(
      gl.TEXTURE_CUBE_MAP_NEGATIVE_Y,
      0, gl.RGBA, gl.RGBA, gl.UNSIGNED_BYTE,
      images[3],
    );
    gl.texImage2D(
      gl.TEXTURE_CUBE_MAP_POSITIVE_Z,
      0, gl.RGBA, gl.RGBA, gl.UNSIGNED_BYTE,
      images[4],
    );
    gl.texImage2D(
      gl.TEXTURE_CUBE_MAP_NEGATIVE_Z,
      0, gl.RGBA, gl.RGBA, gl.UNSIGNED_BYTE,
      images[5],
    );
  }
  // ...
}
```

最後建立好縮圖，並設定輸入好的 texture 到 app.textures.skybox 上：

```
async function setup() {
  // ...
  { // skybox texture
    // images, gl.createTexture(), gl.texImage2D() ...
    gl.generateMipmap(gl.TEXTURE_CUBE_MAP);
    textures.skybox = texture;
  }
  // ...
}
```

回到瀏覽器，如擷圖 6-19 可以在開發者工具的 Network tab 中看到這六張圖：

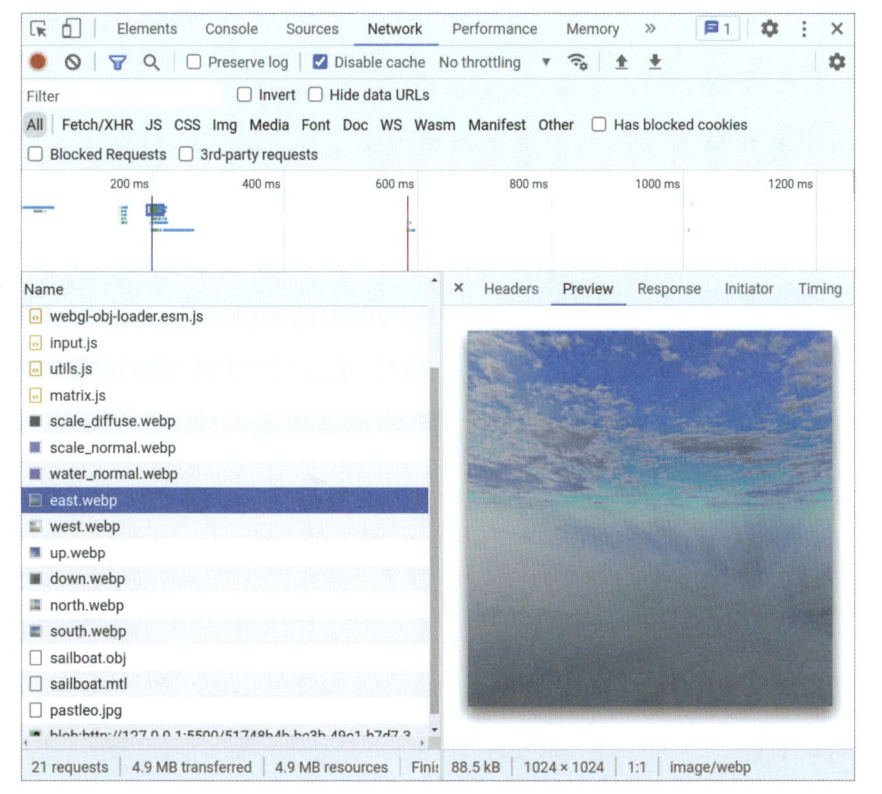

▲ 圖 6-19　開發者工具 Network tab 顯示 skybox 圖片的讀取

6 帆船與海

skybox 的 cube texture 準備好了，對於 3D 物件的部份，skybox 在前面提到事實上是一個背景，運作方式跟前面理解的 3D 物件不太一樣，得要為 skybox 建立屬於他的 shader、bufferInfo、VAO，同時還得依據視角產生正確的方向向量以從 samplerCube 型態的 skybox texture 進行取樣。

⌘ Skybox Shader 與物件

因為 skybox 是『背景』，因此 skybox 這個『物件』最終到達 clip space 時應該落在距離觀察位置最遠但是還看得到的地方，在 CH2 Perspective 3D 這邊提到成像時投影到 z = -1 平面、看著 +z 方向，那麼離觀察最遠的平面為 z = 1，而且為了填滿整個畫面，skybox 物件在 clip space 中即為 x、y 範圍於 -1 ~ +1 的 z = 1 平面，剛好 `twgl.primitives.createXYQuadVertices()` 就是產生一個 z = 0、xy 由 -1 ~ +1 的平面，幾乎可以直接做出我們需要的這個平面，就差在他沒有 z 的值。

這樣聽起來物件的頂點應該是不需要 transform，就只是輸出到 `gl_Position` 的 z 需要設定成 1，比較需要操心的是對 cube texture 取樣時的 normal 法向量，我們將從 clip space 頂點位置出發，透過『某種 transform』指向使用者觀看區域的邊界頂點，接下來就跟一般 texture 一樣利用 varying 補間得到每個 pixel 取用 cube texture 的 normal 法向量。

綜合以上，建立 `skyboxVertexShaderSource` 放置 skybox 之 vertex shader 實做：

```
const skyboxVertexShaderSource = `#version 300 es
in vec2 a_position;
uniform mat4 u_matrix;

out vec3 v_normal;

void main() {
  gl_Position = vec4(a_position, 1, 1);
  v_normal = (u_matrix * gl_Position).xyz;
```

```
}
`;
```

待會將透過 `twgl.primitives.createXYQuadVertices()` 產生頂點資料，將頂點位置輸入到 `a_position`，我們只需要其 x、y 資料就好因此設定成 `vec2`，`gl_Position` 照著上面所說直接輸出並設定 z 為 1；`u_matrix` 變成轉換成 normal 的那個『某種 transform』矩陣，轉換好透過 `v_normal` 給 fragment shader 使用。

fragment shader 的部份就變得很簡單，純粹透過 `v_normal` 把顏色從 cube texture 中取出即可：

```
const skyboxFragmentShaderSource = `#version 300 es
precision highp float;

in vec3 v_normal;

out vec4 outColor;

uniform samplerCube u_skyboxMap;

void main() {
  outColor = texture(u_skyboxMap, normalize(v_normal));
}
`;
```

`skyboxVertexShaderSource`、`skyboxFragmentShaderSource` 原始碼字串寫好，接著建立對應的 programInfo 並放在 `app.skyboxProgramInfo`：

```
async function setup() {
  // canvas, gl ...
  // programInfo, depthProgramInfo, oceanProgramInfo
  const skyboxProgramInfo = twgl.createProgramInfo(gl, [
    skyboxVertexShaderSource,
    skyboxFragmentShaderSource,
  ]);
```

6 帆船與海

```javascript
// textures, framebuffers, objects ...

return {
  gl,
  programInfo, depthProgramInfo,
  oceanProgramInfo, skyboxProgramInfo,
  textures, framebuffers, objects,
  state: { /* ... */ },
  time: 0,
};
}
```

最後把 skybox『物件』建立好，別忘了其 VAO 要使 buffer 與新的 `skyboxProgramInfo` 綁定：

```javascript
async function setup() {
  // canvas, gl ...
  // objects, objects.sphere, objects.plane

  { // skybox
    const attribs = twgl.primitives.createXYQuadVertices();
    const bufferInfo = twgl.createBufferInfoFromArrays(gl, attribs);
    const vao = twgl.createVAOFromBufferInfo(
      gl, skyboxProgramInfo, bufferInfo,
    );

    objects.skybox = {
      attribs,
      bufferInfo,
      vao,
    };
  }

  // ...
}
```

⌘ Cube texture normal 之『某種 transform』

法向量將從 clip space 頂點位置出發，從上方 vertex shader 實做可以看到直接從 `gl_Position` 出發，也就是這四個位置（實際上為兩個三角形六個頂點）：

```
[-1, -1,  1]
[ 1, -1,  1]
[-1,  1,  1]
[ 1,  1,  1]
```

理論上我們是可以透過 `app.state.cameraRotationXY`（`cameraViewing`、`cameraDistance` 是對平移的控制，可以無視）算出對應的 transform，但是另一個方式是透過現成的 `viewMatrix`，這樣的話視角不使用 `app.state.cameraRotationXY` 時也可以通用。

圖 6-20 為場景位置到 clip space 轉換從正上方俯瞰的示意圖，原本使用 `viewMatrix` 是要把場景中透過 `worldMatrix` 轉換的物件投影到 clip space，也就是圖 6-20 中橘色箭頭方向，但是看著上方四個點的座標，現在的出發點是 clip space 中的位置（圖 6-20 黑點），如果轉換成觀察者所能看到最遠平面的四個角落（圖 6-20 藍綠色點），這樣一來再單位矩陣化便成為 cube texture 取樣時所需的 normal 法向量，而這樣的轉換在圖 6-20 中為同黑色箭頭，稍微想一下，這個動作其實是把 clip space 轉換回場景位置，那麼說也就是 `viewMatrix` 的『反向』—它的反矩陣。

6 帆船與海

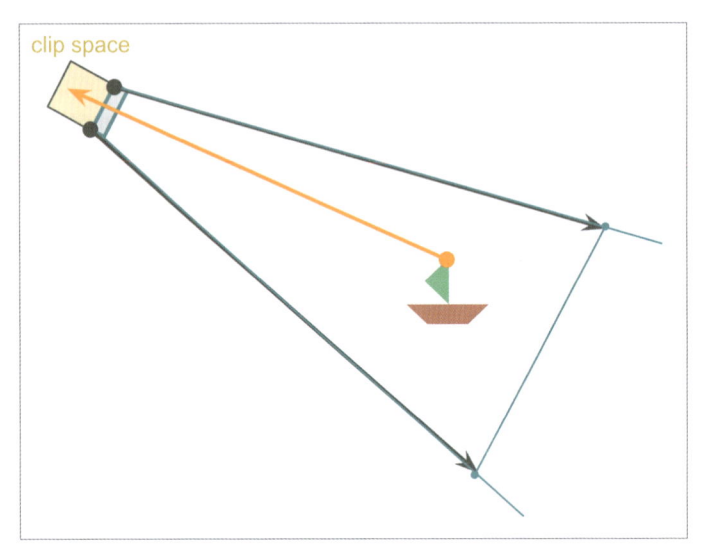

▲ 圖 6-20　場景位置到 clip space 轉換示意圖

不過 `viewMatrix` 會包含平移，需要將平移效果移除，我們把 `viewMatrix` 拆開來看：

```
viewMatrix =
  matrix4.perspective(...) * matrix4.inverse(cameraMatrix)
```

平移會來自於 `matrix4.inverse(cameraMatrix)`，而平移為 4x4 矩陣中最後一行的前三個元素，只要將之設定為 0 即可。綜合以上，轉換成 normal 所需要的矩陣 `u_matrix` 計算公式為：

```
inversedCameraMatrix = matrix4.inverse(cameraMatrix)
u_matrix = inverse(
  matrix4.perspective(...) *
    [
      ...inversedCameraMatrix[0..3],
      ...inversedCameraMatrix[4..7],
      ...inversedCameraMatrix[8..11],
      0, 0, 0, inversedCameraMatrix[15]
    ]
)
```

這段只是公式的 psudo code，不是合法的 Javascript，這邊 `inversedCameraMatrix[0..3]` 表示取得 `inversedCameraMatrix` 的 index 0、1、2、3 的元素。

將計算公式寫成程式碼之前，在程式碼中將 `viewMatrix` 的投影、相機視角兩個矩陣獨立成 `projectionMatrix` 以及 `inversedCameraMatrix`：

```javascript
function render(app) {
  // ...
  const cameraMatrix = matrix4.multiply( /* ... */ );

  const projectionMatrix = matrix4.perspective(
    state.fieldOfView,
    gl.canvas.width / gl.canvas.height,
    0.1,
    2000,
  );
  const inversedCameraMatrix = matrix4.inverse(cameraMatrix);

  const viewMatrix = matrix4.multiply(
    projectionMatrix,
    inversedCameraMatrix,
  );
  // ...
}
```

⌘ Skybox 的繪製

與其他物件一樣，建立一個 function 來實做 skybox 的繪製，並接收拆分出來的 `projectionMatrix` 以及 `inversedCameraMatrix`：

```javascript
// ...
// renderOcean ...

function renderSkybox(app, projectionMatrix, inversedCameraMatrix) {
```

6 帆船與海

```
  const { gl, skyboxProgramInfo, objects, textures } = app;
}
// render ...
```

我們將在 render() 這層進行 program、shader 的切換來統一管理 render 流程，照著上方所描述的 u_matrix 計算公式實做，並且把 skybox cube texture 設定到 skybox shader 之 uniform：

```javascript
function renderSkybox(app, projectionMatrix, inversedCameraMatrix) {
  const { gl, skyboxProgramInfo, objects, textures } = app;

  twgl.setUniforms(skyboxProgramInfo, {
    u_skyboxMap: textures.skybox,
    u_matrix: matrix4.inverse(
      matrix4.multiply(
        projectionMatrix,
        [
          ...inversedCameraMatrix.slice(0, 12),
          0, 0, 0, inversedCameraMatrix[15], // remove translation
        ],
      ),
    ),
  });
}
```

uniform 輸入完成，進行繪製動作，並且要記得切換好 VAO：

```javascript
function renderSkybox(app, projectionMatrix, inversedCameraMatrix) {
  const { gl, skyboxProgramInfo, objects, textures } = app;

  gl.bindVertexArray(objects.skybox.vao);

  twgl.setUniforms(skyboxProgramInfo, { /* ... */ });

  twgl.drawBufferInfo(gl, objects.skybox.bufferInfo);
}
```

有一件事需要注意，在 clip space z = 1 因為沒有『小於』最遠深度 z = 1 而不會被判定在 clip space，所以需要設定成『小於等於』，並在繪製完之後設定回來避免影響到其他物件的繪製：

```javascript
function renderSkybox(app, projectionMatrix, inversedCameraMatrix) {
  const { gl, skyboxProgramInfo, objects, textures } = app;

  gl.bindVertexArray(objects.skybox.vao);

  twgl.setUniforms(skyboxProgramInfo, { /* ... */ });

  gl.depthFunc(gl.LEQUAL);
  twgl.drawBufferInfo(gl, objects.skybox.bufferInfo);
  gl.depthFunc(gl.LESS); // reset to default
}
```

最後在 `render()` 中切換好 program、shader 並呼叫 `renderSkybox()`：

```javascript
function render(app) {
  const {
    gl,
    programInfo,
    depthProgramInfo,
    oceanProgramInfo,
    skyboxProgramInfo,
    framebuffers, textures,
    state,
  } = app;

  // cameraMatrix, projectionMatrix, inversedCameraMatrix ...
  // ... renderSailboat ... renderOcean

  { // skybox
    gl.useProgram(skyboxProgramInfo.program);
    renderSkybox(app, projectionMatrix, inversedCameraMatrix);
  }
}
```

6-37

6 帆船與海

拉低視角,就可以看到天空囉,如圖 6-21、圖 6-22,到此進度的完整程式碼可以在這邊找到:

</>	ch6/03/index.html
</>	ch6/03/main.js

也可以參考上線版本,透過瀏覽器打開此頁面:
https://webgl-book.pastleo.me/ch6/03/index.html

▲ 圖 6-21　拉低視角,可以看到背景繪製了天空

▲ 圖 6-22　旋轉視角到另外一邊，轉動的過程中可感受到 skybox 作為 360 度相片的『Google 街景』效果

⌘ 使海面反射 skybox

看著圖 6-22，顯得海面有點單調，現在有了天空，那不如就讓海面反射天空如何？而且說到底就是要在繪製鏡像世界時繪製 skybox，只是在鏡像世界有個自己的 `viewMatrix` 叫做 `reflectionMatrix`，我們也必須把他拆開來：

```
function render(app) {
  // ... cameraMatrix, projectionMatrix ... viewMatrix

  const reflectionCameraMatrix = matrix4.multiply(
    // ...
  );

  const inversedReflectionCameraMatrix = matrix4.inverse(
    reflectionCameraMatrix,
  );

  const reflectionViewMatrix = matrix4.multiply(
```

6-39

```
    projectionMatrix,
    inversedReflectionCameraMatrix,
  );

  // lightProjectionViewMatrix, globalUniforms ...
}
```

這邊也同時讓 `reflectionViewMatrix` 與 `viewMatrix` 共用 `projectionMatrix`；接下來就是切換 shader 並且在繪製鏡像世界時呼叫 `renderSkybox()`，同時也因為海面會反射整個天空，就不需要自帶顏色了：

```
function render(app) {
  // ... inversedReflectionCameraMatrix, reflectionViewMatrix ...
  { // reflection
    twgl.bindFramebufferInfo(gl, framebuffers.reflection);

    gl.clear(gl.COLOR_BUFFER_BIT | gl.DEPTH_BUFFER_BIT);

    renderSailboat(app, reflectionViewMatrix, programInfo);

    gl.useProgram(skyboxProgramInfo.program);
    renderSkybox(app, projectionMatrix, inversedReflectionCameraMatrix);
  }

  gl.bindFramebuffer(gl.FRAMEBUFFER, null);

  twgl.resizeCanvasToDisplaySize(gl.canvas, state.resolutionRatio);
  gl.viewport(0, 0, canvas.width, canvas.height);

  gl.useProgram(programInfo.program);
  // renderSailboat ...
}
```

因為在 reflection 程式區塊繪製鏡像世界時切換了 program、shader，因此回到主要內容的渲染之前記得要切回來；同時也因為海面會反射整個天空，就不需要自帶顏色了：

```
function renderOcean(app, viewMatrix, reflectionViewMatrix, programInfo) {
  // ...
  twgl.setUniforms(programInfo, {
    // ...
    u_normalMap: textures.oceanNormal,
    u_diffuse: [0, 0, 0],
    u_diffuseMap: textures.reflection,
    // ...
  });

  twgl.drawBufferInfo(gl, objects.plane.bufferInfo);
}
```

海面變成天空的鏡子，晴朗天氣的部份也就完成了，如圖 6-23：

▲ 圖 6-23　海面變成天空的鏡子，天氣晴朗

結束 skybox 之前，回頭看看程式碼會發現讀取 skybox texture 的程式碼我們使用 WebGL 內建 API，也是蠻冗長的，看一下 `twgl.createTextures()` [8]

8　https://twgljs.org/docs/module-twgl.html#.createTextures

6 帆船與海

的文件可以知道此 function 也可以幫忙載入 texture cube map，既然現有實做中已經使用這個 function 載入 `app.textures`，那麼只要加上這幾行就相當於 `setup()` 讀取 skybox texture 的整段程式碼區塊了：

```js
async function setup() {
  // canvas, gl ...

  const textures = twgl.createTextures(gl, {
    scale: { /* ... */ },
    scaleNormal: { /* ... */ },
    oceanNormal: { /* ... */ },
    skybox: {
      target: gl.TEXTURE_CUBE_MAP,
      src: [
        '/assets/skybox/east.webp',
        '/assets/skybox/west.webp',
        '/assets/skybox/up.webp',
        '/assets/skybox/down.webp',
        '/assets/skybox/north.webp',
        '/assets/skybox/south.webp',
      ],
    },
    null: { src: [0, 0, 0, 255] },
    nullNormal: { src: [127, 127, 255, 255] },
  });

  // framebuffers, objects ...
}
```

到此進度的完整程式碼可以在這邊找到：

</>	ch6/04/index.html
</>	ch6/04/main.js

也可以參考上線版本，透過瀏覽器打開此頁面：

https://webgl-book.pastleo.me/ch6/04/index.html

6-42

6-3　半透明的文字看板

　　加上天空以及海面上的天空倒影如圖 6-24，本章的目標『帆船與海』幾乎可以算是完成了：

▲ 圖 6-24　讀取完初始視角所看到的帆船與海

　　但是初次乍到的使用者，除了觀賞畫面之外應該很難知道操作視角的方式，當然我們可以用 HTML 把說明文字加在畫面中，但是這樣的話就太沒挑戰性了，本篇將在場景中加入一段文字簡單說明移動視角的方法：

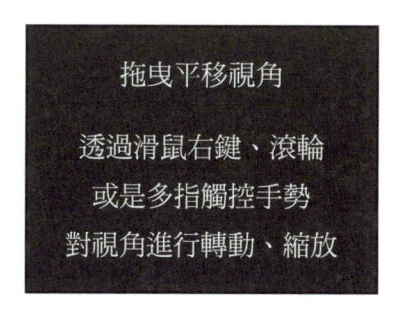

拖曳平移視角

透過滑鼠右鍵、滾輪
或是多指觸控手勢
對視角進行轉動、縮放

6 帆船與海

⌘ 如何在 WebGL 場景中顯示文字？

不論是英文、中文還是任何語言，其實顯示在畫面上時只是一些符號組合在一起形成一幅圖，在 WebGL 中並沒有『透過某個 function 並輸入一個字串，在畫面上就會繪製出該文字』這樣的事情，但是在 `<canvas />` 的另外一組繪製 API canvas2D（`CanvasRenderingContext2D`）[9] 有，同時，輸入 texture 資料的 `gl.texImage2D()` 其實接受 `<canvas />` HTML 元素，允許我們把畫在該 `<canvas />` 中的圖傳送到 texture 上。

這麼一來，我們可以：

1. 建立另一個暫時用的 `<canvas />`
2. 透過 canvas2d 繪製文字到 `<canvas />` 中
3. 建立並將暫時的 `<canvas />` 輸入到 texture
4. 渲染場景物件時，就當成一般的圖片 texture 進行繪製

⌘ 建立文字 Texture

建立一個 function 叫做 `createTextTexture`，實做完成時會回傳 WebGL texture，在 `setup()` 中呼叫並接收放在 `app.textures.text` 中：

```
async function setup() {
  // canvas, gl ...
  // textures

  textures.text = createTextTexture(gl);

  // framebuffers, objects ...
}

function createTextTexture(gl) {
}
```

---

[9] https://developer.mozilla.org/en-US/docs/Web/API/CanvasRenderingContext2D

照著上面的第一步：建立一個暫時用的 `<canvas />`：

```
function createTextTexture(gl) {
  const canvas = document.createElement('canvas');
  canvas.width = 1024;
  canvas.height = 1024;
}
```

長寬設定成 1024，這樣的大小應該可以繪製足夠細緻的文字；接著從 `<canvas />` 取得 canvas2D API，在 WebGL 這邊也是透過一樣的 function，只是傳入不同的 context 名稱字串 `'2d'` 指定要 canvas2D API：

```
function createTextTexture(gl) {
  // const canvas = document.createElement('canvas') ...

  const ctx = canvas.getContext('2d');
}
```

那麼這邊的 `ctx` 變數就類似於 WebGL 的 `gl` 變數，提供一整組繪製用的 API、context，詳細的 API 文件請見註 9 中的 MDN 連結；接著執行一系列的程式進行設定、繪製，canvas2D 的 API 相較 WebGL 單純許多，讀者可以嘗試從其屬性、function 的英文名稱了解所代表的意思：

```
function createTextTexture(gl) {
  // canvas, ctx ...

  ctx.clearRect(0, 0, canvas.width, canvas.height);

  ctx.fillStyle = 'white';
  ctx.textAlign = 'center';
  ctx.textBaseline = 'middle';

  ctx.font = 'bold 80px serif';
  ctx.fillText('拖曳平移視角', canvas.width / 2, canvas.height / 5);

  const secondBaseLine = 3 * canvas.height / 5;
  const secondLineHeight = canvas.height / 7;
```

6-45

6 帆船與海

```
ctx.font = 'bold 70px serif';
ctx.fillText(
  '透過滑鼠右鍵、滾輪',
  canvas.width / 2, secondBaseLine - secondLineHeight,
);
ctx.fillText(
  '或是多指觸控手勢',
  canvas.width / 2, secondBaseLine,
);
ctx.fillText(
  '對視角進行轉動、縮放',
  canvas.width / 2, secondBaseLine + secondLineHeight,
);
}
```

canvas2D 其實也有許多有趣的功能，有 `.bezierCurveTo()` 繪製貝茲曲線、路徑繪製等功能，也可以用來繪製幾何圖形，不過本文的重點是文字，因此只有使用到相關的功能：

- 雖然原本就應該是乾淨的，不過還是先呼叫 `.clearRect()` 確保整個畫布都是透明黑色 `rgba(0, 0, 0, 0)`。
- 繪製文字前，`.fillStyle` 設定文字顏色，而 `.textAlign = 'center'`、`.textBaseline = 'middle'` 使待會繪製時以從下筆的點進行水平垂直置中。
- `.font` 設定字型、字體大小。
- `.fillText(string, x, y)` 如同本文一開始說的『透過某個 API 並輸入一個字串，在畫面上就會繪製出該文字』，此處的 x、y 為下筆的位置。

筆者讓 拖曳平移視角 這行字的大小是 80px，比接下來的文字 70px 來的大，而且有不少『下筆』位置的計算，總之繪製完畢之後，這個暫時性的 canvas 如圖 6-25 所示：

6-3　半透明的文字看板

▲ 圖 6-25　讀取完初始視角所看到的帆船與海

　　有後面的方格是筆者為了避免在書上什麼都看不到而加上，表示該區域是透明的。canvas 準備好了，如同一般 2d texture 一樣建立、輸入 texture 內容，只是先前輸入圖片的參數改成繪製好的 canvas：

```javascript
function createTextTexture(gl) {
  // canvas, ctx ... draw texts using canvas2D ctx ...

  const texture = gl.createTexture();
  gl.bindTexture(gl.TEXTURE_2D, texture);

  gl.texImage2D(
    gl.TEXTURE_2D,
    0, // level
    gl.RGBA, // internalFormat
    gl.RGBA, // format
    gl.UNSIGNED_BYTE, // type
    canvas, // data
  );
  gl.generateMipmap(gl.TEXTURE_2D);

  return texture;
}
```

6-47

6 帆船與海

好的，這麼一來有說明文字的看板 texture 就準備好放在 `app.textures.text` 當中。

⌘ 繪製文字 texture 到場景中

要繪製文字 texture 到場景上，需要一個 3D 物件來當成此 texture 的載體繪製在其表面，最適合的就是一個平面了，而這樣的平面物件也已經有，那就是海面所使用的 `objects.plane.vao`，而且這就是一個單純的物件，與其他物件一樣建立一個 function 進行繪製：

```javascript
function renderText(app, viewMatrix, programInfo) {
  const { gl, textures, objects } = app;

  gl.bindVertexArray(objects.plane.vao);

  const textLeftShift =
    gl.canvas.width / gl.canvas.height < 1.4 ? 0 : -0.9;
  const worldMatrix = matrix4.multiply(
    matrix4.translate(textLeftShift, 0, 0),
    matrix4.xRotate(degToRad(45)),
    matrix4.translate(0, 12.5, 0),
  );

  twgl.setUniforms(programInfo, {
    u_matrix: matrix4.multiply(viewMatrix, worldMatrix),
    u_diffuse: [0, 0, 0],
    u_diffuseMap: textures.text,
  });

  twgl.drawBufferInfo(gl, objects.plane.bufferInfo);
}
// render ...
```

這邊建立的 `renderText()` 與海面、球體的 render function 結構類似：切換 VAO、計算 transform 矩陣、設定 uniform 再呼叫繪製指令。

6-48

對於此看板位置的控制，`worldMatrix` 的 `matrix4.translate(0, 12.5, 0)` 與 `matrix4.xRotate(degToRad(45))` 是為了讓此文字出現在使用者初始時視角位置的前面，另外 `matrix4.translate(textLeftShift, 0, 0)` 的 `textLeftShift` 則是有點 Responsive design[10] 概念，如果裝置為寬螢幕則讓文字面板往左偏移一點使得船可以在一開始不被文字遮到；並且設定 `u_diffuseMap` 使得 texture 為剛才建立的文字 `textures.text`，同時為了避免其他物件設定過的 `u_diffuse`，這邊將之設定成黑色。

最後在 `render()` 中呼叫 `renderText()`，可以看到黑底白字的說明出現如圖 6-26：

```
function render(app) {
  // globalUniforms, lightProjection, reflection

  gl.bindFramebuffer(gl.FRAMEBUFFER, null);

  twgl.resizeCanvasToDisplaySize(gl.canvas, state.resolutionRatio);
  gl.viewport(0, 0, canvas.width, canvas.height);

  gl.useProgram(programInfo.program);

  renderSailboat(app, viewMatrix, programInfo);
  renderText(app, viewMatrix, programInfo);

  // render ocean, skybox
}
```

10　https://developer.mozilla.org/en-US/docs/Learn/CSS/CSS_layout/Responsive_Design

6 帆船與海

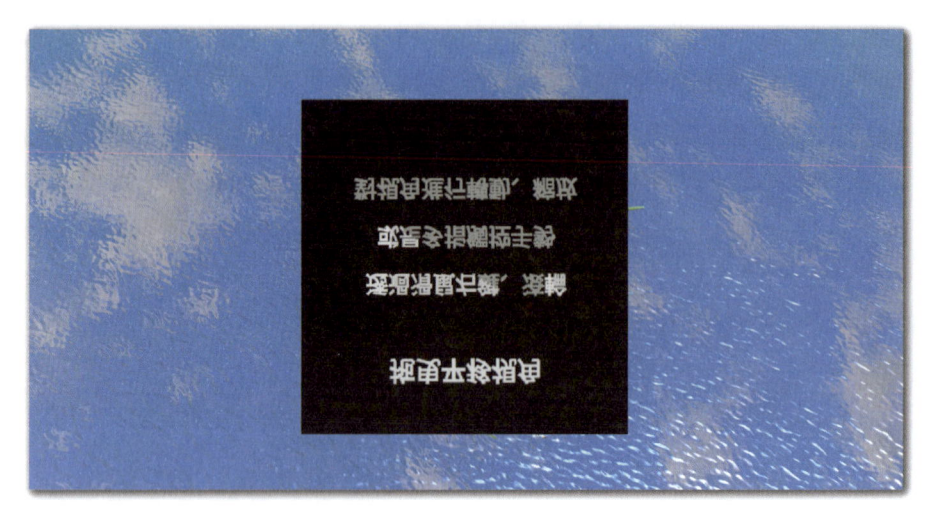

▲ 圖 6-26　繪製文字 texture 到場景中─1st try

圖 6-26 的文字看板顯然上下顛倒了，為什麼呢？因為讀取 canvas 到 texture 的時候，y 軸方向與讀取圖片是相反的，為了修正這個問題，我們在 `gl.texImage2D()` 輸入文字 texture 之前要設定請 WebGL 把輸入資料的 Y 軸顛倒：

```
function createTextTexture(gl) {
  // canvas, ctx, texture

  gl.pixelStorei(gl.UNPACK_FLIP_Y_WEBGL, true);

  gl.texImage2D(
    // ...
  );
  gl.generateMipmap(gl.TEXTURE_2D);

  gl.pixelStorei(gl.UNPACK_FLIP_Y_WEBGL, false);

  return texture;
}
```

6-50

為了避免影響到別的 texture 的載入，用完要設定回來。加入這兩行之後文字就正常囉，如圖 6-27：

▲ 圖 6-27　啟用 gl.UNPACK_FLIP_Y_WEBGL 修正上下顛倒的文字繪製結果

到此進度的完整程式碼可以在這邊找到：

</>	ch6/05/index.html
</>	ch6/05/main.js

也可以參考上線版本，透過瀏覽器打開此頁面：

https://webgl-book.pastleo.me/ch6/05/index.html

6-51

6 帆船與海

⌘ 半透明的渲染與常見的問題

原本半透明的文字 texture 透過 3D 物件渲染到場景變成不透明的了，因為使用的 fragment shader 輸出的 `gl_FragColor` 的第四個元素，也就是 alpha、透明度，固定是 1：

```
// fragmentShaderSource ...
void main() {
  // ...

  outColor = vec4(
    clamp(
      diffuse * diffuseLight +
      u_specular * specularBrightness +
      u_emissive,
      ambient, vec3(1, 1, 1)
    ),
    1
  );
}
```

這使得窄板螢幕一開始文字會遮住導致看不到主角，是不是有辦法可以讓這個說明看板變成半透明的呢？有的，首先當然是要有一個願意根據 texture 輸出 alpha 值的 fragment shader，因為這個文字看板物件不會需要有光影效果，另外寫一個新的簡單 fragment shader `textFragmentShaderSource` 給它用：

```
const textFragmentShaderSource = `#version 300 es
precision highp float;

in vec2 v_texcoord;

uniform vec4 u_bgColor;
uniform sampler2D u_texture;

out vec4 outColor;
```

```
void main() {
  outColor = u_bgColor + texture(u_texture, v_texcoord);
}
`;
```

可以看到除了 `u_texture` 用來輸入文字 texture 之外還有 `u_bgColor`，用來輸入整體的底色。與原本的 vertex shader 連結建立 `textProgramInfo` 並讓看板物件使用：

```
async function setup() {
  // canvas, gl ...

  // programInfo, depthProgramInfo, oceanProgramInfo,
  const textProgramInfo = twgl.createProgramInfo(
    gl, [vertexShaderSource, textFragmentShaderSource],
  );
  const skyboxProgramInfo = twgl.createProgramInfo(gl, [ /* ... */ ]);

  // texture, framebuffers, objects ...

  return {
    gl, /* ... */
    oceanProgramInfo, textProgramInfo,
    skyboxProgramInfo,
    // ...
  };
}
// ...
function render(app) {
  const {
    gl, /* ... */
    oceanProgramInfo,
    textProgramInfo,
    skyboxProgramInfo, /* ... */
  } = app;

  // ...
```

6-53

6 帆船與海

```
  renderSailboat(app, viewMatrix, programInfo);
  gl.useProgram(textProgramInfo.program);
  renderText(app, viewMatrix, textProgramInfo);
}
```

`renderText()` 內設定 uniform 的地方也記得要改一改：

```
function renderText(app, viewMatrix, programInfo) {
  // vao, worldMatrix ...

  twgl.setUniforms(programInfo, {
    u_matrix: matrix4.multiply(viewMatrix, worldMatrix),
    u_bgColor: [0, 0, 0, 0.1],
    u_texture: textures.text,
  });

  twgl.drawBufferInfo(gl, objects.plane.bufferInfo);
}
```

存檔回到瀏覽器，會看到圖 6-28 的樣子，是有變化，但是依然不是半透明的：

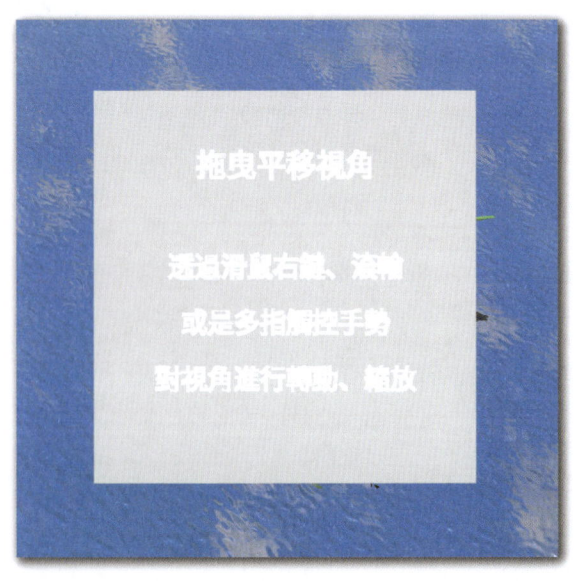

▲ 圖 6-28 改用能輸出半透明的文字專用 shader 渲染的結果

輸出 alpha、透明度之後，底色從黑色變成灰色了，事實上這時 `<canvas />` 的文字看板區塊在『網頁』上是透明的，讀者可以想像成一個有部份是透明的 png `<img />` 元素，如果嘗試使用開發者工具為 `<canvas />` 或是 body 設定一個背景顏色，可以看到該顏色直接透上來，如圖 6-29 筆者使用 CSS 加上綠色背景：

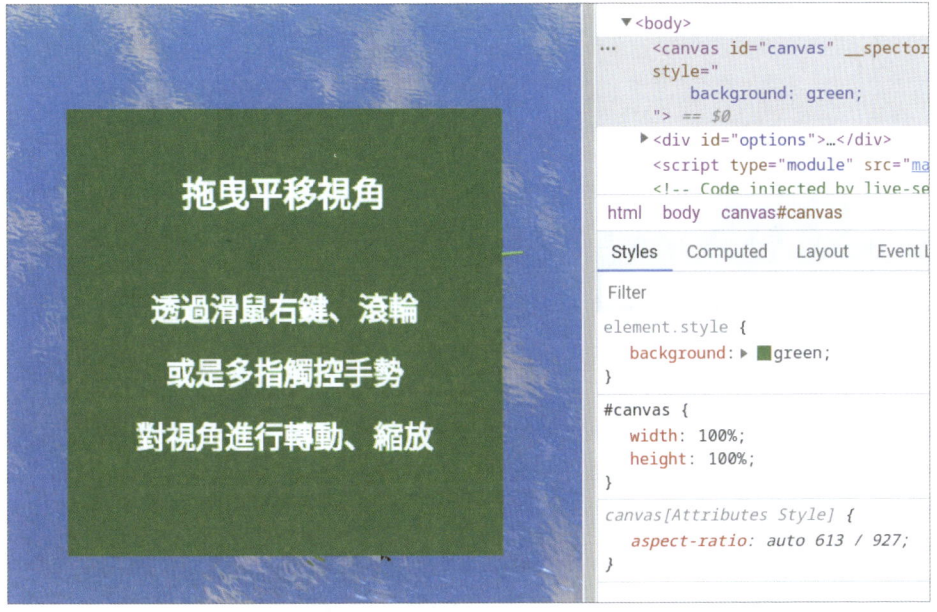

▲ 圖 6-29　因為 canvas 本身在文字看板區塊是透明，因此 CSS 的綠色背景透上來了

先把 HTML `<canvas />` 的背景設定成黑色，以符合 WebGL 渲染這邊的底色：

```
<!DOCTYPE html>
<html lang="en">
<head>
  <!-- ... -->
  <style>
    html, body { /* ... */ }
    #canvas {
      width: 100%;
```

6-55

```
      height: 100%;
      background: black;
    }
    #options { /* ... */ }
  </style>
</head>
<body>
  <!-- ... -->
</body>
</html>
```

文字看板的區域會回復成黑底。回到場景渲染半透明的問題，也就是說，前面寫 `textFragmentShaderSource` 所輸出的 alpha 值是有在作用，只可惜 WebGL 沒有拿來用，事實上 WebGL 有個 `gl.BLEND` 的選項，啟用才會有顏色混合功能：

```
async function setup() {
  // canvas, gl ...
  // textures, framebuffers, objects ...

  gl.enable(gl.CULL_FACE);
  gl.enable(gl.DEPTH_TEST);
  gl.enable(gl.BLEND);
  gl.blendFunc(gl.SRC_ALPHA, gl.ONE_MINUS_SRC_ALPHA);
  gl.clearColor(0, 0, 0, 1);

  return { /* ... */ };
}
```

精確地說，`gl.BLEND` 啟用的行為是讓要畫上去的顏色與畫布上現有的顏色進行運算後相加，而運算的方式由 `gl.blendFunc(sfactor, dfactor)` 設定：

- `sfactor` 表示要畫上去的顏色要乘以什麼，預設值為 `gl.ONE`，設定成 `gl.SRC_ALPHA` 乘以自身之透明度。
- `dfactor` 表示畫布上原本的顏色要乘以什麼，預設值為 `gl.ZERO`，這個顯然也要修改，要不然底下的顏色就等於完全被覆蓋掉，`gl.ONE_MINUS_`

`SRC_ALPHA` 如同字面上的意思，畫上去的 alpha 值越高原本畫布上的顏色被覆蓋程度的越高。

既然是『與畫布上現有的顏色進行運算』，半透明的物件繪製時會需要畫布上已經有繪製好其他物件，我們來看看如果就使用目前的實做，按照『帆船、文字看板、海洋、天空』的順序繪製會變成什麼樣子：

▲ 圖 6-30　啟用 `gl.BLEND` 設定混色之後，按照『帆船、文字看板、海洋、天空』的順序繪製

如圖 6-30 所見，海洋在文字看板的區域是黑色的，為什麼會這樣呢？首先這個黑色是 `<canvas />` 的 CSS 黑色背景透上來所導致，而主要的原因是 WebGL 沒有圖層的概念，而且用來判斷前後的深度（depth）圖並不知道物體是半透明的，每次繪製物體時物體時深度資訊都會被寫入（合併）到畫布上，請參考下圖 6-31 示意以『帆船、文字看板、海洋』繪製步驟的顏色與深度圖：

6-57

6 帆船與海

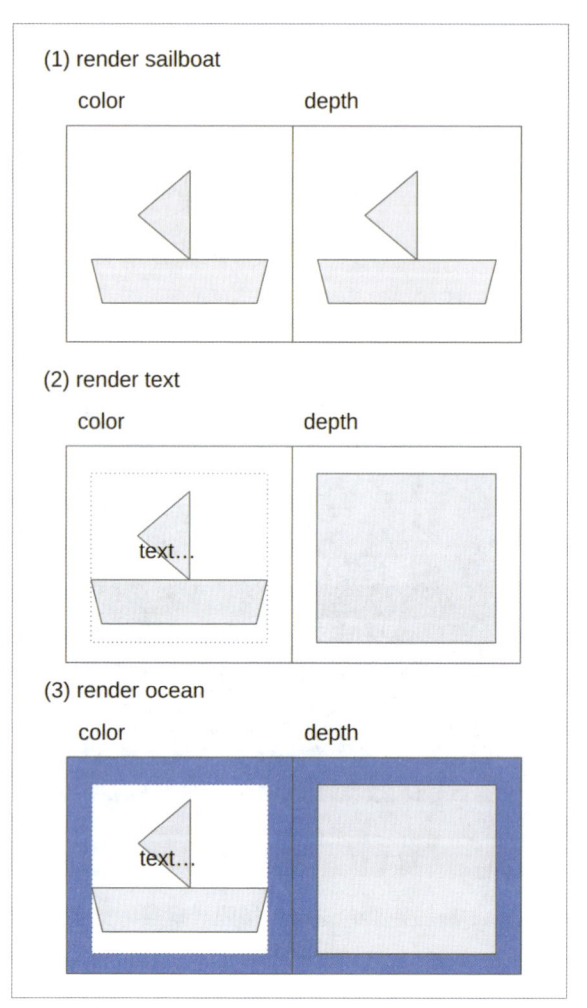

(此處為圖示，內容為三個繪製步驟的 color 與 depth 對照)

(1) render sailboat
 color　　depth

(2) render text
 color　　depth

(3) render ocean
 color　　depth

▲ 圖 6-31　『帆船、文字看板、海洋』繪製步驟的顏色與深度圖

　　在繪製完文字看板之後，儘管 `gl.BLEND` 啟用的狀況下讓文字看板可以以半透明的方式繪製、混合在帆船之上，但是一整塊的文字看板區域還是蓋到了深度圖上，往後在繪製海洋時就被判定成在後面而沒有畫上去，最後形成圖 6-30 的現象。

　　這個問題的解方是『調整繪製順序』，請見圖 6-32 以『帆船、海洋、文字看板』繪製步驟進行的結果，這時半透明物件的深度就不會影響去擋到其他東西了；事實上，半透明物件應該要在實體物件之後繪製，要不然就是要以物件們到相機之間的距離排序，最遠的先畫，一路畫到最近的。

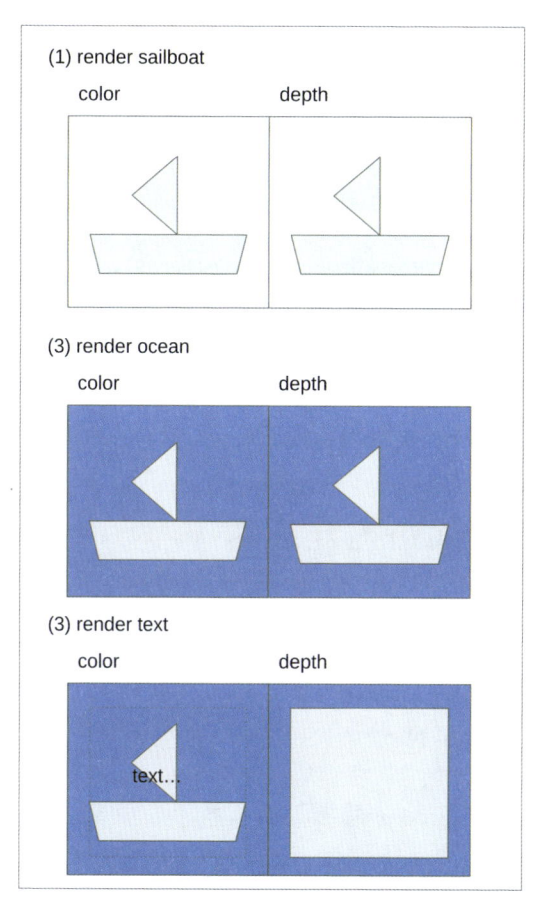

▲ 圖 6-32　『帆船、看板、海洋』繪製步驟的顏色與深度圖

由於目前半透明物件也只有一個，因此把文字看板的繪製順序調整到最後即可：

```
function render(app) {
  // globalUniforms, lightProjection, reflection ...

  gl.bindFramebuffer(gl.FRAMEBUFFER, null);
```

6-59

```
twgl.resizeCanvasToDisplaySize(gl.canvas, state.resolutionRatio);
gl.viewport(0, 0, canvas.width, canvas.height);

gl.useProgram(programInfo.program);

renderSailboat(app, viewMatrix, programInfo);
gl.useProgram(textProgramInfo.program);
renderText(app, viewMatrix, textProgramInfo);

gl.useProgram(oceanProgramInfo.program);
twgl.setUniforms(oceanProgramInfo, globalUniforms);
renderOcean(app, viewMatrix, reflectionViewMatrix, oceanProgramInfo);

{ // skybox
  gl.useProgram(skyboxProgramInfo.program);
  renderSkybox(app, projectionMatrix, inversedCameraMatrix);
}

gl.useProgram(textProgramInfo.program);
renderText(app, viewMatrix, textProgramInfo);
}
```

這樣修改完之後就能正確繪製出正確的半透明效果，終於大功告成，成果可以參考圖 6-33、圖 6-34。

在 WebGL 或甚至 three.js 想要渲染半透明物件時，常常需要去調整渲染的順序、半透明顏色的混合方式等才能達到正確的效果，就像是本次範例刻意示範的這些問題，可說是有許多可能的陷阱需要小心。

▲ 圖 6-33　半透明的文字看板，在手機板界面避免擋住帆船

▲ 圖 6-34　在平板、電腦等寬螢幕下看到的半透明成果

到此進度的完整程式碼可以在這邊找到：

</>	ch6/06/index.html
</>	ch6/06/main.js

也可以參考上線版本，透過瀏覽器打開此頁面：
https://webgl-book.pastleo.me/ch6/06/index.html

6-4 使用 Shader 即時渲染波光粼粼的海面

有了帆船、天空、反射天空的海面以及簡易的操作說明，要說這個是成品應該是沒什麼問題，不過看著那固定的海面以及帆船，如果能讓他們動起來是不是更好？

⌘ Procedural Generation（程序化生成 [11]）

在多媒體的範疇中，圖片 texture、3D model 甚至音樂、遊戲關卡等通常是由人類運用創意創作出來的，而相對而言 procedural generation 是指使用程式產生這些『創作』出來的資料，在產生的過程中為了避免明顯的 pattern，通常會在產生的過程中加入隨機的要素，使用 procedural generation 的好處是應用程式可以不用讀取通常體積龐大的『創作』，並且只要改變隨機要素便可以獲得不同的結果，使得應用程式獲得自動化內容產生的可能性。

有一個名稱叫做 Shadertoy 的網站，像是 Codepen、Jsfiddle、Jsbin 那樣，在網頁中寫程式，然後在旁邊跑起來呈現結果，只是 Shadertoy 是專門寫 WebGL fragment shader 的，同時網站也是分享的平台，在網站上有許多使用了 procedural generation 技巧的 shader：https://www.shadertoy.com/

---

11 https://zh.wikipedia.org/wiki/程序化生成

6-4 使用 Shader 即時渲染波光粼粼的海面

在 Shadertoy 網站中使用 procedural generation shader 的主要資料來源只有繪製的 pixel 位置以及時間，剩下的就是發揮使用者的想像力（以及數學）來畫出絢麗的畫面，可以在這邊看到許多別人寫好的 shader，像是：

- 雲層：https://www.shadertoy.com/view/3l23Rh
- 火焰：https://www.shadertoy.com/view/MdX3zr
- 海洋：https://www.shadertoy.com/view/Ms2SD1
- 水面：https://www.shadertoy.com/view/4dBcRD

可以看到這些 shader 的實做蠻複雜的，究竟是什麼演算法使得只透過簡單的資料來源就能得到這麼漂亮的效果？有一個 Youtube 頻道『The Art of Code』[12] 上面有許多 procedural generation shader 相關的影片，這些 shader 多少運用到了 hash、noise 的技巧來產生上述的隨機性，這些偽隨機函數[13] 接收二或三維度的座標，接著回傳看似隨機的數值，通常介於 0～1，那麼我們就可以利用這個數字當成一個地方的雲層密度、火焰強度、海浪高度、方向等；回到想要讓海面動起來的問題，接下來就看要怎麼實做在 `oceanFragmentShaderSource` 吧。

⌘ 動態的海面法向量 `oceanNormal`

海面反光計算、倒影陰影的計算都會利用到海面的法向量，目前 texture 作為 normal map 使得表面可以利用法向量的變化產生凹凸細節，只可惜這是一張靜態的圖不會動：

```
vec3 normal = texture(
  u_normalMap,
  v_worldPosition.xz / u_normalMapSize
).xyz * 2.0 - 1.0;
```

---

12 https://www.youtube.com/channel/UCcAlTqd9zID6aNX3TzwxJXg
13 https://zh.wikipedia.org/wiki/偽隨機性

6 帆船與海

要讓海面動起來，也就是這個 `normal` 要可以根據時間改變，對於這個我們需要加入一個浮點數 uniform `u_time`，我們使之為 `app.time` 乘以 0.001，每一秒加一：

```
const oceanFragmentShaderSource = `#version 300 es
// ...
uniform sampler2D u_lightProjectionMap;
uniform vec2 u_normalMapSize;
uniform float u_time;
// ...
`;
// ...
function render(app) {
  const {
    // ...
    state, time,
  } = app;
  // ...
  const globalUniforms = {
    // ...
    u_time: time * 0.001,
  }
  // ...
}
```

接著設計一個做 procedural generation 的 function 稱為 `oceanNormal()`，透過海面的 xz 座標來產生偽隨機的法向量：

```
// oceanFragmentShaderSource ...
out vec4 outColor;

vec3 oceanNormal(vec2 pos);

void main() {
  vec2 texcoord =
    (v_reflectionTexcoord.xy / v_reflectionTexcoord.w) * 0.5 + 0.5;
  vec3 normal = oceanNormal(v_worldPosition.xz);
  vec2 distortion = normalize(normal).xy;
  normal = normalize(v_normalMatrix * normal);
  // ...
```

6-64

```
}
vec3 oceanNormal(vec2 position) {
}
```

> 不知不覺這篇可能會變成以 GLSL 基於的 C 語法為主，在 `main()` 之前要先宣告 `vec3 oceanNormal(vec2 pos)` 的存在，要不然編譯會失敗。

接下來假設會實做一個 function，帶入一組 xz 座標會得到 xyz 填上高度，把這個 function 叫做 `vec3 oceanSurfacePosition(vec2 position)`，那麼 `oceanNormal()` 就可以呼叫 `oceanSurfacePosition()` 三次，第一次帶入原始的 `p0 = [x, z]`、第二次 `p1 = [x + 0.01, z]`、第三次 `p2 = [x, z + 0.01]`，拿 `p0 -> p1` 以及 `p0 -> p2` 兩個向量做外積就可以得到一定準度的法向量，`oceanNormal()` 實做如下：

```
// oceanFragmentShaderSource ...
// main ...
#define OCEAN_SAMPLE_DISTANCE 0.01
vec3 oceanNormal(vec2 position) {
  vec3 p1 = oceanSurfacePosition(position);
  vec3 p2 = oceanSurfacePosition(
    position + vec2(OCEAN_SAMPLE_DISTANCE, 0)
  );
  vec3 p3 = oceanSurfacePosition(
    position + vec2(0, OCEAN_SAMPLE_DISTANCE)
  );

  return normalize(cross(
    normalize(p2 - p1), normalize(p3 - p1)
  ));
}
```

這邊出現了 C 語言的 `#define`，雖然實際意義是 macro、程式碼替換，不過讀者們可以理解成常數定義就好，定義 OCEAN_SAMPLE_DISTANCE 為 0.01 表示外積的另外兩個點與 position 的取樣距離。

6 帆船與海

⌘ 波浪函數

讓假設會有的 `oceanSurfacePosition()` function 成為事實：

```
// oceanFragmentShaderSource ...
// main ...
vec3 oceanSurfacePosition(vec2 position) {
  float height = 0.0;
  return vec3(position, height);
}

// oceanNormal ...
```

為了符合 normal 以 `[0, 0, 1]` 為上面，`height` 輸出到 z 的位置，也就跟 normal map 取得的 `vec3` 排列一致；現在要求的是運算 `position` 來取得 `height`，注意到在 Shadertoy 水面效果這個作品[14]的程式碼有一行：

```
float wave = exp(sin(x) - 1.0);
```

將此數學式使用 Desmos 網站畫出來[15]如圖 6-35：

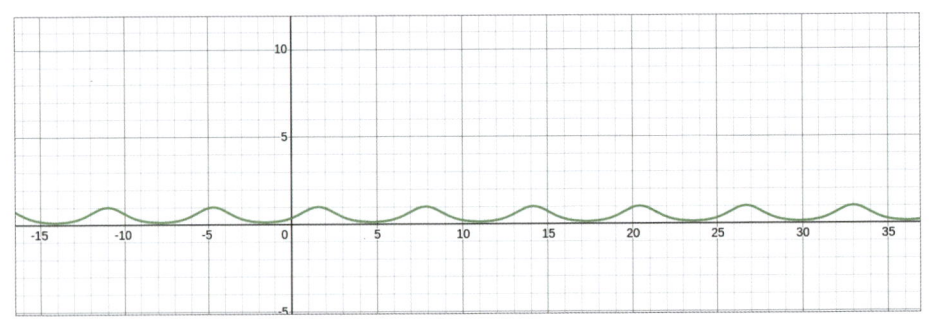

▲ 圖 6-35　exp(sin(x) - 1.0) 呈現漂亮的波浪形狀

14　https://www.shadertoy.com/view/4dBcRD
15　https://www.desmos.com/calculator/fhfmlynqtn

這個函式中，`sin()` 的波形本身就是於 +1 ~ -1 之間不停來回，減 1 套 `exp()` 指數函數[16] 得到最大值為 1 的波浪函數，看起來很適合，實做給 `height` 試試看：

```
vec3 oceanSurfacePosition(vec2 position) {
  float height = exp(sin(position.x) - 1.0);
  return vec3(position, height);
}
```

縮小預覽，可以看到如圖 6-36 規律的波紋與 z 軸平行：

▲ 圖 6-36　以 x 為輸入，產生與 z 軸平行的波紋

很好，這樣可以產生波紋效果；如果想要讓帆船的方向與海浪垂直、並且使之跟著時間而波動，定義一個角度以及向量 `direction` 與位置進行內積，加上時間成為波浪函數的輸入值 `waveX`：

16　https://zh.wikipedia.org/wiki/指數函數

6-67

6 帆船與海

```glsl
vec3 oceanSurfacePosition(vec2 position) {
  float directionRad = radians(-135.0);
  vec2 direction = vec2(cos(directionRad), sin(directionRad));

  float waveX = dot(position, direction) * 2.5 + u_time * 5.0;
  float height = exp(sin(waveX) - 1.0);

  return vec3(position, height);
}
```

　　GLSL 內建 function `radians()`[17] 可以把角度轉換成弧度，`radians(-135.0)` 也就是反向旋轉 135 度；將 `dot(position, direction)` 乘上 2.5 可以縮小波長、時間乘上 5.0 加快波浪速度，波浪就隨著時間動起來囉，如圖 6-37：

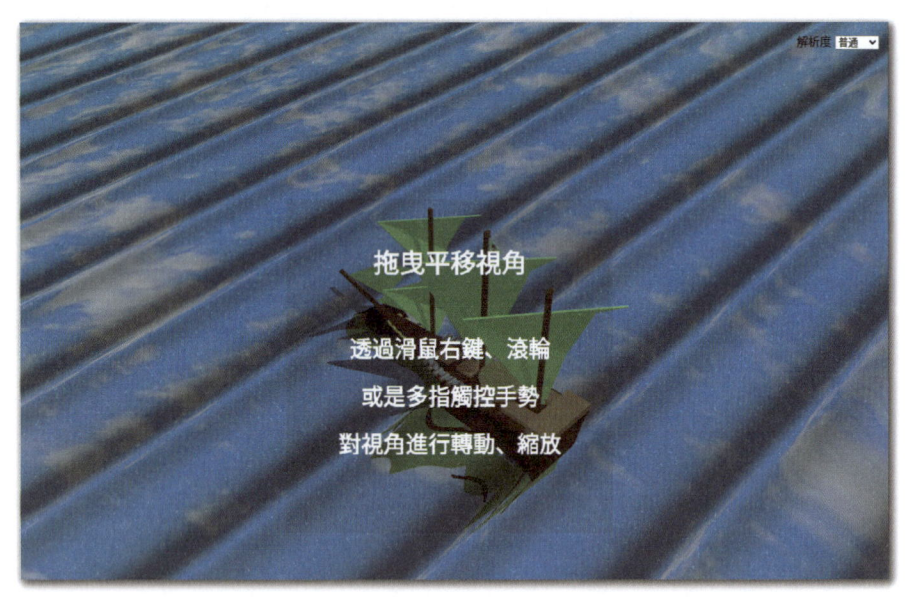

▲ 圖 6-37　以 directionRad 旋轉波浪方向、並使波浪隨著 u_time 流動

17　https://registry.khronos.org/OpenGL-Refpages/es3.0/html/radians.xhtml

6-68

⌘ 實做 hash() 並分格子產生偽隨機值

但是這樣的波浪未免也太規律，這時來實做上方說到的偽隨機技巧，稱為 hash()：

```
// oceanFragmentShaderSource ...

float hash(vec2 p) {
  return fract(sin(mod(dot(p, vec2(13, 17)), radians(180.0))) * 4801.0);
}

// oceanSurfacePosition, oceanNormal ...
```

這個 function 接受一個二維的位置，所謂 hash() 就是傳入相同的值的時候可以得到一樣的結果，但是回傳結果類似亂數難以回推，大致上的原理是利用 sin() 函數製造 -1 ~ +1 之間的數，乘一個大的數字 4801 之後以 fract()[18] 取小數的部份，**最後獲得 0 ~ 1 之間的偽隨機值**，在 sin() 內部的運算則是用 dot() 內積把二維輸入降到一維，13、17 使得兩個維度以不同的速度改變；13、17、4801 這些數字類似於亂數種子，需要稍微反覆實驗看哪些數字效果最好，通常設定成質數時較不容易出現 pattern；另外可以注意到程式碼中還有 mod()[19] 取除以 radians(180.0) 的餘數，radians(180.0) 其實只是要取得一個較為精準的 π，這是因為筆者發現在部份 GPU 傳入 sin() 的輸入值過大時會使得回傳值為 0，sin() 原本就是每個 π 的整數倍一個循環，因此輸入 sin() 之前除以 π 的餘數來保護住避免出現錯誤的結果。

接著使用 The Art of Code 的 Youtube 影片『Shader Coding: Making a starfield - Part 1』[20] 的技巧，利用 floor()[21] 把海面以每個整數分成一格一格，例

18　https://registry.khronos.org/OpenGL-Refpages/es3.0/html/fract.xhtml
19　https://registry.khronos.org/OpenGL-Refpages/es3.0/html/mod.xhtml
20　https://youtu.be/rvDo9LvfoVE?t=789
21　取得小於輸入值的最大整數：
　　https://registry.khronos.org/OpenGL-Refpages/es3.0/html/floor.xhtml

6 帆船與海

如 1~1.99999... 為一格，用 `id` 表示當前的格子，同個格子內的點都會得到一樣的 `id`，使用 `id` 做 `hash()` 出來的偽隨機值當成一個格子的 `directionRad` 角度偏移量：

```
vec3 oceanSurfacePosition(vec2 position) {
  vec2 id = floor(position);

  float directionRad = radians((hash(id) - 0.5) * 45.0 - 135.0);
  vec2 direction = vec2(cos(directionRad), sin(directionRad));

  float waveX = dot(position, direction) * 2.5 + u_time * 5.0;
  float height = exp(sin(waveX) - 1.0);

  return vec3(position, height);
}
```

因為 `hash(id)` 得到 0~1 之間的數值，減掉 0.5 成為 -0.5 ~ +0.5，乘以 45 減掉 135 再傳入 `radians()`，這麼一來角度將為 -135 + (-22.5 ~ 22.5)，採用 id 分格子之後海面變成如圖 6-38 這樣：

▲ 圖 6-38　分格子以格子的 hash(id) 偽隨機影響旋轉波浪方向

6-70

對鄰近格子取海浪高度

每格是有不同的方向，但是格子之間都有明顯的一條線，解決這個問題的方法是每個格子去計算鄰近 1 格的海浪，並且海浪的強度會隨著距離來源格子越遠而越弱，為此再建立一個 function 叫做 `localWaveHeight()` 計算一個位置（`position`）能從一個格子（`id`）得到多少的海浪高度：

```
// oceanFragmentShaderSource ...
// hash ...

float localWaveHeight(vec2 id, vec2 position) {
  float directionRad = (hash(id) - 0.5) * 0.785 - 2.355;
  vec2 direction = vec2(cos(directionRad), sin(directionRad));

  float distance = length(id + 0.5 - position);
  float strength = smoothstep(1.5, 0.0, distance);

  float waveX = dot(position, direction) * 2.5 + u_time * 5.0;
  return exp(sin(waveX) - 1.0) * strength;
}
// oceanSurfacePosition ...
```

- `directionRad`、`direction`、`waveX` 以及 `exp(sin(waveX) - 1.0)` 與先前在 `oceanSurfacePosition()` 的實做相同。
- `distance` 透過 `length()`[22] 計算 `position` 與 `id` 格子中央的距離。
- `strength` 表示海浪的強度，如果距離為 0 則強度最強為 1，而且我們只打算取到鄰近 1 格，距離到達 1.5 時表示已經到達影響力的邊緣，這時強度為 0，並且使用 `smoothstep(edge0, edge1, x)`[23] 平滑化強度差異。

---

22 取得 vector 的長度：
https://registry.khronos.org/OpenGL-Refpages/es3.0/html/length.xhtml
23 https://registry.khronos.org/OpenGL-Refpages/es3.0/html/smoothstep.xhtml

回到 `oceanSurfacePosition()`，這時它的任務便是蒐集鄰近 `id.xy` 相差 -1~+1、共 9 個格子對於當下位置的波浪高度，並且加在一起：

```
vec3 oceanSurfacePosition(vec2 position) {
  vec2 id = floor(position);

  float height = 0.0;

  for (int i = -1; i <= 1; i++) {
    for (int j = -1; j <= 1; j++) {
      height += localWaveHeight(id + vec2(i, j), position);
    }
  }

  return vec3(position, height);
}
```

分成格子產生偽隨機的波浪角度、對鄰近格子進行採樣後，如圖 6-39 看起來真的像是海浪了：

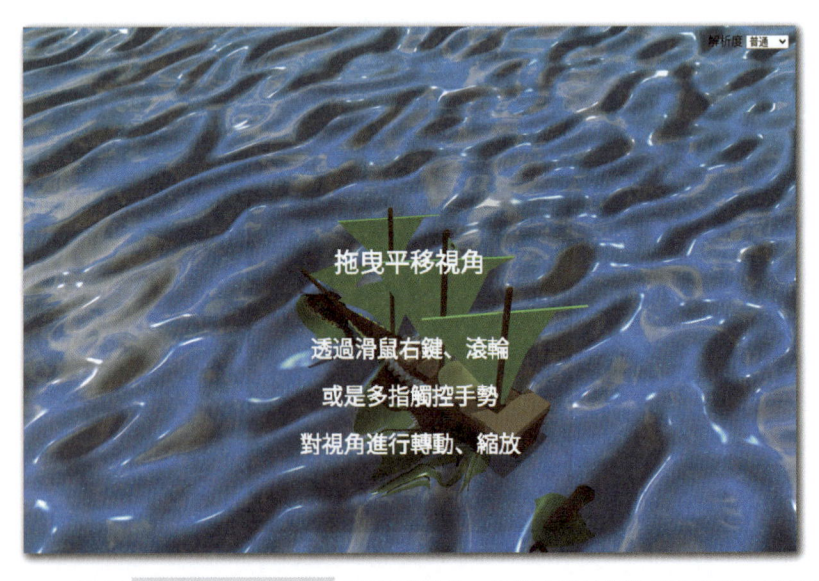

▲ 圖 6-39　使用 `localWaveHeight()` 蒐集鄰近共 9 個格子對於當下位置產生的波浪高度

不過海浪似乎有點太高了，而且希望可以更細緻一點，因此在 oceanSurfacePosition() 加入這兩行調整一下數值：

```
vec3 oceanSurfacePosition(vec2 position) {
  position *= 6.2;
  vec2 id = floor(position);

  float height = 0.0;

  for (int i = -1; i <= 1; i++) {
    for (int j = -1; j <= 1; j++) {
      height += localWaveHeight(id + vec2(i, j), position);
    }
  }

  height *= 0.15;

  return vec3(position, height);
}
```

把 position 乘以 6.2 可以使格子更小，height 則是很直覺地乘以 0.15 降低高度，海面就平靜許多了，如圖 6-40、圖 6-41：

▲ 圖 6-40　調整數值使海浪較為細緻、平靜

6 帆船與海

▲ 圖 6-41　轉一下視角觀察反射光的反應

好的，海面的部份就到這邊，希望讀者覺得這樣的海面有足夠的說服力。既然海面的 normal map 以及相關的兩個 uniform `oceanNormal`、`u_normalMapSize` 已經沒有用到，筆者就順手移除來避免下載不必要的檔案。

事實上，這邊實做的 procedural generation 是於每個 frame 即時運算的，這樣的做法對於實際應用程式上的效能是不利的，理論上應該可以利用原本的 normal map 以更省力的方式達到類似的效果，即時做在一些裝置上可能會有點跑不動，也是因此筆者在實做本篇範例時有稍微避免過度複雜的 shader，最後的成果經過幾個行動裝置的測試覺得效能還不錯，又有提供解析度的選擇，因此海面部份就實做到此；如果要實做的 procedural generation shader 更為複雜會導致裝置跑不動，或許可以思考在初始化的時候使用 framebuffer 先產生好，不過會需要調整整個產生 texture 與渲染之間的策略就是了。

⌘ One more thing — 帆船的晃動

既然有了波浪，那帆船是不是也該隨著時間前後、上下擺動？類似剛剛使用的海浪技巧，只要把時間套 `Math.sin()` 函數，剩下的就是 3D 物件的 transform：

6-4 使用 Shader 即時渲染波光粼粼的海面

```javascript
function renderSailboat(app, viewMatrix, programInfo) {
  const { gl, textures, objects, time } = app;

  const worldMatrix = matrix4.multiply(
    matrix4.yRotate(degToRad(45)),
    matrix4.translate(0, 0, 0),
    matrix4.scale(1, 1, 1),
    matrix4.xRotate(Math.sin(time * 0.0011) * 0.03 + 0.03),
    matrix4.translate(0, Math.sin(time * 0.0017) * 0.05, 0),
  );

  twgl.setUniforms(programInfo, {
    // u_matrix, u_worldMatrix, u_normalMatrix ...
  });

  // objects.sailboat.forEach ...
}
```

這時帆船便會隨著時間慢慢的以 x 軸前後微微旋轉、上下微微晃動。

CH6『帆船與海』的最終成品也就如圖 6-42 完成囉！

▲ 圖 6-42　CH6『帆船與海』

6 帆船與海

到此進度的完整程式碼可以在這邊找到：

</>	ch6/07/index.html
</>	ch6/07/main.js

也可以參考上線版本，透過瀏覽器打開此頁面：
https://webgl-book.pastleo.me/ch6/07/index.html

『帆船與海』這個場景的建構過程中，從讀取及渲染 3D obj 檔案、skybox、文字、半透明到 procedural generation，算是筆者挑選一些 WebGL 除了光影以外的幾個課題，WebGL 作為底層的技術，懂得活用其功能，尤其是 shader（以及數學）的話，能製作出的效果肯定是不勝枚舉的，還有許多概念是沒有提到，像是 raycast 得到滑鼠或觸控位置的 3D 物件[24]、迷霧效果[25]等，甚至透過 WebGL 把 GPU 當成無情的運算機器[26]，舉 Conway's Game of Life[27] 為例，一個棋盤的下一回合每個格子的開關由當前格子周圍的開關決定，現在也可以想像的到使用 WebGL 實做的方向：運用 framebuffer，在 fragment shader 讀取上回合地圖 texture 相關的格子來繪製下回合的地圖。

Shader 的應用非常廣泛且強大，讀者們不妨多多咀嚼其中奧妙之處吧！

---

24 https://webgl2fundamentals.org/webgl/lessons/webgl-picking.html
25 https://webgl2fundamentals.org/webgl/lessons/webgl-fog.html
26 https://webgl2fundamentals.org/webgl/lessons/webgl-gpgpu.html
27 https://zh.wikipedia.org/wiki/康威生命遊戲

7

Catch The Wind 小遊戲

『帆船與海』的場景建構完成，不過就只是一個場景，只能移動視角，像是模型觀賞這樣，在 CH7 這個章節，我們將從做好的場景繼續往下製作，做出一款小遊戲— Catch the Wind。

7　Catch The Wind 小遊戲

在這款小遊戲，玩家將操作帆船，只能控制風帆，打開向右或是向左來乘風前進並控制方向，在航行的過程中避開島嶼以免擱淺，看玩家能航行多遠。

讀者們可以直接玩玩成品：https://webgl-book.pastleo.me/ch7/07/

▲ 圖 7-1　Catch The Wind 標題畫面 (update needed)

▲ 圖 7-2　將會製作適合手機操作與遊玩的界面

本章節的起始點與『帆船與海』的成品功能大致相同，主要差異有：

- `programInfo`、`depthProgramInfo` 等 `shader`、`program` 使用一個 `programInfos`（注意有 s 結尾表示多數）的 Javascript 物件裝起來，主要的 `programInfo` 之後改成了 `programInfos.main`、`depthProgramInfo` 改成 `programInfos.depth`。
- `skyboxVertexShaderSource` 改名為 `simpleVertexShaderSource`、`objects.skybox` 改名為 `objects.xyQuad`，在本章將會用於 skybox 以外的全畫布渲染，因此改名成通用一點的名字。
- 平行光方向改用符合 skybox 中太陽照射下來的方向，而非隨著時間改變方向。
- 網頁初始 HTML 顯示『Loading…』，讀取、初始化完成後移除此文字，若出現錯誤則顯示瀏覽器可能不支援以及錯誤訊息。

起始點完整程式碼可以在這邊找到：

</>	ch7/00/index.html
</>	ch7/00/main.js

也可以參考上線版本，透過瀏覽器打開此頁面：

https://webgl-book.pastleo.me/ch7/00/index.html

▲ 圖 7-3　CH7 起始點，固定平行光方向與天空太陽照射方向一致

7 Catch The Wind 小遊戲

7-1 地形高度圖的產生

對於『Catch The Wind』小遊戲來說，場景部份缺少的即為作為障礙物的島嶼，這些島嶼將以隨機的方式產生，並且隨著玩家前進產生更多島嶼，實做方向其實有很多，筆者列出兩個：

- 購買、下載可免費使用或是自己用軟體製作島嶼的 3D 模型，遊戲進行時產生一些隨機位置、大小，把島嶼 3D 模型當成一個個的物件放置在場景中，到時候開船的同時比較各個島嶼的位置與船隻的距離，小於一定程度判斷為觸礁。

- 實做一個 shader，使用 procedural generation 的方式繪製地形高度圖，並利用 framebuffer 把地圖寫入 texture 中，繪製島嶼時使用一個有大量三角形的平面，讓頂點從地圖取得地形高度作為該點的高度，航行時對地形高度圖的 texture 取樣，如果船體位置在地圖中地形高度高於水面則判定觸礁。

第一種應為正常遊戲開發所使用的方式，一來是島嶼位置產生的方式比較容易實做，二來是島嶼的樣貌可以根據 3D 模型而更有更豐富、更有創意的細節，也可以使用多組不同的 3D 模型；不過本書專注在 WebGL 這項技術的研討，第二種方式可以繼續應用這個底層 API 以及 GPU 的功能，其中對地形高度圖 texture 的取樣這件事，有點像是讓 CPU 讀取 GPU 運算結果，這個功能也可以應用在滑鼠點選 3D 物體做 raycast，讀取結果回 CPU 來判斷點擊到哪一個物件。

不過可能就只是筆者覺得使用 GPU 做 procedural generation 的方式很有趣吧，因此接下來要請讀著們忍耐一下，我們使用 shader 繪製地形圖作為島嶼的產生吧！

⌘ 地形高度圖之 texture、framebuffer

先前使用 framebuffer 渲染陰影所需的深度圖時，使用了一種叫做 `gl.DEPTH_COMPONENT32F` 作為每個 pixel 的資料，除了可以與 WebGL 的深度寫入整合之外，這也使得該 texture 不是放置一般的 RGBA 顏色資料，在這邊作為高度圖的『高度』數值也不適合使用 RGBA，所有 WebGL2 支援的 texture 型別組合在 WebGL2 規格中有一張表[1]，但在 MDN 對於 `gl.texImage2D()` 的文件中又有一張表[2]，表示不是所有的組合都是『可渲染』的，我們要從可渲染的組合中，找到盡量不浪費空間，也就是只需要一個 channel R，同時也提供盡量大的 space，也就是希望高度的數值可以盡量精細，筆者最後選擇 R32I，也就是每個 pixel 為一個 32bit 的整數，可表示 2 的 32 次方個可能性，並且是『可渲染』的。

建立一個稱為 `landMap` 的 framebuffer、texture，其型別為 `gl.R32I`：

```
async function setup() {
  // framebuffers ...
  framebuffers.landMap = twgl.createFramebufferInfo(gl, [{
    attachmentPoint: gl.COLOR_ATTACHMENT0,
    internalFormat: gl.R32I,
    minMag: gl.NEAREST,
  }], 2048, 2048);
  textures.landMap = framebuffers.landMap.attachments[0];
  // ...
}
```

1　https://registry.khronos.org/webgl/specs/latest/2.0/#TEXTURE_TYPES_FORMATS_FROM_DOM_ELEMENTS_TABLE

2　https://developer.mozilla.org/en-US/docs/Web/API/WebGLRenderingContext/texImage2D#internalformat

7 Catch The Wind 小遊戲

⌘ 地圖的 Infinite Scroll

　　Infinite scroll — 無限捲軸，相信有寫過一些前端案例的讀者應該知道，也就是在網頁讀取時載入一定數量的內容，使用者往下捲動接近到底時載入更多內容，並且繼續重複此『捲動到底時載入』的行為，使捲軸有種無限展開的感覺；在這款遊戲中的島嶼、地圖也將採用這樣的行為，請見下頁圖 7-4，表示初始化時應產生出來的俯視地圖、可視範圍以及座標，往上為 vec2 location 的 -y 方向，對應三維場景內 -z 方向，藍色點表示原點 (0, 0)，在原點與之重疊的三角形表示帆船的位置，一個正方形區域為一個 chunk，並定義 `LAND_CHUNK_SIZE` 為 96 表示一個 chunk 正方形區域的邊長，灰色（包含被半透明黃色蓋住）區域表示『地圖』`textures.landMap` 的範圍，地圖為 3 個 chunk，同時也表示了 3 個陸地物件，半透明黃色區域則表示地表『可視』範圍，精確一點說是島嶼允許浮出水面的範圍；在圖 7-4 中可以看到 landMapOffset 為 (0, -96)，我們將會利用這個數值來控制地圖 shader 渲染範圍中心點的平移。

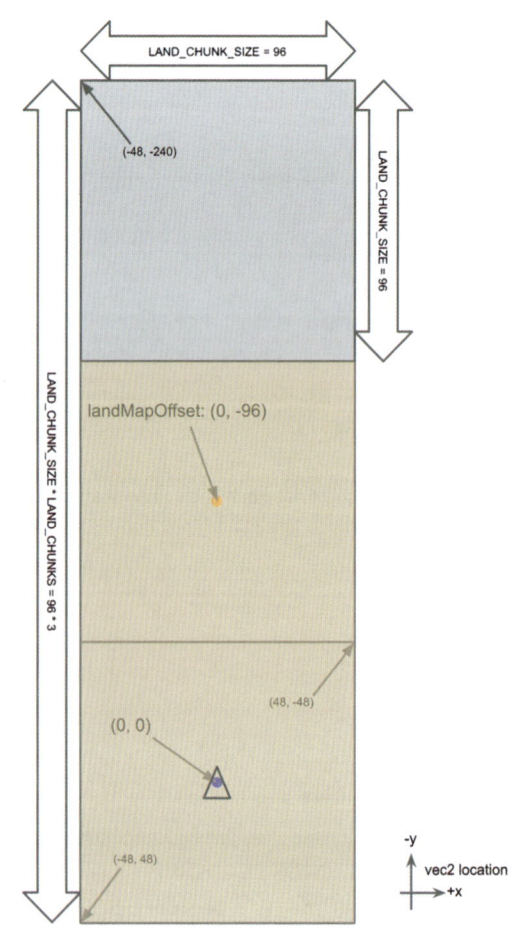

▲ 圖 7-4　地圖、陸地物件、可視範圍初始化預想俯視示意圖

7-1 地形高度圖的產生

接著遊戲開始，半透明黃色區域的島嶼可視範圍將會隨著帆船移動，這個可視範圍前緣為帆船位置往前推 1.5 倍的 `LAND_CHUNK_SIZE`，當這個前緣位置超出渲染好的地圖範圍，則移動 landMapOffset 到 (0, -192) 並再執行一次地圖渲染，這麼一來就把捲軸接下來的捲紙產生好囉，如圖 7-5：

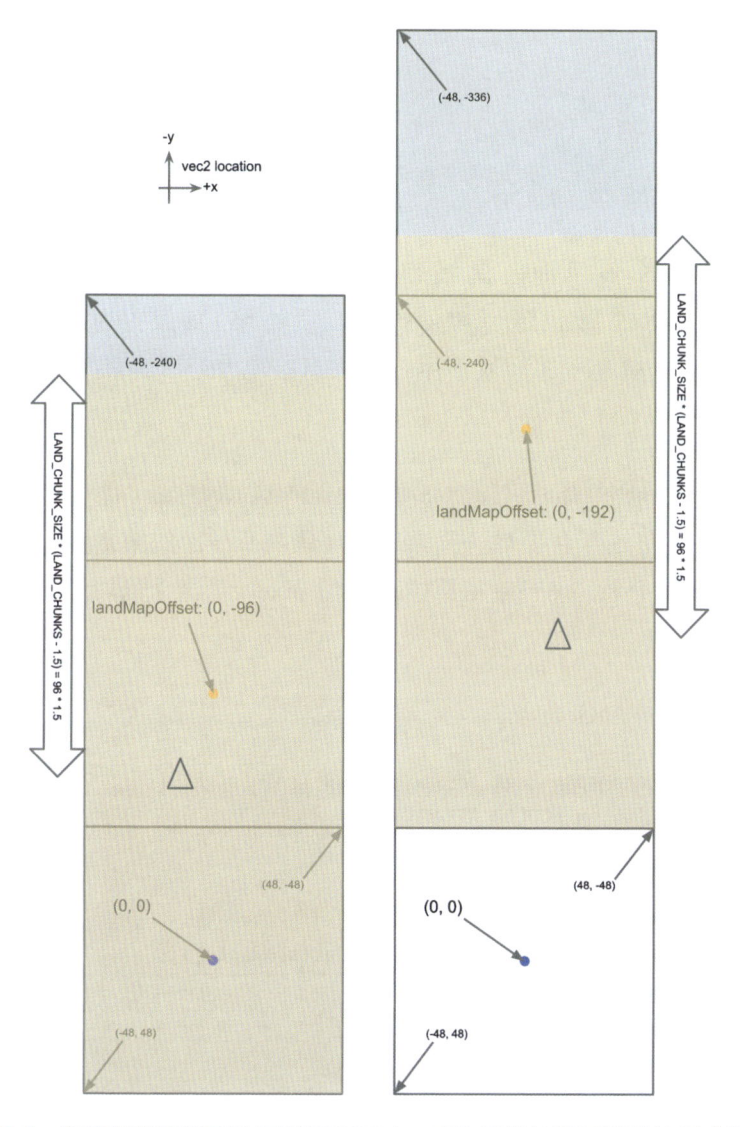

▲ 圖 7-5　隨著帆船推進到地圖範圍不足時，產生捲軸接著要捲開的新『捲紙』

7 Catch The Wind 小遊戲

這邊 `LAND_CHUNK_SIZE` 以及圖 7-4、圖 7-5 內座標位置的長度單位皆指的是場景內的，應與 worldPosition 一致；大概了解了這個遊戲地圖的運作方式，回到程式碼，先把提到的幾個應用程式常數寫到 Javascript 的上方：

```javascript
// import ...

const LAND_CHUNK_SIZE = 96;
const LAND_CHUNKS = 3;
const LAND_MAP_SIZE = [LAND_CHUNK_SIZE, LAND_CHUNK_SIZE * LAND_CHUNKS];

// vertexShaderSource ...
```

`LAND_CHUNKS` 表示地圖區域包含的 chunk 數量，並依據 `LAND_CHUNK_SIZE`、`LAND_CHUNKS` 算出 `LAND_MAP_SIZE` 表示地圖在場景中的大小，這些常數先寫好，接下來撰寫程式時就不用在各處直接寫 96、3 等不容易直接看出意義的數值。

再回到 `framebuffers.landMap` 與 `textures.landMap`，他們的大小就應該對應圖 7-4、圖 7-5 灰色區域，又希望在場景內有一定的細緻度，場景 1 單位的大小可以參考下圖 7-6，每 1 單位給 8 個 pixel，因此把解析度設定成 `LAND_MAP_SIZE` 的 8 倍：

```javascript
async function setup() {
  // framebuffers ...
  framebuffers.landMap = twgl.createFramebufferInfo(gl, [{
    attachmentPoint: gl.COLOR_ATTACHMENT0,
    internalFormat: gl.R32I,
    minMag: gl.NEAREST,
  }], LAND_MAP_SIZE[0] * 8, LAND_MAP_SIZE[1] * 8);
  textures.landMap = framebuffers.landMap.attachments[0];
  // ...
}
```

▲ 圖 7-6 把海面的 scale 調整成 1、稍微平移到旁邊，與帆船比較來感受 1x1 長度的大小

⌘ 建立繪製地圖高低圖用的 shader

建立繪製地圖專用的 shader、program，此 shader 與 skybox 共用同一個 simpleVertexShaderSource，因為他們都只是使用 objects.xyQuad 把整個畫布填滿，而 fragment shader 則是到時候 procedural generation 發生的地方，不過現階段先把整個架構建設好，寫上這個 fragment shader 原始碼：

```
// skyboxFragmentShaderSource ...

const landMapFragmentShaderSource = `#version 300 es
precision highp float;

in vec2 v_position;

out int outLandAltitude;

uniform float u_seed;
uniform vec2 u_landMapSize;
uniform vec2 u_landMapOffset;

void main() {
```

7-9

7 Catch The Wind 小遊戲

```
  vec2 location =
    v_position * u_landMapSize * vec2(0.5, -0.5) +
    u_landMapOffset;
  outLandAltitude = 0;
}
`;
```

與先前的 fragment shader 有一個很不一樣的地方,那就是輸出顏色的 `outColor` 在這邊稱為 `outLandAltitude`,並且型別為一個整數,這是要對應 `framebuffers.landMap` 與 `textures.landMap` 所設定的型別,輸出的值先設定為 0,待會再回來處理;此 shader 接收三個 uniforms 以及一個 varying:

- `u_seed`:為了讓每次遊戲都產生不同的地圖,這個 uniform 在實做 procedural generation 時會當成偽隨機的其中一個變因。
- `u_landMapSize`:等於已經設定好的 `LAND_MAP_SIZE` 應用程式常數,應為 [96, 96 * 3] = [96, 288]。
- `u_landMapOffset`:等於圖 7-4、圖 7-5 的 landMapOffset,用來控制地圖渲染範圍中心點的平移,初始時應為 [0, -96]。
- `v_position`:`objects.xyQuad` 的頂點位置補間而來,表示該點於 clip space 的 xy 位置,介於 (-1, -1) 到 (1, 1)。

由這些輸入資料,我們可以由 `v_position` 乘上 `u_landMapSize` 得到介於 (-96, -288) 到 (96, 288),再乘上 `vec2(0.5, -0.5)` 得到介於 (-48, -144) 到 (48, 144),最後加上 `u_landMapOffset` 就得到介於 (-48, -240) 到 (48, 48) 的 `location`,表示俯瞰場景中的 xz 位置,同時也符合圖 7-4 灰色區域範圍。

不過現有的 simpleVertexShaderSource 不會輸出 `v_position`,在此補上:

```
// simpleVertexShaderSource ...

in vec2 a_position;
uniform mat4 u_matrix;
```

```
out vec2 v_position;
out vec3 v_normal;

void main() {
  gl_Position = vec4(a_position, 1, 1);
  v_position = a_position;
  v_normal = (u_matrix * gl_Position).xyz;
}
```

然後就可以把繪製地圖高低圖用的 shader、program 編譯、連結放好於 `app.programInfos.landMap` 囉：

```
async function setup() {
  // ...
  const programInfos = {
    // main, depth ...
    skybox: twgl.createProgramInfo(gl, [
      simpleVertexShaderSource,
      skyboxFragmentShaderSource,
    ]),
    landMap: twgl.createProgramInfo(gl, [
      simpleVertexShaderSource,
      landMapFragmentShaderSource,
    ]),
  };
  // ...
}
```

⌘ 遊戲初始化 initGame()

先前鏡面、陰影的所需要的鏡面世界、深度圖需要每個 frame 更新，這次地形繪製不用，是等到帆船航行捲動到不夠時再移動 landMapOffset 並繪製，不過當然在遊戲一開始時還是要繪製一次，因此就有了遊戲初始化的概念，建立 `initGame()` function，除了在 `main()` 中接續 `setup()` 後呼叫，筆者同時也把初始狀態的設定也移動到此 `initGame()` 中進行：

7 Catch The Wind 小遊戲

```
async function setup() {
  // ...
  return {
    // ...
    state: {
      fieldOfView: degToRad(45),
      cameraRotationXY: [degToRad(-45), 0],
      cameraDistance: 15,
      cameraViewing: [0, 0, 0],
      lightRotationXY: [degToRad(20), degToRad(-60)],
      resolutionRatio: 1,
    },
    time: 0,
  };
}
// ... render ...
function initGame(app) {
  app.state = {
    fieldOfView: degToRad(45),
    cameraRotationXY: [degToRad(-45), 0],
    cameraDistance: 15,
    cameraViewing: [0, 0, 0],
    lightRotationXY: [degToRad(20), degToRad(-60)],
    resolutionRatio: 1,

    seed: Math.random() * 20000 + 5000,
    level: 0,
  };

  renderLandMap(app);
}
// startLoop ...
async function main() {
  // ...
  try {
    const app = await setup();
    window.app = app;
    window.gl = app.gl;
```

7-12

```
    initGame(app);
    app.input = listenToInputs(app.gl.canvas, app.state);

    // ...
  } catch (error) { /* ... */ }
}
```

稍微看一下 `initGame()`，會發現多了 `seed` 與 `level` 這兩個狀態，seed 對應前面 landMap shader 中的 u_seed，作為偽隨機的其中一個變因，而 `level` 這個狀態有點像是遊戲的『第幾關』，表示 landMapOffset 切換了幾次，以圖 7-5 為例，圖中左邊表示 landMapOffset 是初始數值 level 為 0，右邊則是經過了一次切換 level 為 1。

在 `initGame()` 最後呼叫 `renderLandMap()` 進行初次的 landMap 地圖高低圖渲染，這個 function 的所使用的 WebGL 功能在前面的章節都有應用過，切換 shader、program、framebuffer、VAO，設定 uniform 並進行 draw call：

```
// render ...

function renderLandMap(app) {
  const { gl, programInfos, framebuffers, objects, state } = app;

  gl.useProgram(programInfos.landMap.program);
  twgl.bindFramebufferInfo(gl, framebuffers.landMap);
  gl.bindVertexArray(objects.xyQuad.vao);

  twgl.setUniforms(programInfos.landMap, {
    u_seed: state.seed,
    u_landMapSize: LAND_MAP_SIZE,
    u_landMapOffset: getLandMapOffset(app),
  });

  twgl.drawBufferInfo(gl, objects.xyQuad.bufferInfo);
}

// initGame ...
```

7　Catch The Wind 小遊戲

其中 `u_landMapOffset` uniform 需要從 `app.state.level` 狀態與 `LAND_CHUNK_SIZE` 等常數來算出，考慮到接下來在島嶼渲染等狀況時也會需要去算得這個座標，因此實做 `getLandMapOffset()` function 使之後可以重複使用：

```
// renderLandMap ...

function getLandMapOffset(app) {
  return [
    0,
    -LAND_CHUNK_SIZE * (app.state.level + (LAND_CHUNKS * 0.5 - 0.5)),
  ];
}
```

這個座標的 y 值計算需要搭配圖 7-5，因為帆船出生在最下方 chunk 的中心點，所以初始時要往 -y 移動地圖高度的一半減去半個 chunk，接下來每次 level 增加時多往 -y 移動一個 chunk，稍微經過數學把 `LAND_CHUNK_SIZE` 提取到最外面就可以得到這樣的公式。

⌘ 繪製一個測試用的整數高度地圖

筆者希望地圖高度單位與 location 一樣可以符合場景所使用的，在某個 function 回傳的浮點數即表示該點的高度在場景中的 y 值，這樣一來設計 procedural generation 時比較好思考，這個『某個 function』就命名為 `landAltitude()`，但是 landMap 因為選擇的 texture 型別是整數，所以需要一個乘法以及一個轉型：

```
// landMapFragmentShaderSource ...
precision highp float;

in vec2 v_position;

out int outLandAltitude;

// uniforms ...
```

7-14

```
float landAltitude(vec2 location);

void main() {
  vec2 location =
    v_position * u_landMapSize * vec2(0.5, -0.5) +
    u_landMapOffset;
  outLandAltitude = int(landAltitude(location) * 65536.0);
}
```

整數不能表示小數點，直接把浮點數轉換成整數會導致小數全部遺失，那麼我們在把浮點數轉換成整數之前先乘以一個數值，到時候在另外一端讀取值的時候先轉換回浮點數再除以這個數值，就可以一定程度的保留小數點了，在這邊這個數值選用 65536，等於 2 的 16 次方，因為選用的型別為 R32I 有 32 bits，最大值為 2 的 31 - 16 = 15 次方，1 bit 給正負、15 bits 給小數點以上、16 bits 給小數點以下，這樣的精準度、範圍應該是非常夠用。

接著來實做簡單測試用的 `landAltitude()`，先一律讓高度為 -1，接著讓一些區域高出水面，之後方便測試與驗證：

- `location.x` 介於 -1 到 +1、`location.y` 任意延展所形成的長條型區域，高度為 1.0。
- `location.y` 介於 -1 到 +1、`location.x` 任意延展所形成的長條型區域，高度為 1.0。
- 介於 (-49, 47) 到 (-47, 49) 的矩形區域，高度為 1.0。
- 介於 (47, 47) 到 (49, 49) 的矩形區域，高度為 1.0。
- 介於 (47, -49) 到 (49, -47) 的矩形區域，高度為 1.0。
- 介於 (-49, -241) 到 (-47, -239) 的矩形區域，高度為 1.0。

到時候場景中 x 與 z 軸應該會突起，並且會有四個突起點，實做如下：

```
// landMapFragmentShaderSource ...
// main() ...
float landAltitude(vec2 loc) {
```

7 Catch The Wind 小遊戲

```
float altitude = -1.0;

if (loc.x >= -1.0 && loc.x <= 1.0) {
  altitude = 1.0;
}
if (loc.y >= -1.0 && loc.y <= 1.0) {
  altitude = 1.0;
}

if (loc.x >= -49.0 && loc.x <= -47.0 && loc.y >= 47.0 && loc.y <= 49.0) {
  altitude = 1.0;
}
if (loc.x >= 47.0 && loc.x <= 49.0 && loc.y >= 47.0 && loc.y <= 49.0) {
  altitude = 1.0;
}
if (loc.x >= 47.0 && loc.x <= 49.0 && loc.y >= -49.0 && loc.y <= -47.0) {
  altitude = 1.0;
}

if (
  loc.x >= -49.0 && loc.x <= -47.0 &&
  loc.y >= -241.0 && loc.y <= -239.0
) {
  altitude = 1.0;
}

return altitude;
}
```

⌘ 把地圖繪製在文字看板上

為了先預覽地圖渲染的結果，並且確保真的能在實際渲染時使用，我們借用一下文字看板，把文字看板使用的 u_texture 從文字內容 texture 改成 landMap：

```
function renderText(app, viewMatrix, programInfo) {
  // vao, textLeftShift ...
  const worldMatrix = matrix4.multiply(
```

```
    matrix4.translate(textLeftShift, 0, 0),
    matrix4.xRotate(degToRad(45)),
    matrix4.translate(0, 10, 0),
    matrix4.scale(1, 1, LAND_CHUNKS),
  );
  twgl.setUniforms(programInfo, {
    u_matrix: matrix4.multiply(viewMatrix, worldMatrix),
    u_bgColor: [0, 0, 0, 0.2],
    u_texture: textures.landMap,
  });
  // ...
}
```

同時筆者也調整了一下文字看板的位置，並上下拉伸符合 `LAND_MAP_SIZE` 的比例。

因為地圖高度 texture 是用整數儲存，並且放大了 65536 倍，在文字看板的 shader 要做點調整：

```
// textFragmentShaderSource ...
precision highp float;
precision highp isampler2D;

in vec2 v_texcoord;

uniform vec4 u_bgColor;
uniform isampler2D u_texture;

out vec4 outColor;

void main() {
  outColor = u_bgColor + float(texture(u_texture, v_texcoord).r) / 65536.0;
}
```

配合資料來源改為整數，代表 texture 的 `sampler2D` 型別改成 `isampler2D`，而且 WebGL 要求設定這個型別的精準度：`precision highp isampler2D;`，最後把放大 65536 倍的整數轉成浮點數並縮小 65536 倍來還原成當初在 `landAltitude()` 輸出的數值。

7 Catch The Wind 小遊戲

到瀏覽器查看地圖的預覽，結果如圖 7-7，可以看到一個十字表示原點位置，並且位於原點 chunk 的三個角落有白點，在地圖左上角也有一個點，這樣一來往後在 `landAltitude()` 就可以用場景中的位置來思考。

到此進度的完整程式碼可以在這邊找到：

</>	ch7/01/index.html
</>	ch7/01/main.js

也可以參考上線版本，透過瀏覽器打開此頁面：

https://webgl-book.pastleo.me/ch7/01/index.html

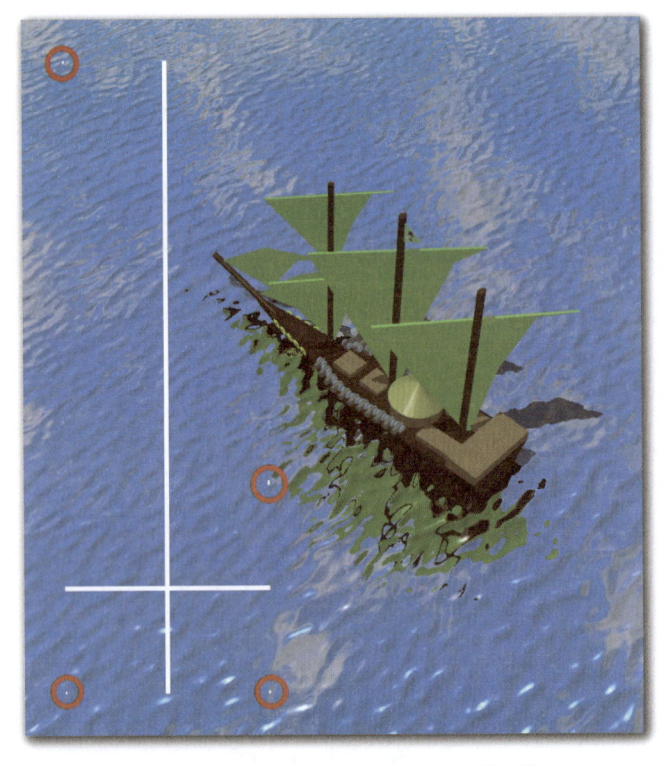

▲ 圖 7-7 借用文字看板預覽地形高度圖渲染的結果

7-18

7-2 依照地形高度圖繪製島嶼

大致架構好了 `textures.landMap` 地形高度圖的渲染，接著加入地面物件，這個地面物件不能用海洋以及文字看板所用的 `objects.plane` 三角形頂點資料，因為 `objects.plane` 只有兩個三角形，地面物件會需要仰賴一定密集程度的三角形頂點，每個頂點根據其 xz 座標值去 `textures.landMap` 地形高度圖取樣當成 y 值高度，來達成地形的高低起伏，為此建立一個三角形數量較多的 plane，稱為 `objects.land`：

```
async function setup() {
  // objects ...
  { // land
    const vertexDataArrays = twgl.primitives.createPlaneVertices(
      LAND_CHUNK_SIZE, LAND_CHUNK_SIZE,
      LAND_CHUNK_SIZE * 2, LAND_CHUNK_SIZE * 2,
    );
    const bufferInfo = twgl.createBufferInfoFromArrays(
      gl, vertexDataArrays,
    );
    const vao = twgl.createVAOFromBufferInfo(
      gl, programInfos.main, bufferInfo,
    );

    objects.land = {
      vertexDataArrays,
      bufferInfo,
      vao,
    };
  }
  // ...
}
```

7 Catch The Wind 小遊戲

使用 TWGL 產生頂點資料時，呼叫 `createPlaneVertices()`[3] 第一、第二參數傳入 `LAND_CHUNK_SIZE`（= 96）作為長寬，第三、第四參數 `subdivisionsWidth`、`subdivisionsDepth` 表示『要分成多少個頂點』，在圖 7-6 可以再感受一次場景中 1 單位的大小，這邊傳入 `LAND_CHUNK_SIZE * 2`，也就是在場景中地面每隔 0.5 一個頂點。

但是這邊怎麼長寬以及要分的頂點數不是根據 `LAND_MAP_SIZE` 的 [96, 96 * 3] = [96, 288]？因為 WebGL 的 indexed element 功能下，indices 指標們的型別[4]在不使用 WebGL extension 的狀況下只能到 `gl.UNSIGNED_SHORT`，indices 是 `Uint16Array`，也就是說一個指標只有 16 bits，最大只能表示 2 的 16 次方減 1 為 65535，如果三角形頂點資料過多，那麼會導致指標需要指到超過 65535 指不上去導致渲染結果不正確，為此我們讓 `objects.land` 一次只繪製一個 chunk（圖 7-4 的一個正方形區域），繪製這個物件 3 次來把整張地圖畫完，就可以在維持每隔 0.5 一個頂點的精細度狀況下，把 `objects.land` 的資料量壓在 WebGL indices 能指到的範圍內，可以打開 Console 輸入 `app.objects.land.vertexDataArrays` 查看產生的資料量，如圖 7-8，position、normal 每組資料有 3 個元素，數量是 111747 在 65535 * 3 = 196608 以下，texcoord 每組 2 個元素，數量是 74498 在 65535 * 2 = 131070 以下。

▲ 圖 7-8　檢查 app.objects.land.vertexDataArrays 的資料量

3　https://twgljs.org/docs/module-twgl_primitives.html#.createPlaneVertices
4　https://developer.mozilla.org/en-US/docs/Web/API/WebGLRenderingContext/drawElements#type

⌘ 渲染地形、島嶼的 Shader

有了 `objects.land` 作為三角形頂點資料，下一步是繪製地形用的 shader、program，在這邊會需要 vertex shader 從高度圖中取樣當成頂點位置 y 值，是一個特別的 vertex shader，到著色時行為倒是與帆船等一般物件相同，也就是說地面物件的 shader、program 將由一個專用的 vertex shader 與一般的 `fragmentShaderSource` 組成，這個專用的 vertex shader 就叫做 `landVertexShaderSource` 吧！從 `vertexShaderSource` 複製出來改：

```
// landMapFragmentShaderSource ...
const landVertexShaderSource = `#version 300 es
in vec4 a_position;
in vec2 a_texcoord;
in vec3 a_normal;

// uniforms, outs ...

void main() {
  gl_Position = u_matrix * a_position;
  v_texcoord = vec2(a_texcoord.x, 1.0 - a_texcoord.y);

  vec3 normal = mat3(u_normalMatrix) * a_normal;
  vec3 normalMatrixI = normal.y >= 1.0 ?
    vec3(1, 0, 0) :
    normalize(cross(vec3(0, 1, 0), normal));
  vec3 normalMatrixJ = normalize(cross(normal, normalMatrixI));

  v_normalMatrix = mat3(
    normalMatrixI,
    normalMatrixJ,
    normal
  );

  v_worldPosition = u_worldMatrix * a_position;
  v_surfaceToViewer = u_worldViewerPosition - v_worldPosition.xyz;
  v_lightProjection = u_lightProjectionMatrix * v_worldPosition;
}
`;
```

7 Catch The Wind 小遊戲

開始對 `a_position` 動手腳，寫一個 function 接收原始 `a_position`（在 function 內為 `pos` 變數），根據 uniform 回傳動完手腳的 position，這個 function 稱為 `getAltitudePosition()`；從 `a_position` 取出 xz 作為 `location`，再從高度圖拿到該 location 對應的高度作為 altitude 高度，這個過程本身就牽扯了 `textures.landMap` 取樣位置與整數 texture 的轉換，同時 `a_position` 是單個 `objects.land` 內的位置，稍早有說會需要渲染 `objects.land` 三次來構成完整的地板，為此需要一個 unform 來指定目前是渲染哪個 chunk，現階段 `getAltitudePosition()` 的完整實做如下：

```
// landVertexShaderSource ...
precision highp isampler2D;
// uniforms ...
uniform isampler2D u_landMap;
uniform vec2 u_landMapSize;
uniform vec2 u_landMapOffset;
uniform vec2 u_landOffset;

// outs ...

vec3 getAltitudePosition(vec4 pos) {
  vec2 location = pos.xz + u_landOffset;

  int altitudeInt = texture(
    u_landMap,
    (location - u_landMapOffset) / u_landMapSize * vec2(1, -1) +
      vec2(0.5, 0.5)
  ).r;
  float altitude = float(altitudeInt) / 65536.0;

  return vec3(location.x, altitude, location.y);
}
// main ...
```

加入 `u_landMap` 整數 texture uniform 並設定 `isampler2D` 精準度，`u_landOffset` 這個 uniform 控制當前繪製 chunk 的中心點，與原始頂點位置 xz

值相加得到 location，接著 u_landMapOffset 為 getLandMapOffset() 目前地形高度圖中心點，拿 location 減去 u_landMapOffset 並除以地圖大小 u_landMapSize 時地圖範圍內的數值為 [-0.5, 0.5] 到 [0.5, -0.5]，需要把 y 值反轉並加 [0.5, 0.5] 便得到符合 textures 採樣範圍 [0, 0] 到 [1, 1]；取樣完畢得到整數高度轉換回浮點數之後，還要還原先前輸出到 texture 為了整數而做的放大，縮小 65536 倍得到 altitude，最後把 location、altitude 組合回新的 position 給 main() 使用：

```
// landVertexShaderSource ...

void main() {
  vec4 position = vec4(getAltitudePosition(a_position), 1);
  gl_Position = u_matrix * position;
  v_texcoord = vec2(a_texcoord.x, 1.0 - a_texcoord.y);

  // normals ...

  v_worldPosition = u_worldMatrix * position;
  v_surfaceToViewer = u_worldViewerPosition - v_worldPosition.xyz;
  v_lightProjection = u_lightProjectionMatrix * v_worldPosition;
}
```

我們修改了 position，那法向量 normal 呢？TWGL 的 createPlaneVertices() 所產生的法向量基本上都是全部指向 +y 的，但是地形已經不是全平的了，也就是會產生法向量的改變，在此可以利用 CH6 海面法向量的技巧：對另外兩個位置取樣得到不同的高度位置，然後用外積求得法向量：

```
// landVertexShaderSource ...
// ...
#define LAND_SAMPLE_DISTANCE 0.5
void main() {
  vec4 position = vec4(getAltitudePosition(a_position), 1);
  // ...

  vec3 p2 = getAltitudePosition(
    a_position + vec4(0, 0, LAND_SAMPLE_DISTANCE, 0)
```

7 Catch The Wind 小遊戲

```
);
vec3 p3 = getAltitudePosition(
  a_position + vec4(LAND_SAMPLE_DISTANCE, 0, 0, 0)
);
vec3 landNormal = normalize(cross(
  normalize(p2 - position.xyz), normalize(p3 - position.xyz)
));

vec3 normal = mat3(u_normalMatrix) * landNormal;
vec3 normalMatrixI = normal.y >= 1.0 ?
  vec3(1, 0, 0) :
  normalize(cross(vec3(0, 1, 0), normal));
vec3 normalMatrixJ = normalize(cross(normal, normalMatrixI));

v_normalMatrix = mat3(
  normalMatrixI,
  normalMatrixJ,
  normal
);
// ...
}
```

定義 LAND_SAMPLE_DISTANCE 表示與另外兩個用來比較高低的點的距離，分別呼叫 getAltitudePosition() 取得另外兩個點，把 position -> p2 與 position -> p3 做外積得到法向量 landNormal，取代原本的 a_normal。

這麼一來 landVertexShaderSource 就準備好囉，與一般 fragmentShaderSource 編譯、連結成 app.programInfos.land：

```
async function setup() {
  // ...
  const programInfos = {
    // main, depth, ocean ...
    land: twgl.createProgramInfo(gl, [
      landVertexShaderSource,
      fragmentShaderSource,
    ]),
    // text, skybox, landMap ...
```

7-24

```
};
// ... objects ...
{ // land
  const vertexDataArrays = twgl.primitives.createPlaneVertices(
    LAND_CHUNK_SIZE, LAND_CHUNK_SIZE,
    LAND_CHUNK_SIZE * 2, LAND_CHUNK_SIZE * 2,
  );
  const bufferInfo = twgl.createBufferInfoFromArrays(
    gl, vertexDataArrays,
  );
  const vao = twgl.createVAOFromBufferInfo(
    gl, programInfos.land, bufferInfo,
  );

  objects.land = { /* ... */ };
}
// ...
}
```

同時也記得把 `objects.land` 之 VAO 工作區域與這個使用新 `landVertexShaderSource` 的 `programInfos.land` 綁定。

⌘ 繪製地形到場景中

建立 `renderLand()` function，把繪製物體常用的前置動作如切換 VAO、設定 uniform 寫上：

```
// renderOcean ...
function renderLand(app, viewMatrix, programInfo) {
  const { gl, textures, objects, state } = app;

  gl.bindVertexArray(objects.land.vao);

  const worldMatrix = matrix4.identity();

  twgl.setUniforms(programInfo, {
    u_matrix: matrix4.multiply(viewMatrix, worldMatrix),
    u_worldMatrix: worldMatrix,
```

7 Catch The Wind 小遊戲

```
    u_normalMatrix: matrix4.transpose(matrix4.inverse(worldMatrix)),
    u_diffuse: [0.97265625, 0.9140625, 0.62890625],
    u_diffuseMap: textures.null,
    u_emissive: [0, 0, 0],
    u_specular: [0, 0, 0],
    u_normalMap: textures.nullNormal,
    u_landMap: textures.landMap,
    u_landMapSize: LAND_MAP_SIZE,
    u_landMapOffset: getLandMapOffset(app),
  });
}
```

transform 部份把 `worldMatrix` 設定成單位矩陣不做任何事，`u_diffuse` 散射光顏色設定成 `#f9eaa1` 沙灘的顏色，把地形高度圖 `textures.landMap` 設定到 `u_landMap`，`u_landMapSize`、`u_landMapOffset` 地形高度圖的大小、中心點位置依照該設定的設定上去，剩下的都設定成 null 或是 0 關閉效果。

因為 `objects.land` 只包含了一個 chunk，需要寫一個迴圈把 3 個 chunk 都繪製上去，`u_landOffset` 的部份有需要可以回去參考圖 7-5，3 個 chunk 的 -y 偏移量分別為 `app.state.level + 0`、`app.state.level + 1`、`app.state.level + 2` 乘以 `LAND_CHUNK_SIZE`（= 96），`app.state.level` 加一會讓整個地板區域移動一個 chunk、圖 7-5 的正方形格子：

```
function renderLand(app, viewMatrix, programInfo) {
  // vao, worldMatrix, setUniforms ...

  for (let i = 0; i < LAND_CHUNKS; i++) {
    twgl.setUniforms(programInfo, {
      u_landOffset: [0, -LAND_CHUNK_SIZE * (state.level + i)],
    });
    twgl.drawBufferInfo(gl, objects.land.bufferInfo);
  }
}
```

7-2 依照地形高度圖繪製島嶼

最後在 `render()` 內呼叫 `renderLand()`，記得要先切換好 program、shader：

```
function render(app) {
  // ... renderSailboat, ocean ...

  gl.useProgram(programInfos.land.program);
  twgl.setUniforms(programInfos.land, globalUniforms);
  renderLand(app, viewMatrix, programInfos.land);

  // skybox, text ...
}
```

完成修改之後，可以看到前面地形高度圖測試用繪製的結果，圖 7-9 清楚看到一個十字突起的陸地，也可以轉到另一邊看到 chunk 邊角的突起陸地，如圖 7-10：

▲ 圖 7-9　測試用的十字陸地繪製結果

7-27

7 Catch The Wind 小遊戲

▲ 圖 7-10　轉向並拉遠一點，可以看到 chunk 邊角的突起陸地

到此進度的完整程式碼可以在這邊找到：

</>	ch7/02/index.html
</>	ch7/02/main.js

也可以參考上線版本，透過瀏覽器打開此頁面：

https://webgl-book.pastleo.me/ch7/02/index.html

⌘ 地圖高度、島嶼之 procedural generation

可以來畫龍點睛一下了，回到 `landMapFragmentShaderSource`，地圖的產生就與遊戲玩法很有關了，筆者有稍微嘗試過經過內插的 noise 方式作為地形起伏，不過玩起來很容易遇到死路，因此最後決定的實做方式是：

- 把地圖每 16 單位分成一個格子，每個格子會產生一個島嶼。
- 這些島嶼會從格子中央隨機偏移，最遠偏移 10 單位。
- 島嶼的大小也使用偽隨機決定，同時有一定程度越接近水道中央島嶼越小。

7-2 依照地形高度圖繪製島嶼

至於島嶼本身的形狀怎麼產生呢？筆者設計一個 function 叫做 `island()`，接收當下 location、島嶼圓心位置與島嶼大小，一開始從純粹 location 與島嶼圓心位置的距離開始，並使用島嶼大小減去這個距離作為地形高度，這樣的話會像圖 7-11 這樣平平尖尖的：

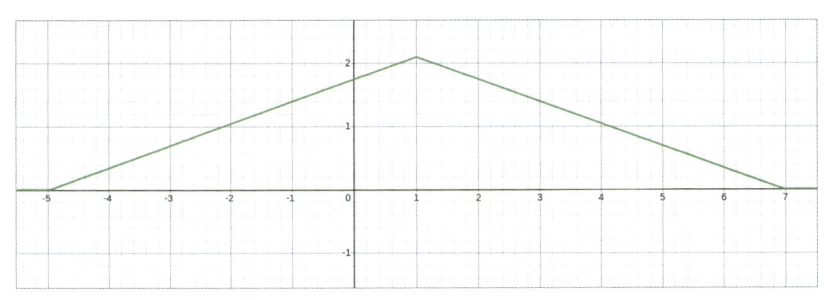

▲ 圖 7-11　假設島嶼圓心在 x = 1，與圓心位置的距離乘以 0.35

這時候套上自然對數，稍微調高島嶼大小，就獲得了圖 7-12 紅色線條較為自然的圖形：

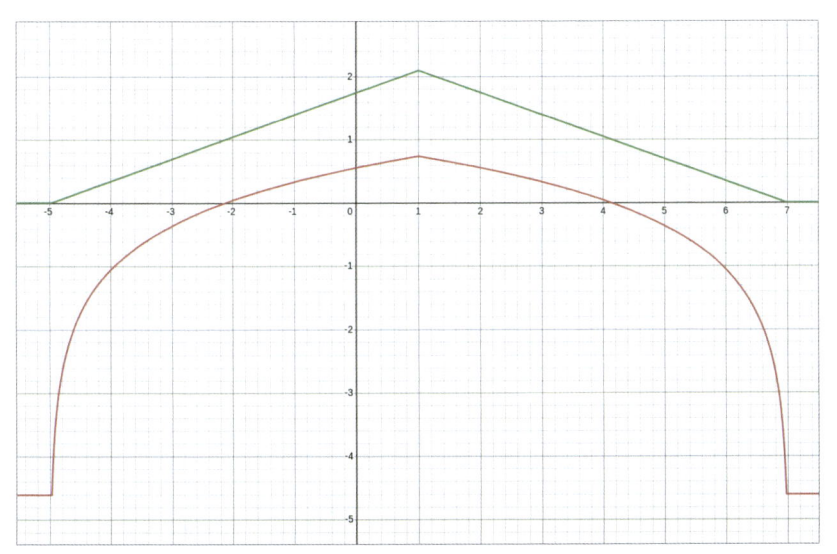

▲ 圖 7-12　由圓心距離套上自然對數之後島嶼曲線如紅色線條所示

數學公式以及圖形可以參考：
https://www.desmos.com/calculator/qufpkvaalv

7-29

7 Catch The Wind 小遊戲

這麼一來，寫成 `island()` function 如下：

```
// landMapFragmentShaderSource ...

float island(vec2 loc, vec2 origin, float size) {
  return log(max((size - length(loc - origin)) * 0.35, 0.01));
}

// landAltitude ...
```

加入一個 `max()` 把代入 `log()` 的最低數值限制在 0.01 是為了避免得到奇怪的結果；接著準備好三個偽隨機函式，基本上與 `oceanFragmentShaderSource` 海面的 `hash()` 一樣，只是這邊讓 `u_seed` 進來參一腳：

```
// landMapFragmentShaderSource ...

float hash0(vec2 p) {
  return fract(sin(mod(dot(p, vec2(101, 107)), radians(180.0))) * u_seed);
}
float hash1(vec2 p) {
  return fract(sin(mod(dot(p, vec2(113, 127)), radians(180.0))) * u_seed);
}
float hash2(vec2 p) {
  return fract(sin(mod(dot(p, vec2(179, 181)), radians(180.0))) * u_seed);
}

// landAltitude ...
```

最後是每個格子內的計算，因為在島嶼偏移量與大小都最大時還是會有部份陸地跑到別的格子去，因此也需要蒐集鄰近格子的島嶼，為此再定義一個 function 稱為 `localLandAltitude()`，接收當下的位置 `loc` 與格子 `id`：

```
// landMapFragmentShaderSource ...

float localLandAltitude(vec2 loc, vec2 id) {
  if (
    id.y >= 2.0 || id.y >= -2.0 && id.x <= 0.0 && id.x >= -1.0
  ) return -1.0;
```

7-30

```
  vec2 origin =
    id * 16.0 + vec2(8, 8) +
    (vec2(hash0(id), hash1(id)) - 0.5) * 10.0;
  float size = hash2(id) * 2.0 + abs(id.x + 0.5) * 1.2 + 4.0;

  return island(loc, origin, size);
}

// landAltitude ...
```

一開始的 `if` 是用來檢查是不是一開始玩家的所在格子,如果是直接回傳 -1。

接著產生 `origin` 作為島嶼圓心位置,`id * 16.0 + vec2(8, 8)` 可以得到該格子的中心點,接著使用 `hash0()`、`hash1()` 分別對 x、y 偏移 -5 ~ +5;`size` 的部份基礎大小是 4,根據格子增加基底,最接近 z 軸(或是 location 的 x = 0 線)的兩排格子加上 0.6,接著每往外一排加上 1.2,最後 `hash2() * 2` 隨機加上 0 ~ 2;最後呼叫前面寫好的 `island()` 產生島嶼弧形。

終於回到 `landAltitude()`,可以先把測試用產生十字以及突起點的一堆 if 移除,將 location 除以 16(等於乘以 0.0625)得到 `id` 並且使用雙層迴圈蒐集 3x3 共九個格子呼叫 `localLandAltitude()`:

```
// landMapFragmentShaderSource ...
float landAltitude(vec2 loc) {
  float altitude = -1.0;

  vec2 id = floor(loc * 0.0625);

  for (int i = -1; i <= 1; i++) {
    for (int j = -1; j <= 1; j++) {
      altitude = max(altitude, localLandAltitude(loc, id + vec2(i, j)));
    }
  }

  return altitude;
}
```

7 Catch The Wind 小遊戲

與海面蒐集波浪不同的是，島嶼高度不是用相加的，而是取最高點，這樣才不會在島嶼重疊的區域因為相加變得異常的高，到這邊島嶼產生就初步完成囉，如圖 7-13：

▲ 圖 7-13　偽隨機產生的島嶼

不過如果只是有這些隨機圓形的島嶼，玩家是有可能可以直接開到外海去，這樣就變得太簡單了，為了避免，我們得在左右側加上『護欄』，這邊的核心是 `abs(loc.x + cos(loc.y) +- 45.0)`，也就是與左右側 x = +-45 的位置，以 cosine 波形隨著 y 值擺動，然後用 `log(max())` 形成左右兩條島嶼，而 `clamp(loc.y * -0.375 - 3.0, 0.0, 3.0)` 的部份則讓這條護欄從玩家出發點漸漸浮出水面：

```
// landMapFragmentShaderSource ...
float landAltitude(vec2 loc) {
  // id, 3x3 calling localLandAltitude() ...

  altitude = max(
    altitude,
    log(max(
      clamp(loc.y * -0.375 - 3.0, 0.0, 3.0) -
        abs(loc.x + cos(loc.y) - 45.0),
      0.01
    ))
```

```
);
altitude = max(
  altitude,
  log(max(
    clamp(loc.y * -0.375 - 3.0, 0.0, 3.0) -
      abs(loc.x + cos(loc.y) + 45.0),
    0.01
  ))
);

return altitude;
}
```

產生的地圖如圖 7-14 所示，可以在文字看板上看到左右側各有一條『護欄』。

▲ 圖 7-13　借用文字看板來觀察，左右兩條避免玩家開到外海的護欄正確產生了

7 Catch The Wind 小遊戲

⌘ 島嶼隨著捲軸捲動浮出水面

現在是一次全部繪製出來,如果就保持這樣,到時候遊玩往前推進到 `app.state.level` 增加並更新完 `textures.landMap` 地圖時會看到一整個 chunk 直接出現;為了製造捲軸效果,再度回到渲染地形、島嶼的 `landVertexShaderSource`,加上 `u_landFarthest` `uniform` 控制最遠可見那條線的位置,並且在 `getAltitudePosition()` 輸出 `altitude` 之前利用 `min()` 取最小值把超過這條線的高度壓下去:

```
// landVertexShaderSource ...
uniform vec2 u_landMapOffset;
uniform vec2 u_landOffset;
uniform float u_landFarthest;

// ...

vec3 getAltitudePosition(vec4 pos) {
  // location, altitude ...

  altitude = min(
    altitude,
    location.y * 0.333333 +
      u_landFarthest * -0.333333 - 1.0
  );

  return vec3(location.x, altitude, location.y);
}
```

因為航行方向是 `location` 的 -y,`location.y` 大於 `u_landFarthest` 的區域才是可見的,這邊同時也給了一些斜率 0.333333 讓島嶼『漸漸浮出水面』。

接著把 uniform `u_landFarthest` 在 `renderLand()` 補上:

```
function renderLand(app, viewMatrix, programInfo) {
  // ...
  twgl.setUniforms(programInfo, {
    // ...
```

7-34

```
    u_landMapSize: LAND_MAP_SIZE,
    u_landMapOffset: getLandMapOffset(app),
    u_landFarthest: (
      state.sailboatLocation[1] -
        LAND_CHUNK_SIZE * (LAND_CHUNKS - 1.5)
    ),
  });
  // ...
}
```

照圖 7-5 所說，由帆船位置往前推進 `(LAND_CHUNKS - 1.5) = 1.5` 倍的 `LAND_CHUNK_SIZE`（= 96），但是這個帆船位置 `state.sailboatLocation[1]` 目前在 state 中還沒加入，順手把它加上：

```
function initGame(app) {
  app.state = {
    // ...
    sailboatLocation: [0, 0],
  }

  renderLandMap(app);
}
```

這麼一來，Catch the wind 小遊戲的地圖部份就大功告成，移動到最遠的一端看回去，可以看到整體產生出來的島嶼、護欄以及浮出水面的樣子，如圖 7-14：

▲ 圖 7-14 Catch the wind 小遊戲的地圖部份完成，可以看見離帆船最遠的護欄漸漸沒入水中

7 Catch The Wind 小遊戲

　　與此同時，遊戲核心邏輯的第一個狀態 — 帆船位置也加入了，接下來會繼續實做遊戲邏輯，到此進度的完整程式碼可以在這邊找到：

</>	ch7/03/index.html
</>	ch7/03/main.js

也可以參考上線版本，透過瀏覽器打開此頁面：
https://webgl-book.pastleo.me/ch7/03/index.html

▲ 圖 7-14　從起始點拉遠觀察地形

7-3　Set Sail! 航行帆船

　　有了 `app.state.sailboatLocation` 狀態，老實講要讓帆船動起來其實只需要把事件串接到 `sailboatLocation` 的變化，接著讓 `renderSailboat()` 使用這個狀態平移帆船，我們先做鍵盤事件就好，畢竟還沒有製作畫面上的按鈕來按。

7-36

鍵盤事件的部份,就是監聽 keydown、keyup 事件,在遊戲初始化時開始監聽:

```
function initGame(app) {
  // app.state, renderLandMap ...
  document.addEventListener('keydown', event => {
    if (event.code === 'KeyA' || event.code === 'ArrowLeft') {
      pressDirection(app, 'left');
    } else if (event.code === 'KeyD' || event.code === 'ArrowRight') {
      pressDirection(app, 'right');
    }
  });
  document.addEventListener('keyup', event => {
    if (event.code === 'KeyA' || event.code === 'ArrowLeft') {
      releaseDirection(app, 'left');
    } else if (event.code === 'KeyD' || event.code === 'ArrowRight') {
      releaseDirection(app, 'right');
    }
  });
}
```

當按下『方向鍵左』、『方向鍵右』、『A』、『D』這幾個按鍵時,執行 `pressDirection()`,放開時執行 `releaseDirection()`,在按下、放開時皆傳入方向字串,因為在之後會加入按鈕觸控等更多輸入,因此在此加入這幾個 function 讓不同的事件監聽可以共用,接著實做這些 function:

```
function pressDirection(app, direction) {
  if (app.state.directionPresseds[direction]) return;
  app.state.sailing = direction;
  app.state.directionPresseds[direction] = true;
}
function releaseDirection(app, direction) {
  app.state.directionPresseds[direction] = false;

  if (app.state.directionPresseds.left) {
    app.state.sailing = 'left';
  } else if (app.state.directionPresseds.right) {
```

```
    app.state.sailing = 'right';
  } else {
    app.state.sailing = false;
  }
}
// initGame ...
```

`app.state.directionPresseds` 將會是一個包含 `.left`、`.right` 兩個 key 的 Javascript 物件，紀錄按下的按鈕，在 `pressDirection()`、`releaseDirection()` 中操作此物件，因為鍵盤按鍵按著時事件會不斷產生，在設定航行方向之前會先檢查是否已經按下，`releaseDirection()` 除了解除 `app.state.directionPresseds` 對應方向的按下狀態，也要查看 `app.state.directionPresseds` 是否還有按下的按鈕，設定到 `app.state.sailing` 狀態上，如果 `.left`、`.right` 皆為 `false`，則表示所有按鈕都已經放開，設定 `app.state.sailing` 為 `false`。

為什麼要設計一個 `app.state.directionPresseds` 呢？如果考慮一個快速切換左右可能會產生的情況：先按下左，再按下右，接著放開右，如果這時候希望方向可以回復到左，會需要一個資料結構紀錄先前按下的按鈕，以便在放開時能夠回到先前的狀態；把這些狀態加入 `app.state` 當中：

```
function initGame(app) {
  app.state = {
    // fieldOfView, cameraRotationXY ...

    seed: Math.random() * 20000 + 5000,
    level: 0,

    directionPresseds: { left: false, right: false },
    sailing: false,
    sailboatLocation: [0, 0],
  };
  // ...
}
```

⌘ 遊戲進行 gameUpdate()

好的，接下來在每個 frame 更新時看 `app.state.sailing` 是往哪邊，修改帆船位置 `app.state.sailboatLocation` ，而這個『每個 frame 更新』的程式就是遊戲的核心邏輯：

```javascript
// initGame ...
function gameUpdate(app, timeDiff, now) {
  const { state } = app;

  if (state.sailing) {

    let direction;
    if (state.sailing === 'left') {
      direction = -SAIL_DIRECTION_RAD;
    } else if (state.sailing === 'right') {
      direction = SAIL_DIRECTION_RAD;
    }

    state.sailboatLocation[0] +=
      timeDiff * MAX_VELOCITY * Math.sin(direction);
    state.sailboatLocation[1] -=
      timeDiff * MAX_VELOCITY * Math.cos(direction);
  }
}
```

`gameUpdate()` 除了 `app` 之外接收 `timeDiff`、`now` 兩個時間相關參數，這邊的實做根據應用程式常數 `SAIL_DIRECTION_RAD` 偏移角度，並先以最大速度 MAX_VELOCITY 移動帆船，加上這兩個常數並且在 `startLoop()` 裡頭呼叫：

```javascript
// import ...
const LAND_CHUNK_SIZE = 96;
const LAND_CHUNKS = 3;
const LAND_MAP_SIZE = [LAND_CHUNK_SIZE, LAND_CHUNK_SIZE * LAND_CHUNKS];
const MAX_VELOCITY = 0.05;
const SAIL_DIRECTION_RAD = degToRad(20);
```

7 Catch The Wind 小遊戲

```
// ...

function startLoop(app, now = 0) {
  const timeDiff = now - app.time;
  app.time = now;

  inputUpdate(app.input, app.state);
  gameUpdate(app, timeDiff, now);

  render(app, timeDiff);
  requestAnimationFrame(now => startLoop(app, now));
}
```

因為 `MAX_VELOCITY` 會乘上 `timeDiff`，意思是說 `MAX_VELOCITY` 的單位是每毫秒移動的單位長，所以 0.05 其實已經很快，在這邊同時也移除 `inputUpdate()` 以免鍵盤操作去移動視角，這樣一來就會隨著按鍵移動 `app.state.sailboatLocation`，把這狀態傳接到帆船繪製上：

```
function renderSailboat(app, viewMatrix, programInfo) {
  const { gl, textures, objects, state, time } = app;

  const worldMatrix = matrix4.multiply(
    matrix4.yRotate(degToRad(45)),
    matrix4.translate(
      state.sailboatLocation[0], 0, state.sailboatLocation[1]
    ),
    matrix4.xRotate(Math.sin(time * 0.0011) * 0.03 + 0.03),
    matrix4.translate(0, Math.sin(time * 0.0017) * 0.05, 0),
  );

  // ...
}
```

帆船要往 -z 的方向航行，把之前為了展示轉向 45 度的 transform 移除，存檔回到瀏覽器按下左右鍵，可以看到帆船很快速的駛離畫面，在帆船還沒到太遠的地方之前稍微拉遠還可以看到奇怪的陰影：

7-40

7-3　Set Sail! 航行帆船

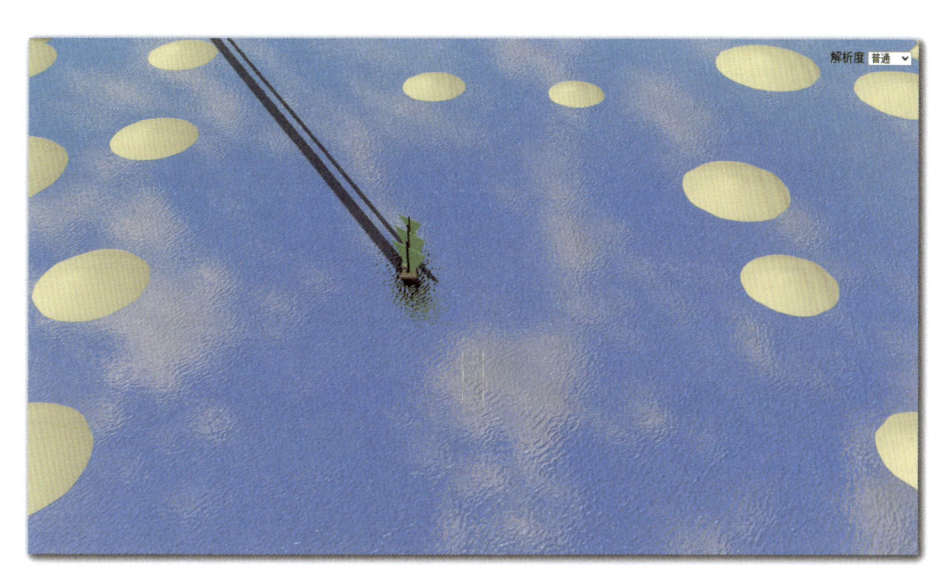

▲ 圖 7-15　帆船產生的陰影不正確

　　第一個問題是鏡頭應該要跟著帆船移動，這個不會很難，把帆船位置同步到 `app.state.cameraViewing`；陰影的部份，則需要讀著回想一下 CH5 陰影的實做，從光源拍攝深度照片時會捕捉場景中 xz 介於 -10 ~ +10，y 介於 -5 ~ +5 的物件，最好的方法就是直接移動拍攝深度照片的相機位置；最後還有一個，那就是海面物件，目前設定成 4000x4000 的大小，但是還是有機會船開一開開出去的，我們可以很簡單讓海面隨著帆船的平移來避免這個問題，對於這三個問題修改程式碼：

```javascript
function renderOcean(app, viewMatrix, reflectionViewMatrix, programInfo) {
  const { gl, textures, objects, state } = app;
  // ...
  const worldMatrix = matrix4.multiply(
    matrix4.translate(
      state.sailboatLocation[0], 0, state.sailboatLocation[1],
    ),
    matrix4.scale(4000, 1, 4000),
  );
  // setUniforms, drawBufferInfo ...
}
```

7-41

7 Catch The Wind 小遊戲

```javascript
// ...
function render(app) {
  // ...
  const lightProjectionViewMatrix = matrix4.multiply(
    // projection, shearing ...
    matrix4.inverse(
      matrix4.multiply(
        matrix4.translate(
          state.sailboatLocation[0], 0, state.sailboatLocation[1],
        ),
        matrix4.yRotate(state.lightRotationXY[1]),
        matrix4.xRotate(degToRad(90)),
      )
    ),
  );
  // ...
}
// ...
function gameUpdate(app, timeDiff, now) {
  const { state } = app;

  if (state.sailing) {
    // direction ...

    state.sailboatLocation[0] +=
      timeDiff * MAX_VELOCITY * Math.sin(direction);
    state.sailboatLocation[1] -=
      timeDiff * MAX_VELOCITY * Math.cos(direction);
    state.cameraViewing =
      [state.sailboatLocation[0], 0, state.sailboatLocation[1]];
  }
}
```

經過這三個修正過後,就可以正式啟航了!如圖 7-16 把帆船開到左側:

7-42

7-3　Set Sail! 航行帆船

▲ 圖 7-16　正式啟航！

⌘ 開、關帆的動畫演出

　　雖然可以駕駛帆船航行，但是帆船顯得很死板，這個小遊戲名稱『Catch the wind』就是要控制風帆乘風前進，當玩家按下向左、向右時打開風帆，放開時關閉，同時也讓開關風帆具有轉換狀態的動畫效果，有了這樣的動畫演出就可以讓帆船航行不會這麼死板，也看起來比較自然。

　　帆船 3D 物件是一個個不同材質的子物件所集成的陣列，這些材質各有一個名稱由 `.mtl` 檔案定義，筆者有調整過帆船 3D 模型與材質檔案使得在 Javascript 可以利用材質名稱來區分特定的子物件，在這邊我們關注的是風帆，而且分成材質名稱為 `sails`、`sails-front` 兩種風帆，如圖 7-17：

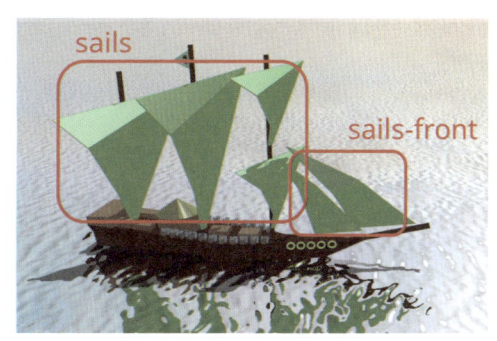

▲ 圖 7-17　風帆的材質分為主風帆 sails 與前風帆 sails-front

7-43

7 Catch The Wind 小遊戲

回去查看 `WebGLObjLoader.downloadModels()` 讀取帆船模型的結果 `boatModel`，可以看到 `boatModel.materialsByIndex` 的各個材質中有一個 `.name` 的屬性可以讓我們判斷，如圖 7-18：

```
▼ n {name: 'boatModel', indicesPerMaterial: Array(10), materialsByIndex: {…}, tangents: Array(0), bitangents: Array(0), …}
  ▶ bitangents: []
  ▶ indices: (34560) [14290, 14291, 14292, 14292, 14293, 14290, 14294, 14295, 14296, 14296,
  ▶ indicesPerMaterial: (10) [Array(168), Array(552), Array(186), Array(1488), Array(240), A
  ▶ materialIndices: {body: 0, building: 1, building_roof: 2, mast: 3, sails: 4, …}
  ▶ materialNames: (10) ['body', 'building', 'building_roof', 'mast', 'sails', 'sails-front'
  ▼ materialsByIndex:
    ▶ 0: l {name: 'body', ambient: Array(3), diffuse: Array(3), specular: Array(3), emissive
    ▶ 1: l {name: 'building', ambient: Array(3), diffuse: Array(3), specular: Array(3), emis
    ▶ 2: l {name: 'building_roof', ambient: Array(3), diffuse: Array(3), specular: Array(3),
    ▶ 3: l {name: 'mast', ambient: Array(3), diffuse: Array(3), specular: Array(3), emissive
    ▶ 4: l {name: 'sails', ambient: Array(3), diffuse: Array(3), specular: Array(3), emissive
    ▶ 5: l {name: 'sails-front', ambient: Array(3), diffuse: Array(3), specular: Array(3), em
    ▶ 6: l {name: 'flag-pastleo', ambient: Array(3), diffuse: Array(3), specular: Array(3),
    ▶ 7: l {name: 'cannon-wheel', ambient: Array(3), diffuse: Array(3), specular: Array(3),
    ▶ 8: l {name: 'cannon-body', ambient: Array(3), diffuse: Array(3), specular: Array(3), em
    ▶ 9: l {name: 'lifebuoy', ambient: Array(3), diffuse: Array(3), specular: Array(3), emis
    ▶ [[Prototype]]: Object
    name: "boatModel"
  ▶ tangents: []
    textureStride: 2
  ▶ textures: (74660) [0.559949, 0, 0.625, 0, 0.625, 0.25, 0.559949, 0.25, 0.559949, 0.25, 0
  ▶ vertexMaterialIndices: (37330) [0, 0, 0, 0, 0, 0, 0, 0, 0, 0, 0, 0, 0, 0, 0, 0, 0, 0,
  ▶ vertexNormals: (111990) [-0.7759, -0.625, 0.0861, -0.7759, -0.625, 0.0861, -0.7759, -0.6
  ▶ vertices: (111990) [-0.374793, 0.372056, 2.750663, -0.551836, 0.552017, 2.746644, -0.551
  ▶ [[Prototype]]: Object
```

▲ 圖 7-18　boatModel.materialsByIndex 中可以看到 name 為 sails、sails-front 的兩個材質

在讀取階段把這個名字在子物件中放好，之後遊戲開始進行時在 `renderSailboat()` 就可以使用 `.name` 來判斷哪個子物件是主風帆、前風帆：

```
async function loadSailboatModel(gl, textures, programInfo) {
  // ...
  const parts = boatModel.indicesPerMaterial.map((indices, mtlIdx) => {
    const material = boatModel.materialsByIndex[mtlIdx];
    // ...

    return {
```

```
      name: material.name,
      bufferInfo,
      vao: twgl.createVAOFromBufferInfo(gl, programInfo, bufferInfo),
      uniforms: { /* ... */ },
    }
  });
  // ...
}
```

要讓遊戲進行階段能夠讓這兩個材質形成的子物件進行開關，並且要有轉換的動畫，最好的方式就是夠過漸漸變化的 3D transform，例如說關閉主風帆的時候，使用 scale 把 y 縮小，就可以達到收帆樣子，同時透過一個存在 state 中的數字來控制這個 y 的 scale，每個 frame 慢慢減少這個數字來達到轉換的動畫效果；來到 `renderSailboat()`，算出帆船本身的 transform 之後，再為主風帆、前風帆算出他們的 transform：

```
function renderSailboat(app, viewMatrix, programInfo) {
  const { gl, textures, objects, state, time } = app;

  const worldMatrix = matrix4.multiply( /* ... */ );
  const sailWorldMatrix = matrix4.multiply(
    worldMatrix,
    matrix4.translate(0, state.sailTranslateY, 0),
    matrix4.scale(state.sailScaleX, state.sailScaleY, 1),
  );
  const sailFrontWorldMatrix = matrix4.multiply(
    worldMatrix,
    matrix4.scale(
      state.sailFrontScaleXY * state.sailScaleX,
      state.sailFrontScaleXY, 1,
    ),
  );

  // ...
}
```

7 Catch The Wind 小遊戲

`sailWorldMatrix` 將套用在主風帆上，scale 中的 `state.sailScaleY` 以 y 方向縮放，即為控制是否收帆，而 translate 的 `state.sailTranslateY` 則是因為模型會以原點進行縮放，所以也要往上（+y）做平移讓風帆收在桅杆偏上的位置，另外 `sailFrontWorldMatrix` 將套用在前風帆上，這個子物件收起來就直接縮小到 0 就好，最後還可以看到有個 `state.sailScaleX`，這個是用來控制向左還是向右開帆，帆船模型的原始樣子是向右開帆，我們只要讓 scale 以 x 方向縮放 -1 倍，那麼就左右相反向左開帆囉！

3D transform 矩陣準備好，定義對應帆船整體與主風帆、前風帆子物件個別的 uniforms，並且在依序渲染子物件時根據 `name` 決定是否是主風帆、前風帆設定對應的 uniform 來做繪製：

```
function renderSailboat(app, viewMatrix, programInfo) {
  // worldMatrix, sailWorldMatrix, sailFrontWorldMatrix ...

  twgl.setUniforms(programInfo, { /* ... */ });
  const objUniforms = {
    u_matrix: matrix4.multiply(viewMatrix, worldMatrix),
    u_worldMatrix: worldMatrix,
    u_normalMatrix: matrix4.transpose(matrix4.inverse(worldMatrix)),
    u_normalMap: textures.nullNormal,
  };
  const sailUniforms = {
    u_matrix: matrix4.multiply(viewMatrix, sailWorldMatrix),
    u_worldMatrix: sailWorldMatrix,
    u_normalMatrix: matrix4.transpose(matrix4.inverse(sailWorldMatrix)),
  };
  const sailFrontUniforms = {
    u_matrix: matrix4.multiply(viewMatrix, sailFrontWorldMatrix),
    u_worldMatrix: sailFrontWorldMatrix,
    u_normalMatrix: matrix4.transpose(
      matrix4.inverse(sailFrontWorldMatrix)
    ),
  };

  gl.disable(gl.CULL_FACE);
```

```
objects.sailboat.forEach(({ name, bufferInfo, vao, uniforms }) => {
  gl.bindVertexArray(vao);
  twgl.setUniforms(programInfo, {
    ...objUniforms,
    ...(name === 'sails' && sailUniforms),
    ...(name === 'sails-front' && sailFrontUniforms),
    ...uniforms,
  });
  twgl.drawBufferInfo(gl, bufferInfo);
});

gl.enable(gl.CULL_FACE);
}
```

原本帆船整體可以共用 3D transform 矩陣，因此進入在依序繪製子物件的迴圈之前設定，這邊改成一律在迴圈內進行，並且依據 name 知道是否要設定對應的 uniforms；同時可以注意到呼叫了 `gl.disable(gl.CULL_FACE)` 取消『只繪製正面』的功能，這是因為向左航行時把風帆整個以 x 方向縮放 -1 會導致三角性的正面不再對外，為了讓背面對外的三角形能夠被繪製，因此暫時關閉這個功能，並且在繪製完帆船之後開啟回來確保其他物件的渲染維持原行為。

這麼一來繪製流程就調整完畢，回到遊戲核心邏輯 `gameUpdate()`，在『遊戲開始後』，如果 state.sailing 沒有值表示使用者放開方向鍵，則每個 frame 讓 `state.sailScaleY`、`state.sailTranslateY` 等狀態改變一點點，並且以 Javascript 內建函式 Math.max() 或是 `Math.min()` 來限制轉換狀態動畫終點的 3D transform 數值，也就是停在風帆收起來的樣子；如果 `state.sailing` 有值表示使用者按下方向鍵，要把狀態一點一點改變回來，重新展開風帆：

```
function gameUpdate(app, timeDiff, now) {
  const { state } = app;

  if (state.started) {
```

7 Catch The Wind 小遊戲

```javascript
    if (state.sailing) {
      state.sailTranslateY = Math.max(
        state.sailTranslateY - timeDiff * 0.0045, 0,
      );
      state.sailScaleY = Math.min(state.sailScaleY + timeDiff * 0.0018, 1);
      state.sailFrontScaleXY = Math.min(
        state.sailFrontScaleXY + timeDiff * 0.002, 1,
      );

      let direction;
      if (state.sailing === 'left') {
        direction = -SAIL_DIRECTION_RAD;
        state.sailScaleX = -1;
      } else if (state.sailing === 'right') {
        direction = SAIL_DIRECTION_RAD;
        state.sailScaleX = 1;
      }

      // state.sailboatLocation, state.cameraViewing ...
    } else {
      state.sailTranslateY = Math.min(
        state.sailTranslateY + timeDiff * 0.0045, 2.25,
      );
      state.sailScaleY = Math.max(
        state.sailScaleY - timeDiff * 0.0018, 0.1,
      );
      state.sailFrontScaleXY = Math.max(
        state.sailFrontScaleXY - timeDiff * 0.002, 0,
      );
    }
  } else if (state.sailing) {
    state.started = true;
  }
}
```

『遊戲開始後』這個狀態就由 `state.started` 來表示，筆者在此希望之後遊戲讀取完畢的標題畫面還是帆船模型的展示，為了避免遊戲還沒開始玩家沒按下按鈕導致風帆收起來，因此把原本邏輯用一個 `if (state.started)`

7-48

`{ ... }` 包起來，當玩家在遊戲尚未開始時按下方向鍵才開始遊戲；而遊戲開始後，如果有按下方向鍵則依照 frame 之間經過的時間（`timeDiff`）漸漸改變 `state.sailTranslateY`、`state.sailScaleY`、`state.sailFrontScaleXY` 直到開、關風帆到達定點；向左、向右開帆的部份如果要做動畫，用 `state.sailScaleX` 一路從 +1 改變成 -1 的過程會看起來有點奇怪，因此這邊就不做動畫，向右時直接設定為 +1、向左時設定成 -1。

開、關帆的動畫演出需要不少狀態，在 `initGame()` 把初始值設定好：

```javascript
function initGame(app) {
  app.state = {
    // ...

    seed: Math.random() * 20000 + 5000,
    started: false,
    level: 0,

    directionPresseds: { left: false, right: false },
    sailing: false,
    sailTranslateY: 0,
    sailScaleX: 1,
    sailScaleY: 1,
    sailFrontScaleXY: 1,
    sailboatLocation: [0, 0],
  };

  // renderLandMap, addEventListener ...
}
```

回到瀏覽器測試，這時初次按下方向鍵開始航行後，放開方向鍵將會有風帆收起來的動畫，如圖 7-18；如果向左航行也可以看到風帆方向的改變，如圖 7-19：

7 Catch The Wind 小遊戲

▲ 圖 7-18　收起風帆的樣子

▲ 圖 7-19　帆船開帆向左航行

⌘ 向前推進產生新地圖、完成 Infinite Scroll

先前已經在地形、島嶼繪製 shader 中加入 `u_landFarthest` 來控制可見區域的邊緣，並且串上帆船位置加上 1.5 倍的 `LAND_CHUNK_SIZE`，現在把視角調整一下往前航行可以看到在往前航行的同時前方島嶼進入可見區域而浮出水面，但是由於目前不會把地形高度圖往前產生新區域，因此往前航行很快就不會有島嶼繼續浮現，如圖 7-20：

▲ 圖 7-20 地形沒有繼續產生，稍微往前島嶼就不會繼續浮現

因此在往前進的同時，要檢查帆船位置是否已經到達當前地形圖的邊緣，可以再回去查看圖 7-5、圖 7-6，當帆船位置到達當前地形圖中心點的位置時，表示已經可見區域已經捲動到當前地圖的邊緣，該是時候增加 `app.state.level` 並且呼叫 `renderLandMap()` 進行『新區域的產生』：

```
function gameUpdate(app, timeDiff, now) {
  const { state } = app;

  if (state.started) {
    if (state.sailing) {
      // sailTranslateY, sailScaleY, sailFrontScaleXY ...
      // direction ...
```

7 Catch The Wind 小遊戲

```
    // sailboatLocation, cameraViewing ...

    if (state.sailboatLocation[1] < getLandMapOffset(app)[1]) {
      state.level++;
      state.windStrength = Math.min(state.windStrength + 0.005, 0.4);
      console.log('next level!', {
        level: state.level,
        windStrength: state.windStrength,
      });
      renderLandMap(app);
    }
  } else {
    // sailTranslateY, sailScaleY, sailFrontScaleXY ...
  }
} else if (state.sailing) {
  state.started = true;
}
}
```

因為往前是往 `-position.z`、`-location.y` 方向，帆船位置的 y 值『小於』地形圖中心點 y 值觸發，在 `renderLandMap()` 內因為呼叫的 `getLandMapOffset()` 會根據 `app.state.level` 算出地形圖中心點位置，因此呼叫之後便可以更新地形圖往前移動一個 chunk；而程式碼中 `state.windStrength` 是什麼意思？這是筆者想要讓風力強度隨著玩家 `level` 的增加而變強，隨著遊戲繼續下去，帆船速度更快、遊戲難度漸漸提昇，初始數值為 0.1，每個 `level` 增加 0.005，最後到達 0.4，先把初始值寫上：

```
function initGame(app) {
  app.state = {
    // ...
    started: false,
    level: 0,
    windStrength: 0.1,
    // ...
  };
  // ...
}
```

至於為什麼這個 `windStrength` 的數值是 0.1 ~ 0.4 呢？因為風力強度可以反應在海面的波浪高度上，回到海面 fragment shader 加上 uniform `u_windStrength` 改變海面高度的數值調整：

```
// oceanFragmentShaderSource ...
uniform vec3 u_ambient;
uniform sampler2D u_lightProjectionMap;
uniform float u_time;
uniform float u_windStrength;

// ...

float localWaveHeight(vec2 id, vec2 position) {
  float directionRad = radians((hash(id) - 0.5) * 45.0 + 90.0);
  vec2 direction = vec2(cos(directionRad), sin(directionRad));
  // ...
}
// ...
vec3 oceanSurfacePosition(vec2 position) {
  // id, height ...

  height *= u_windStrength;

  return vec3(position, height);
}
```

原本海浪高度在彙整加總之後會乘以 0.15，現在這個數值從 0.1 出發，隨著 `level` 增加最後到 0.5 封頂；順便把海浪主要方向改成與風向一致向著 -z 方向。

最後把 `app.state.windStrength` 串到 uniform `u_windStrength` 上：

```
function renderOcean(app, viewMatrix, reflectionViewMatrix, programInfo) {
  // ...

  twgl.setUniforms(programInfo, {
    // ...
```

7-53

7　Catch The Wind 小遊戲

```
    u_ambient: [0.4, 0.4, 0.4],
    u_reflectionMatrix: reflectionViewMatrix,
    u_windStrength: state.windStrength,
  });

  twgl.drawBufferInfo(gl, objects.plane.bufferInfo);
}
```

現在帆船能夠開關風帆、前進，同時地形、島嶼的 infinite scroll 也就完成囉，在觸發新區域的產生時也會使用 `console.log()` 在 Console 上印出目前的 `level`、`windStrength`，如圖 7-21 一路從 level 0 航行到 level 5：

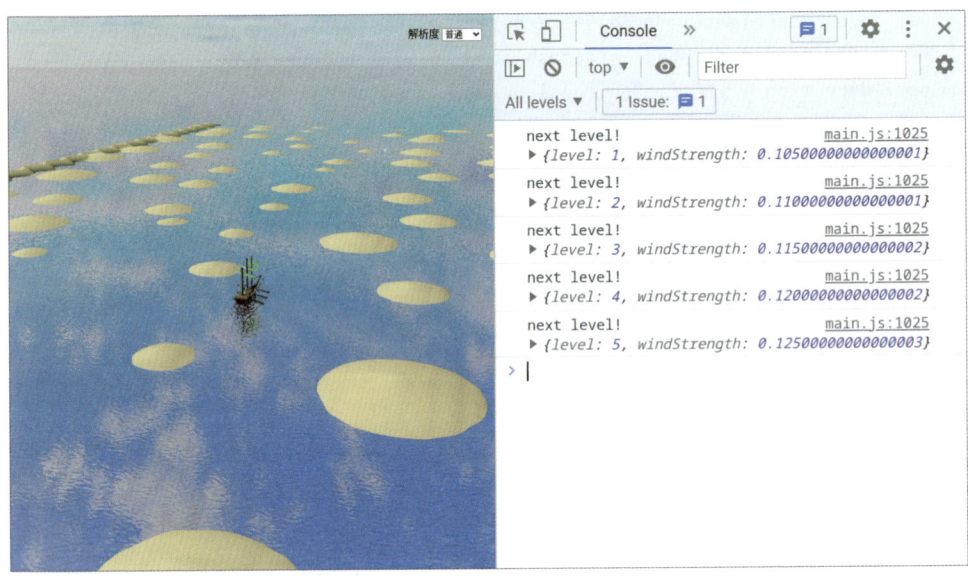

▲ 圖 7-21　地形、島嶼隨著航行持續產生，並印出 level、windStrength

到此進度的完整程式碼可以在這邊找到：

</>	ch7/04/index.html
</>	ch7/04/main.js

7-54

也可以參考上線版本，透過瀏覽器打開此頁面：
https://webgl-book.pastleo.me/ch7/04/index.html

帆船前進觸發新區域的產生時，走過的區域也會因為往前推進而消失，讀者可以試著視角轉往後看並航行帆船，觀察地形圖隨著 chunk 推進的區域範圍，如圖 7-22：

▲ 圖 7-22　往前航行觸發新區域產生，舊區域也就跟著消失

⌘ 加速度、慣性與摩擦力

現在帆船的移動方式很不真實，不是按鍵按下後 state.sailing 有值，每毫秒讓帆船移動 MAX_VELOCITY 的距離，就是放開按鍵，帆船直接靜止，比較真實的狀況應該是開帆之後速度慢慢增加並維持慣性，在放開按鍵收帆時因為摩擦力慢慢降低速度。

先把帆船速度 sailboatVelocity 加到初始狀態中：

```
function initGame(app) {
  app.state = {
    // ...
```

7 Catch The Wind 小遊戲

```
    sailFrontScaleXY: 1,
    sailboatLocation: [0, 0],
    sailboatVelocity: [0, 0],
  };
  // ...
}
```

來到 `gameUpdate()`，有按下方向鍵時增加速度：

```
function gameUpdate(app, timeDiff, now) {
  const { state } = app;

  if (state.started) {

    if (state.sailing) {
      // ...
      let direction;
      if (state.sailing === 'left') {
        direction = -SAIL_DIRECTION_RAD;
        state.sailScaleX = -1;
      } else if (state.sailing === 'right') {
        direction = SAIL_DIRECTION_RAD;
        state.sailScaleX = 1;
      }
      state.sailboatVelocity[0] += (
        state.windStrength * timeDiff * ACCELERATION * Math.sin(direction)
      );
      state.sailboatVelocity[1] -= (
        state.windStrength * timeDiff * ACCELERATION * Math.cos(direction)
      );

      // ...
    } else {
      // ...
    }
  } // ...
}
```

7-56

增加速度的量，可以看到除了由加速度 `ACCELERATION` 應用程式常數決定之外，也會受到 `state.windStrength` 影響，使得航行越遠、加速度越快、難度越高。

原本 `state.sailboatLocation` 只有在按下方向按鈕時會有變化，現在因為慣性的緣故，無論是否有按下按鈕都要依據速度 `state.sailboatVelocity` 算出新的帆船位置 `state.sailboatLocation`：

```javascript
function gameUpdate(app, timeDiff, now) {
  const { state } = app;

  if (state.started) {

    if (state.sailing) {
     // ... direction ...
      state.sailboatVelocity[0] += (
        state.windStrength * timeDiff * ACCELERATION * Math.sin(direction)
      );
      state.sailboatVelocity[1] -= (
        state.windStrength * timeDiff * ACCELERATION * Math.cos(direction)
      );

      state.sailboatLocation[0] +=
        timeDiff * MAX_VELOCITY * Math.sin(direction);
      state.sailboatLocation[1] -=
        timeDiff * MAX_VELOCITY * Math.cos(direction);
      state.cameraViewing =
        [state.sailboatLocation[0], 0, state.sailboatLocation[1]];

      if (state.sailboatLocation[1] < getLandMapOffset(app)[1]) {
       // level, windStrength, renderLandMap() ...
      }
    } else {
     // ...
    }

    state.sailboatLocation[0] += state.sailboatVelocity[0] * timeDiff;
    state.sailboatLocation[1] += state.sailboatVelocity[1] * timeDiff;
```

7 Catch The Wind 小遊戲

```
    state.cameraViewing =
      [state.sailboatLocation[0], 0, state.sailboatLocation[1]];

    if (state.sailboatLocation[1] < getLandMapOffset(app)[1]) {
      // level, windStrength, renderLandMap() ...
    }
  } // ...
}
```

  `state.sailboatLocation` 的移動量為每毫秒 `state.sailboatVelocity`，除此之外，相機視角的同步與是否要產生新地形區域的檢查也得跟著 `state.sailboatLocation` 的變化移動出來。

  再來是摩擦力減速的部份，筆者設計一個 `calcSlowDownRatio()` function，傳入帆船速度，回傳一個小於等於 1 的倍率來調整帆船速度的 xy：

```
// gameUpdate ...
function calcSlowDownFactor(velocity, timeDiff) {
  const velocityLength = Math.sqrt(
    velocity[0] * velocity[0] + velocity[1] * velocity[1]
  );
  if (velocityLength > MAX_VELOCITY) {
    return MAX_VELOCITY / velocityLength;
  }
  const slowDownLength = timeDiff * DEACCELERATION;
  if (velocityLength < slowDownLength) return 0;
  return (velocityLength - slowDownLength) / velocityLength;
}
```

  透過畢氏定理[5]算出二維 `velocity` 的長度 `velocityLength`，如果速度已經大於最大速度 `MAX_VELOCITY`，回傳把速度限制在最大速度的倍率，否則算出這個 frame 要減少的速度量 `slowDownLength`，如果速度已經小於這個要減少的速度量，則回傳 0 讓速度歸零，要不然就是速度不會太快也不會太慢的狀況，使用 `(velocityLength - slowDownLength) / velocityLength` 算出當

---

[5] https://zh.wikipedia.org/zh-tw/畢氏定理

前速度減少之後與傳入速度 `velocityLength` 的比例作為降速的倍率；回到 `gameUpdate()` 在使用速度改變帆船位置之前呼叫並調整速度：

```
function gameUpdate(app, timeDiff, now) {
  // ...
  if (state.started) {
    if (state.sailing) {
      // ... direction ...
      // sailboatVelocity ...
    } else {
      // ...
    }

    const slowDownFactor = calcSlowDownFactor(
      state.sailboatVelocity, timeDiff,
    );
    state.sailboatVelocity[0] *= slowDownFactor;
    state.sailboatVelocity[1] *= slowDownFactor;

    state.sailboatLocation[0] += state.sailboatVelocity[0] * timeDiff;
    state.sailboatLocation[1] += state.sailboatVelocity[1] * timeDiff;

    // cameraViewing ...
    // getLandMapOffset(), renderLandMap() ...
  } // ...
}
```

最後把加速度、減速度這兩個應用程式常數寫上：

```
// imports ...
const MAX_VELOCITY = 0.05;
const ACCELERATION = 0.00025;
const SAIL_DIRECTION_RAD = degToRad(20);
const DEACCELERATION = 0.00002;
```

這麼一來，航行帆船的功能就完成了！按下方向鍵打開風帆乘風開始加速，放開收帆時具有慣性並減速，到此進度的完整程式碼可以在這邊找到：

7 Catch The Wind 小遊戲

</>	ch7/05/index.html
</>	ch7/05/main.js

也可以參考上線版本，透過瀏覽器打開此頁面：
https://webgl-book.pastleo.me/ch7/05/index.html

7-4 遊戲標題與 UI

稍早我們拿文字看板來顯示地形、島嶼圖是為了開發、測試用，現在地形功能已經完成，也是時候讓遊戲看起來更『完整』一些了，而不是網頁載入後顯示如圖 7-23 這樣，除了為了測試目的的地圖顯示，畫面中的帆船也是屁股對著玩家：

▲ 圖 7-23　目前進度網頁載入後所顯現的畫面

7-60

7-4 遊戲標題與 UI

為了讓遊戲看起來更完整，接下來對程式碼修改，讓網頁載入後，映入眼簾的標題畫面以展示的角度看著帆船，並且配上文字看板顯示遊戲規則說明，在遊戲開始時做一個進場效果，從展示的角度旋轉鏡頭換成看著航行的方向；同時用 HTML 製作 UI（使用者界面），顯示目前航行的距離、時間，以及觸控裝置的左右按鈕。

⌘ 標題、遊戲說明文字看板

遊戲標題文字看板內容大致設計如下，第一行『Catch the Wind!』為遊戲名稱、標題，剩下的則是遊戲玩法簡單說明：

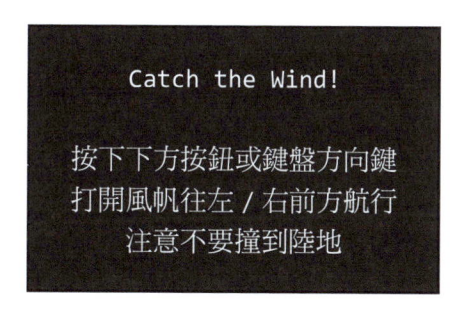

CH7-1 為了將地形圖繪製在文字看板上，修改了文字看板的 shader 使用整數型態的 texture，先把這個 shader 還原成一般顏色型態的 texture：

```
// textFragmentShaderSource ...
precision highp float;
precision highp isampler2D;

in vec2 v_texcoord;

uniform vec4 u_bgColor;
uniform sampler2D u_texture;

out vec4 outColor;

void main() {
  outColor = u_bgColor + texture(u_texture, v_texcoord);
}
```

7-61

7 Catch The Wind 小遊戲

考慮到之後遊戲結束時也會需要用這個文字看板來顯示結算，因此把 `createTextTexture()` 的文字內容產生部份拆分出去，改接收一個繪製好的 `canvas` 參數為其建立 WebGL texture：

```
function createTextTexture(gl, canvas) {
  const canvas = document.createElement('canvas');
  // ...

  const ctx = canvas.getContext('2d');
  // ctx.clearRect(), ctx.fillText( ... ) ...

  const texture = gl.createTexture();
  gl.bindTexture(gl.TEXTURE_2D, texture);

  gl.pixelStorei(gl.UNPACK_FLIP_Y_WEBGL, true);

  gl.texImage2D(
    gl.TEXTURE_2D,
    0, // level
    gl.RGBA, // internalFormat
    gl.RGBA, // format
    gl.UNSIGNED_BYTE, // type
    canvas, // data
  );
  gl.generateMipmap(gl.TEXTURE_2D);

  gl.pixelStorei(gl.UNPACK_FLIP_Y_WEBGL, false);

  return texture;
}
```

獨立寫一個回傳畫好標題文字看板內容 canvas 的 function，原理與 CH6-3 相同，使用 canvas2D[6] API：

```
// createTextTexture ...
function createWelcomeCanvas() {
```

6　https://developer.mozilla.org/en-US/docs/Web/API/CanvasRenderingContext2D

```js
const canvas = document.createElement('canvas');
canvas.width = 1024;
canvas.height = 1024;

const ctx = canvas.getContext('2d');
ctx.fillStyle = 'white';
ctx.textAlign = 'center';
ctx.textBaseline = 'middle';

ctx.font = 'bold 100px serif';
ctx.fillText('Catch The Wind!', canvas.width / 2, canvas.height / 5);

const secondBaseLine = 3 * canvas.height / 5;
const secondLineHeight = canvas.height / 7;
ctx.font = 'bold 70px serif';
ctx.fillText(
  '按下下方按鈕或鍵盤方向鍵',
  canvas.width / 2, secondBaseLine - secondLineHeight,
);
ctx.fillText(
  '打開風帆往左/右前方航行',
  canvas.width / 2, secondBaseLine,
);
ctx.fillText(
  '注意不要撞到陸地',
  canvas.width / 2, secondBaseLine + secondLineHeight,
);

return canvas;
}
```

在 setup() 中組裝起來：

```js
async function setup() {
  // textures ...

  const welcomeCanvas = createWelcomeCanvas();
  textures.text = createTextTexture(gl, welcomeCanvas);
  // ...
}
```

7 Catch The Wind 小遊戲

並且在 renderText() 內把改回 textures.text：

```javascript
function renderText(app, viewMatrix, programInfo) {
  // ... textLeftShift ...
  const worldMatrix = matrix4.multiply(
    matrix4.translate(textLeftShift, 0, 0),
    matrix4.xRotate(degToRad(45)),
    matrix4.translate(0, 10, 0),
    matrix4.scale(1, 1, LAND_CHUNKS),
  );

  twgl.setUniforms(programInfo, {
    u_matrix: matrix4.multiply(viewMatrix, worldMatrix),
    u_bgColor: [0, 0, 0, 0.2],
    u_texture: textures.text,
  });

  twgl.drawBufferInfo(gl, objects.plane.bufferInfo);
}
```

在此也取消先前為了符合地形高低圖比例的縮放，這麼一來文字看板就有做為『遊戲標題』的功能了，如圖 7-24：

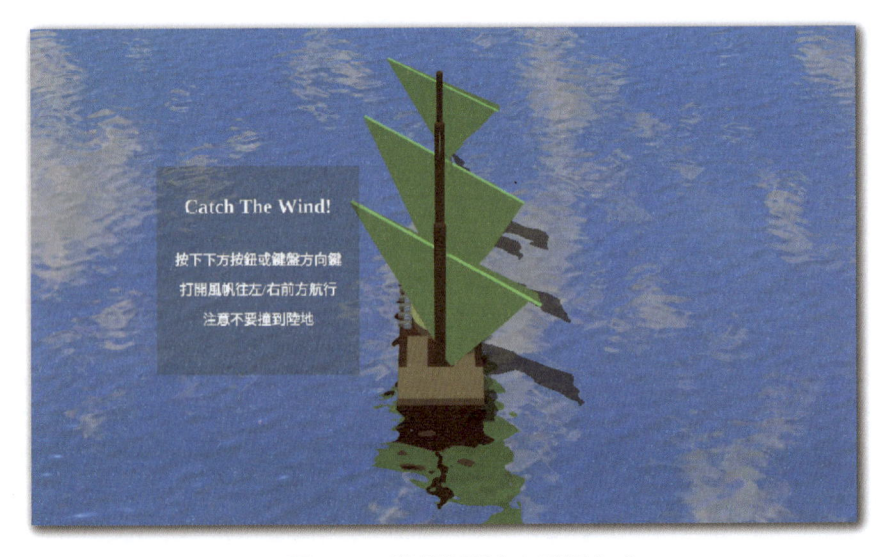

▲ 圖 7-24 遊戲標題文字看板完成

7-64

7-4 遊戲標題與 UI

⌘ 開場展示帆船、遊戲開始進場效果

筆者稍微使用現有的視角操作試了幾個角度，找到這個角度可以看到帆船風帆的正面、背景也可以有幾個小島：

```
function initGame(app) {
  app.state = {
    fieldOfView: degToRad(45),
    cameraRotationXY: [degToRad(-15), degToRad(135)],
    cameraDistance: 15,
    cameraViewing: [0, 0, 0],
    lightRotationXY: [degToRad(20), degToRad(-60)],
    resolutionRatio: 1,
    // ...
  };
  // ...
}
```

使用這個角度設定的開場畫面看起來如下圖 7-25：

▲ 圖 7-25　開場以展示角度看著帆船

7-65

7 Catch The Wind 小遊戲

但是總不能這樣航行帆船吧，根本沒有看著航行方向，在此也運用到每個 frame 慢慢改變狀態數值的方式，在玩家按下方向鍵開始遊戲後，也就是 `app.state.started` 為 true 時，把鏡頭轉向看著 -z 方向：

```javascript
function gameUpdate(app, timeDiff, now) {
  const { state } = app;

  if (state.started) {
    state.cameraRotationXY[1] = (
      Math.max(state.cameraRotationXY[1] - timeDiff * 0.0045, 0)
    );
    state.cameraDistance = Math.min(
      state.cameraDistance + timeDiff * 0.01, 25,
    );

    if (state.sailing) {
      // ... direction, sailboatVelocity ...
    } // ...
    // slowDownFactor, sailboatLocation ...
    // level, renderLandMap() ...
  } else if (state.sailing) {
    state.started = true;
  }
}
```

除了轉向從 135 轉到 0 度之外，也把鏡頭從 15 拉遠到 25 使得玩家可以看到更寬廣的地圖區域，如圖 7-26：

▲ 圖 7-26　開始遊戲、進場後的視角

想必讀者也已經注意到了，那就是文字看板應該要符合開場標題畫面展示的角度，再次回到 renderText() 調整角度：

```javascript
function renderText(app, viewMatrix, programInfo) {
  const { gl, textures, objects } = app;

  gl.bindVertexArray(objects.plane.vao);

  const textLeftShift =
    gl.canvas.width / gl.canvas.height < 1.4 ? 0 : -1.6;
  const worldMatrix = matrix4.multiply(
    matrix4.translate(textLeftShift, 0, 0),
    matrix4.yRotate(degToRad(135)),
    matrix4.xRotate(degToRad(75)),
    matrix4.translate(0, 12.5, 0),
  );

  // setUniforms(), drawBufferInfo() ...
}
```

7-67

7 Catch The Wind 小遊戲

這麼一來標題畫面就更完整了，如圖 7-27；遊戲開始後文字看板也不會擋到玩家視線，如圖 7-28：

▲ 圖 7-27　修正文字看板的位置後的遊戲標題畫面

▲ 圖 7-28　開始遊戲、進場後文字看板不再擋住視線

⌘ 遊戲 UI 與觸控裝置的按鈕

觸控裝置通常沒有鍵盤可以按,因此在畫面上增加兩個按鈕使得手機、平板裝置可以透過畫面上的觸控按鈕來操作帆船風帆的開關、左右,這部份就使用 HTML 來做就可以了,使用系統 emoji 來表示左前、右前:

```html
<!DOCTYPE html>
<html lang="en">
<head>
  <!-- ... -->
</head>
<body>
  <!-- canvas, loading ... -->
  <div id="options">
    <label for="resolution-ratio">解析度</label>
    <select id="resolution-ratio" name="resolution-ratio">
      <!-- 低/普通/Retina -->
    </select>
  </div>
  <div id="ui">
    <button class="sail-btn btn" id="sail-left">↖</button>
    <button class="sail-btn btn" id="sail-right">↗</button>
  </div>
  <script type="module" src="main.js"></script>
</body>
</html>
```

並搭配簡單的 CSS 使這兩個按鈕的容器 `<div id="ui" />` 位置在畫面下方:

```html
<!DOCTYPE html>
<html lang="en">
<head>
  <!-- ... -->
  <style>
    /* html, body, #canvas, #loading, #options */
    #ui {
      position: fixed; bottom: 0; left: 0; right: 0;
      margin: 1rem;
```

7 Catch The Wind 小遊戲

```
      height: 10%;
      display: flex; gap: 1rem;
    }
    @media (max-aspect-ratio: 1/1) {
      #ui { height: 20%; }
    }
    #ui .sail-btn {
      flex: 1;
      font-size: 2rem;
    }
    </style>
</head>
<!-- ... -->
```

`<div id="ui" />` 使用 CSS flexbox[7] 排版，兩個 `sail-btn` 按鈕使用 `flex: 1` 左右平分 50% 的寬度，可以注意到這邊也有使用到 Responsive design[8]，裝置比例較接近直立螢幕時使 `<div id="ui" />` 高度佔螢幕 20%，使按鈕較大、比較好按；客製化左、右按鈕的樣式為半透明：

```
<!DOCTYPE html>
<html lang="en">
<head>
  <!-- ... -->
  <style>
    /* html, body, #canvas */
    .btn {
      background: #00000033;
      border: none; border-radius: 1rem; outline: none;
      cursor: pointer;
    }
    .btn:hover {
      background: #00000040;
    }
```

---

7 https://developer.mozilla.org/en-US/docs/Learn/CSS/CSS_layout/Flexbox
8 https://developer.mozilla.org/en-US/docs/Learn/CSS/CSS_layout/Responsive_Design

```
    .btn.active {
      background: #00000050;
    }
    /* #status, #ui */
  </style>
</head>
<!-- ... -->
```

這邊客製化的是 `btn` 這個 CSS class，讓之後要加入的按鈕可以共用這個 CSS 規則，並且達到風格一致的效果，滑鼠進入按鈕範圍時（hover）降低透明度提醒玩家，按下方向時在按鈕加上 `active` CSS class 來使玩家知道按鈕已經按下，HTML 修改完成後如圖 7-29、7-30：

▲ 圖 7-29　加入向左、向右開帆 UI 按鈕

7 Catch The Wind 小遊戲

▲ 圖 7-30　手機界面的向左、向右開帆 UI 按鈕

　　向左向右的 emoji 圖示可能因為作業系統不同而有所不同；把這些 UI 的按下、放開事件串到之前寫好的 `pressDirection()`、`releaseDirection()`：

```js
function initGame(app) {
  // app.state, renderLandMap() ...
  const sailLeftBtn = document.getElementById('sail-left');
  const sailRightBtn = document.getElementById('sail-right');

  [sailLeftBtn, sailRightBtn].forEach(element => {
    element.addEventListener('touchstart', event => {
      event.preventDefault();
    });
  });
```

```javascript
sailLeftBtn.addEventListener('pointerdown', () => {
  pressDirection(app, 'left');
});
sailRightBtn.addEventListener('pointerdown', () => {
  pressDirection(app, 'right');
});
sailLeftBtn.addEventListener('pointerup', () => {
  releaseDirection(app, 'left')
});
sailRightBtn.addEventListener('pointerup', () => {
  releaseDirection(app, 'right')
});

document.addEventListener('keydown', event => { /* ... */ });
document.addEventListener('keyup', event => { /* ... */ });
}
```

使用 `pointerdown`、`pointerup` 同時兼顧滑鼠、觸控事件,同時對兩個按鈕監聽 `touchstart` 事件並且避免預設行為,例如 android 長按按鈕有可能會模擬滑鼠右鍵選單。現在使用手機、平板按下按鈕已經可以打開風帆航行帆船囉,只是按鈕沒有 CSS class `active` 樣式的『按下』狀態提示,增加一個 `updateDirection()` 在 `pressDirection()`、`releaseDirection()` 之後呼叫,使得使用鍵盤時也可以得到視覺上的回饋:

```javascript
function pressDirection(app, direction) {
  // sailing, directionPresseds ...

  updateDirection(app);
}
function releaseDirection(app, direction) {
  // sailing, directionPresseds ...

  updateDirection(app);
}
```

`updateDirection()` 的實做如下,負責把當下航行方向反應在 HTML 按鈕上:

7 Catch The Wind 小遊戲

```javascript
function updateDirection(app) {
  if (app.state.sailing === 'left') {
    document.getElementById('sail-left').classList.add('active');
    document.getElementById('sail-right').classList.remove('active');
  } else if (app.state.sailing === 'right') {
    document.getElementById('sail-left').classList.remove('active');
    document.getElementById('sail-right').classList.add('active');
  } else {
    document.getElementById('sail-left').classList.remove('active');
    document.getElementById('sail-right').classList.remove('active');
  }
}
```

這麼一來在手機、平板上的按鈕就完成囉，桌上型裝置想要也可以使用這些按鈕，如圖 7-31：

▲ 圖 7-31　從手機等觸控裝置遊玩 Catch the Wind 小遊戲

遊戲最後的計分首先當然要參考航行的距離，同時如果玩家技術高超，同樣的距離使用較短的時間完成，那麼分數也會比較高，這樣的話可以在左

7-74

7-4 遊戲標題與 UI

上角加入兩個數值使玩家知道目前航行的距離以及使用的時間，這部份 UI 也是使用 HTML：

```html
<!DOCTYPE html>
<html lang="en">
<head>
  <!-- ... -->
  <style>
    /* html, body, #canvas, .btn ... */
    #status {
      position: fixed; top: 0; left: 0;
      margin: 1rem;
    }
    /* #options, #ui ... */
  </style>
</head>
<body>
  <canvas id="canvas"></canvas>
  <div id="status">
    <table>
      <tr>
        <td> 距離：</td>
        <td id="status-distance">0</td>
      </tr>
      <tr>
        <td> 時間：</td>
        <td id="status-time">0 秒 </td>
      </tr>
    </table>
  </div>
  <div id="options">
    <!-- 低 / 普通 /Retina -->
  </div>
  <div id="ui">
    <!-- sail-left, sail-right btns -->
  </div>
  <script type="module" src="main.js"></script>
</body>
```

7-75

7 Catch The Wind 小遊戲

建立一個 `updateStatus()` function 把最新的遊戲進度更新到 UI 上：

```javascript
function updateStatus(app) {
  document.getElementById('status-distance').textContent = (
    (-app.state.sailboatLocation[1]).toFixed(2)
  );
  document.getElementById('status-time').textContent = (
    `${((app.time - app.state.startedTime) / 1000).toFixed(1)} 秒`
  );
}
// startLoop ...
```

在 `gameUpdate()` 更新完帆船位置之後，呼叫 `updateStatus()`，同時因為 `updateStatus()` 需要一個遊戲開始的時間，在遊戲開始時把當時的時間（`now`）設定到 `app.state.startedTime`：

```javascript
function initGame(app) {
  app.state = {
    // ...
    seed: Math.random() * 20000 + 5000,
    started: false,
    startedTime: 0,
    level: 0,
    // ...
  };
  // ...
}
// ...
function gameUpdate(app, timeDiff, now) {
  const { state } = app;
  if (state.started) {
    // ...
    if (state.sailing) { /* ... */ }
    // slowDownFactor, sailboatVelocity ...

    state.sailboatLocation[0] += state.sailboatVelocity[0] * timeDiff;
    state.sailboatLocation[1] += state.sailboatVelocity[1] * timeDiff;
    state.cameraViewing =
```

```
      [state.sailboatLocation[0], 0, state.sailboatLocation[1]];

    updateStatus(app);

    // level, renderLandMap() ...
  } else if (state.sailing) {
    state.started = true;
    state.startedTime = now;
  }
}
```

左上角就有個 UI 顯示當前的距離與時間囉，如圖 7-32：

▲ 圖 7-32　簡易的狀態列顯示當前航行距離與時間

　　加上 UI、標題畫面、進場動畫與觸控裝置支援之後，『Catch the Wind!』遊戲現在已經看起來有模有樣了，如圖 7-33、7-34，到此進度的完整程式碼可以在這邊找到：

</>	ch7/06/index.html
</>	ch7/06/main.js

也可以參考上線版本，透過瀏覽器打開此頁面：

https://webgl-book.pastleo.me/ch7/06/index.html

7 Catch The Wind 小遊戲

▲ 圖 7-32 桌上型裝置的標題畫面

▲ 圖 7-33 智慧型手機遊玩擷圖

7-5 碰撞島嶼判定、結束遊戲

遊戲之所以可以稱為遊戲，就是要有一點挑戰，現在根本可以直接開進島裡面：

▲ 圖 7-34　沒有島嶼碰撞判定機制，遊戲不會結束

把遊戲完成還需要一個島嶼碰撞的判斷機制，同時在碰撞島嶼之後進入遊戲結束的狀態，使用文字看板來顯示玩家得分，並且這時候才啟用視角控制功能讓玩家可以環顧四周。

⌘ 判定碰撞島嶼

地形高地圖已經繪製好放置在 `app.textures.landMap`，並且也有 `app.framebuffers.landMap` 作為其畫布，要得知帆船的位置是否高低圖高於水面，需要讀取高低圖上的資料回到 CPU、Javascript，這件事可以透過 WebGL `readPixels()`[9] function 來做到，它會對『當下對著的 framebuffer』讀取資料，且它接收的參數如下：

9　https://developer.mozilla.org/en-US/docs/Web/API/WebGLRenderingContext/readPixels

7 Catch The Wind 小遊戲

```
readPixels(x, y, width, height, format, type, pixels)
```

`x`、`y`、`width`、`height` 這四個參數想當然爾對應了想要取出畫布上什麼範圍的矩形區域，`format`、`type` 應對應 texture 上的型別設定，`pixels` 是一個 Javascript 型別陣列[10]，表示一個記憶體區塊傳入讓 GPU 把讀取好的資料寫回，因此 `readPixels()` 讀取好的資料並不是用回傳值傳送，它沒有回傳值。

開始實做一個專門檢查碰撞的 function，稱為 `checkSailboatCollision()`：

```javascript
// renderLandMap ...
function checkSailboatCollision(app) {
  const { gl, framebuffers, state } = app;

  twgl.bindFramebufferInfo(gl, framebuffers.landMap);

  const landMapOffset = getLandMapOffset(app);

  const sailboatLocInLandMap = [
    state.sailboatLocation[0] - landMapOffset[0],
    state.sailboatLocation[1] - landMapOffset[1],
  ];
}
```

把 framebuffer 對準好 `framebuffers.landMap`，使用 `getLandMapOffset()` 取得當前高低圖中央位置，並且計算帆船與其相對座標用變數 `sailboatLocInLandMap` 存起來；接著定義一組帆船的四個碰撞邊框與帆船的距離，命名為 `SAILBOAT_COLLISION_BORDERS` 做為應用程式常數：

```javascript
// imports ...
const ACCELERATION = 0.00025;
const SAIL_DIRECTION_RAD = degToRad(20);
const DEACCELERATION = 0.00002;
const SAILBOAT_COLLISION_BORDERS =
  [0.5, 0.5, 2.75, 1.4]; // [+x, -x, +z, -z]
```

---

10 https://developer.mozilla.org/zh-TW/docs/Web/JavaScript/Typed_arrays

7-5 碰撞島嶼判定、結束遊戲

　　四個元素分別為從帆船位置出發，往 +x、-x、+z、-z 方向的長度，往船身的左右都是 0.5，往前（-z）只有 1.4，因為帆船模型的前面有部份是在水面上的，往後則是延伸 2.75 單位，把這些碰撞區域繪製出來的話如圖 7-35、圖 7-36 的橘色區域：

▲ 圖 7-35　從側邊看的碰撞邊框　　▲ 圖 7-36　從上方看的碰撞邊框

　　為了算出 `readPixels()` 所需要的 `x`、`y`、`width`、`height`，把邊框對應到 texture、framebuffer 的上下左右算出來：

```javascript
function checkSailboatCollision(app) {
  // twgl.bindFramebufferInfo() ...
  // landMapOffset, sailboatLocInLandMap ...

  const left = Math.floor(
    (
      (sailboatLocInLandMap[0] - SAILBOAT_COLLISION_BORDERS[1]) /
        LAND_MAP_SIZE[0] * 1 + 0.5
    ) * framebuffers.landMap.width
  );
  const right = Math.ceil(
    (
      (sailboatLocInLandMap[0] + SAILBOAT_COLLISION_BORDERS[0]) /
        LAND_MAP_SIZE[0] * 1 + 0.5
    ) * framebuffers.landMap.width
```

```
  );
  const top = Math.floor(
    (
      (sailboatLocInLandMap[1] + SAILBOAT_COLLISION_BORDERS[2]) /
        LAND_MAP_SIZE[1] * -1 + 0.5
    ) * framebuffers.landMap.height
  );
  const bottom = Math.ceil(
    (
      (sailboatLocInLandMap[1] - SAILBOAT_COLLISION_BORDERS[3]) /
        LAND_MAP_SIZE[1] * -1 + 0.5
    ) * framebuffers.landMap.height
  );
}
```

從帆船於地形圖中位置 `sailboatLocInLandMap` 往邊框方向出發之後，除以 `LAND_MAP_SIZE` 得到 -0.5~0.5 的數值，同時因為 z 軸是往 -z 方向，要乘以 -1 反轉，加上 0.5 得到在地圖中的小數位置，最後乘上 texture、framebuffer 原生的解析度並無條件捨去、進位後得到 `left`、`right`、`top`、`bottom`。

`readPixels()` 的 x、y 即為 `left`、`top`，`width`、`height` 也不難算出來：

```
function checkSailboatCollision(app) {
  // twgl.bindFramebufferInfo() ...
  // landMapOffset, sailboatLocInLandMap ...
  // left, right, top, bottom

  const width = Math.max(right - left + 1, 1);
  const height = Math.max(bottom - top + 1, 1);
}
```

接著準備一個 Javascript 型別陣列、記憶體區塊，因為 texture 使用的型別是 32 bits 整數，Javascript 接收端使用 `Int32Array`：

```
function checkSailboatCollision(app) {
  // twgl.bindFramebufferInfo() ...
  // landMapOffset, sailboatLocInLandMap ...
```

```
  // left, right, top, bottom, width, height

  if (!app.sailboatAltitudes) {
    app.sailboatAltitudes =
      new Int32Array(width * height * 2);
  }
}
```

遊戲運作的時候，此 `checkSailboatCollision()` 將會在每個 frame 檢查，筆者希望可以重複使用建立好的記憶體區塊 `Int32Array`，因此這邊有個邏輯：如果 `app.sailboatAltitudes` 尚未建立，建立一個長度為 `width * height * 2` 的 `Int32Array` 到 `app.sailboatAltitudes`；為什麼長度除了長乘以寬之外要乘以 2 呢？因為不同位置的上下左右無條件捨去或是進位時，長寬可能會有所不同，因而加大空間作為緩衝。

好的，給 `gl.readPixels()` 所需的參數都準備好了，請 WebGL、GPU 把帆船碰撞範圍內的高度資料取回：

```javascript
function checkSailboatCollision(app) {
  // twgl.bindFramebufferInfo() ...
  // landMapOffset, sailboatLocInLandMap ...
  // left, right, top, bottom, width, height
  // app.sailboatAltitudes

  gl.readPixels(
    left, top, width, height,
    /* format: */ gl.RED_INTEGER,
    /* type: */ gl.INT,
    app.sailboatAltitudes,
  );
}
```

雖然 `checkSailboatCollision()` 還沒辦法回傳『是否碰撞』結果，不過會把從 GPU 讀取出來的資料放到 `app.sailboatAltitudes` 當中，回到 `gameUpdate()` 中『遊戲開始後』的每個 frame 呼叫：

7 Catch The Wind 小遊戲

```
function gameUpdate(app, timeDiff, now) {
  const { state } = app;

  if (state.started) {
    // ...
    if (state.sailing) { /* ... */ }
    // sailboatVelocity, sailboatLocation, updateStatus() ...

    checkSailboatCollision(app);

    // level, renderLandMap() ...
  } // ...
}
```

把遊戲跑起來試試看，比較 `app.sailboatAltitudes` 的數值變化，圖 7-37 把帆船保持在海面上，可以看到數值皆為負數，表示帆船位置對應高低圖中的位置皆小於 0：

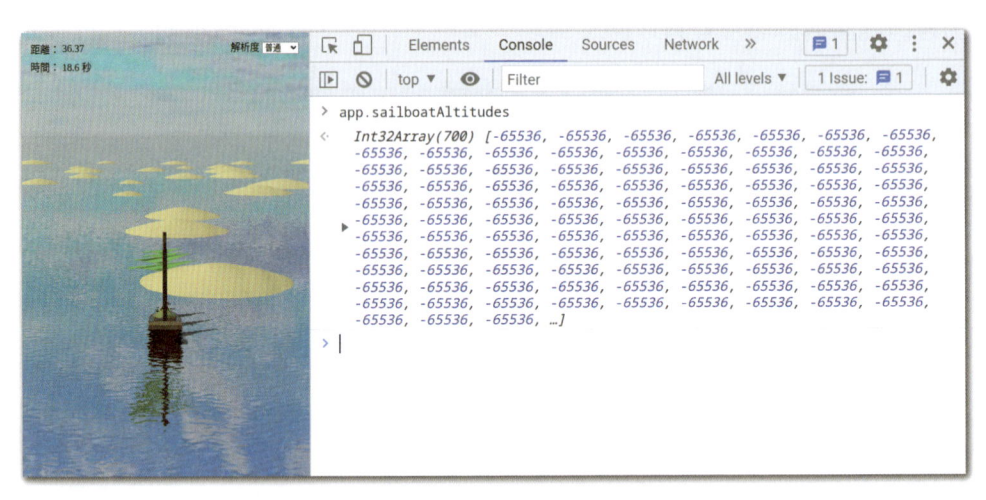

▲ 圖 7-37　在海面上時，地形圖中的高低皆小於 0

7-84

7-5 碰撞島嶼判定、結束遊戲

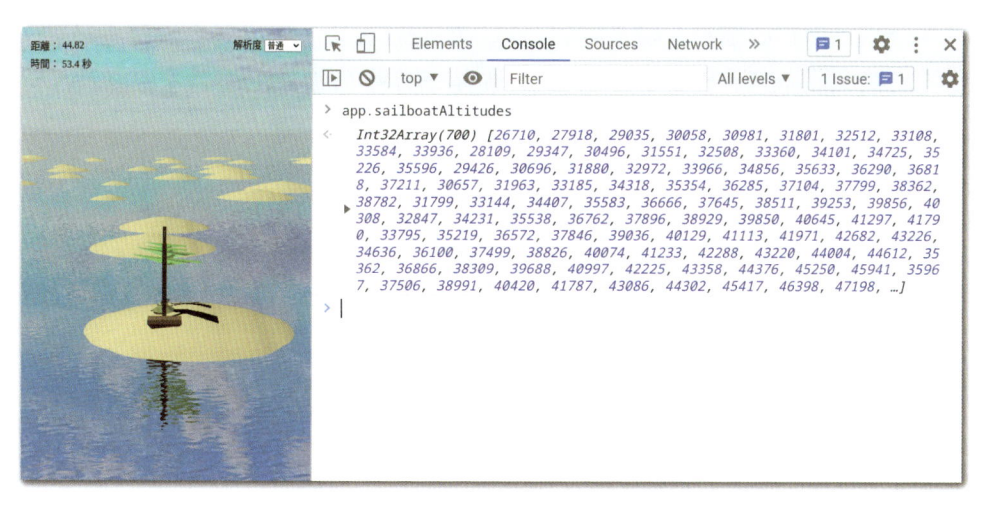

▲ 圖 7-38　開到島嶼中時，數值轉正

當帆船開到島嶼中，如圖 7-38，則可以看到表示高低高於水面的正值出現，這樣一來，在 `checkSailboatCollision()` 可以透過一個迴圈檢查 `app.sailboatAltitudes` 在長寬範圍內的資料是否大於 0，只要有大於 0 則表示碰撞到陸地，回傳 `true` 讓 `gameUpdate()` 可以做後續遊戲結束的處理，如果整個迴圈跑完都沒有沒有大於零，那麼玩家是安全的，回傳 `false` 表示沒有觸碰到島嶼：

```javascript
function checkSailboatCollision(app) {
  // twgl.bindFramebufferInfo() ...
  // landMapOffset, sailboatLocInLandMap ...
  // left, right, top, bottom, width, height
  // app.sailboatAltitudes
  // gl.readPixels()

  for (let i = 0, len = width * height; i < len; i += 1) {
    if (app.sailboatAltitudes[i] > 0) return true;
  }
  return false;
}
```

7-85

7　Catch The Wind 小遊戲

　　回到遊戲核心邏輯，遊戲開始、每次帆船移動後，檢查是否碰撞到島嶼，是的話把 `app.state.gameOver` 設定為 `ture` 表示遊戲結束，否則才檢查是否需要新區域：

```javascript
function gameUpdate(app, timeDiff, now) {
  const { state } = app;

  if (state.started) {
    // ...
    if (state.sailing) { /* ... */ }
    // sailboatVelocity, sailboatLocation, updateStatus() ...

    if (checkSailboatCollision(app)) {
      state.gameOver = true;
      state.score = Math.max(
        Math.floor(
          -state.sailboatLocation[1] -
            4 * (app.time - state.startedTime) / 1000
        ),
        0
      );
    } else if (state.sailboatLocation[1] < getLandMapOffset(app)[1]) {
      // level, renderLandMap() ...
    }
  } // ...
}
```

　　遊戲結束的同時，為玩家計算分數，分數的計算主體是航行距離，減掉時間秒數乘以 4，並且最少有 0 分；不過只有這樣的話遊戲還是會繼續進行，因此把整個遊戲運作時期的邏輯用『沒有遊戲結束』的檢查包起來，確保遊戲結束時真的『結束』：

```javascript
function initGame(app) {
  app.state = {
    // ...
    seed: Math.random() * 20000 + 5000,
    started: false,
```

```
    gameOver: false,
    startedTime: 0,
    level: 0,
    // ...
  };
  // ...
}

function gameUpdate(app, timeDiff, now) {
  const { state } = app;

  if (state.started) {
    if (!state.gameOver) {
      // cameraRotationXY, cameraDistance ...
      if (state.sailing) { /* ... */ }
      // sailboatVelocity, sailboatLocation, updateStatus() ...

      if (checkSailboatCollision(app)) {
        state.gameOver = true;
        // state.score = ...
      } else if (state.sailboatLocation[1] < getLandMapOffset(app)[1]) {
        // level, renderLandMap() ...
      }
    }

  } // ...
}
```

這邊也順手把 `app.state.gameOver` 初始值 false 補上。

上機試玩，可以看到帆船撞到島嶼之後停下來了，遊戲結束，不能再繼續航行，使用 Console 可以看到此時 `app.state.gameOver` 為 `true`，如圖 7-39：

7　Catch The Wind 小遊戲

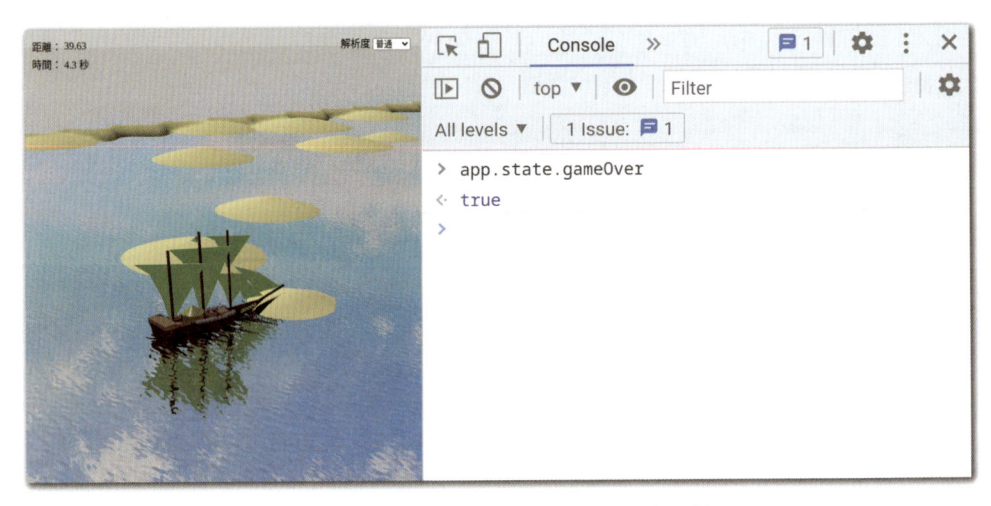

▲ 圖 7-39　撞到島嶼後，遊戲結束狀態

⌘ 沈船動畫

撞到島嶼之後會發生什麼事？筆者第一個直覺反應就是船身破裂、進水，然後沈船，沈船的動畫也不難做，把船頭朝上，船身整體向下（-y）平移，並且如同先前的動畫轉場、演出，每個 frame 慢慢調整，配合原本的微微前後、上下擺動，就可以作到沈船的動畫效果。

來到 `initGame()` 加上沈船動畫所需的狀態：

```javascript
function initGame(app) {
  app.state = {
    // ...
    sailScaleX: 1,
    sailScaleY: 1,
    sailFrontScaleXY: 1,
    sailboatRotationX: degToRad(0),
    sailboatTranslateY: 0,

    sailboatLocation: [0, 0],
    sailboatVelocity: [0, 0],
  };
```

```
  // ...
}
```

　　船頭朝上為 x 軸的轉動由 `app.state.sailboatRotationX` 控制、船體向下的平移由 `app.state.sailboatTranslateY` 控制；在 `gameUpdate()` 中加入遊戲結束之後每個 frame 要做的事：慢慢改變這兩個狀態演出沈船動畫。

```javascript
function gameUpdate(app, timeDiff, now) {
  const { state } = app;

  if (state.started) {
    if (!state.gameOver) {
      // game running logic ...
    } else {
      state.sailboatRotationX =
        Math.min(
          state.sailboatRotationX + timeDiff * 0.00005,
          degToRad(60),
        );
      state.sailboatTranslateY = Math.max(
        state.sailboatTranslateY - timeDiff * 0.00005,
        -1,
      );
    }
  } // ...
}
```

　　最後串到 `renderSailboat()` 上：

```javascript
function renderSailboat(app, viewMatrix, programInfo) {
  const { gl, textures, objects, state, time } = app;

  const worldMatrix = matrix4.multiply(
    matrix4.translate(
      state.sailboatLocation[0],
      0,
      state.sailboatLocation[1],
    ),
```

7　Catch The Wind 小遊戲

```
  matrix4.xRotate(
    Math.sin(time * 0.0011) * 0.03 + 0.03 + state.sailboatRotationX
  ),
  matrix4.translate(
    0,
    Math.sin(time * 0.0017) * 0.05 + state.sailboatTranslateY,
    0,
  ),
);
// ...
}
```

在撞到島嶼、遊戲結束之後會開始播放沈船動畫，圖 7-40 為播放完畢之後的樣子：

▲ 圖 7-40　撞到島嶼、遊戲結束，帆船沈到水裡

筆者故意不讓整個模型沈到水裡，畢竟這個海域這麼多沙丘島嶼應該是個淺水區，水其實沒有非常深，一方面讓玩家可以觀察模型與沈船地點。

7-90

⌘ Game Over 文字看板與 UI

除了沈船之外，把文字看板帶到玩家面前顯示『Game Over』與分數，使得玩家清楚知道自己撞到島嶼導致遊戲結束了，這個 Game Over 文字看板設計如下：

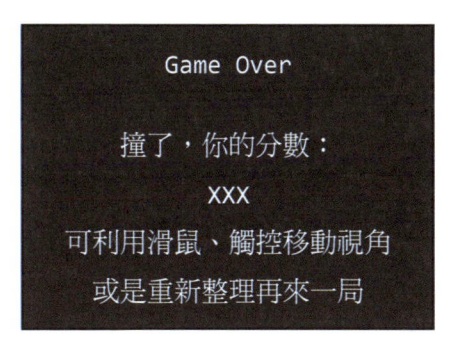

方法在本書已經示範兩次了，使用 canvas2D API 繪製 Game Over 文字看板內容：

```
function createGameOverCanvas(app) {
  const canvas = document.createElement('canvas');
  canvas.width = 1024;
  canvas.height = 1024;

  const ctx = canvas.getContext('2d');
  ctx.fillStyle = 'white';
  ctx.textAlign = 'center';
  ctx.textBaseline = 'middle';

  ctx.font = 'bold 100px serif';
  ctx.fillText('Game Over', canvas.width / 2, canvas.height / 5);

  const secondBaseLine = 2 * canvas.height / 3;
  const secondLineHeight = canvas.height / 8;
  ctx.font = 'bold 70px serif';
  ctx.fillText(
    '撞了，你的分數：',
    canvas.width / 2, secondBaseLine - secondLineHeight * 2,
  );
```

7　Catch The Wind 小遊戲

```
  ctx.fillText(
    '可利用滑鼠、觸控移動視角',
    canvas.width / 2, secondBaseLine,
  );
  ctx.fillText(
    '或是重新整理再來一局',
    canvas.width / 2, secondBaseLine + secondLineHeight,
  );

  ctx.font = 'bold 90px serif';
  ctx.fillText(
    app.state.score,
    canvas.width / 2, secondBaseLine - secondLineHeight,
  );
  return canvas;
}
```

比較特別的是 `createGameOverCanvas()` 接收 app 參數，並且根據分數 `app.state.score` 不同會有不一樣的文字看板內容，先前的文字看板內容實務上應該會用製圖軟體製作好作為遊戲材質的一部分。

為遊戲結束的界面調整加入一個 `showGameOver()` function：

```
// gameUpdate ...
function showGameOver(app) {
  app.input = listenToInputs(app.gl.canvas, app.state);
  const gameOverCanvas = createGameOverCanvas(app);
  app.textures.text = createTextTexture(gl, gameOverCanvas);
}
```

這邊呼叫來自 `lib/input.js` 的 `listenToInputs()` 啟用視角控制，並且將文字看板所使用之 texture 改成新建立的 Game Over 結算內容。

文字看板的位置也需要移動到玩家面前，在 `renderText()` 使用 `app.state.gameOver` 判斷要用那一組 transform 即可：

```
function renderText(app, viewMatrix, programInfo) {
```

7-5 碰撞島嶼判定、結束遊戲

```
// ... textLeftShift
const worldMatrix = !app.state.gameOver ?
  matrix4.multiply(
    matrix4.translate(textLeftShift, 0, 0),
    matrix4.yRotate(degToRad(135)),
    matrix4.xRotate(degToRad(75)),
    matrix4.translate(0, 12.5, 0),
  ) : matrix4.multiply(
    matrix4.translate(
      app.state.sailboatLocation[0],
      0,
      app.state.sailboatLocation[1],
    ),
    matrix4.xRotate(degToRad(75)),
    matrix4.translate(0, 22.5, 0),
  );

  // twgl.setUniforms(), twgl.drawBufferInfo() ...
}
```

主要是利用 `app.state.sailboatLocation` 平移到帆船位置，並且最後移動 22.5，距離玩家比較近，文字看板更大。

文字看板的更動準備完成，回到 `gameUpdate()` 在遊戲結束結算之後呼叫 `showGameOver()`：

```
function gameUpdate(app, timeDiff, now) {
  const { state } = app;

  if (state.started) {
    if (!state.gameOver) {
      // cameraRotationXY, cameraDistance ...
      if (state.sailing) { /* ... */ }
      // sailboatVelocity, sailboatLocation, updateStatus() ...

      if (checkSailboatCollision(app)) {
        state.gameOver = true;
        // state.score = ...
        showGameOver(app);
      } // level, renderLandMap() if needed ...
```

7-93

7 Catch The Wind 小遊戲

```
    }
  } // ...
}
// ...
async function main() {
  // ...
  try {
    // ...

    initGame(app);
    app.input = listenToInputs(app.gl.canvas, app.state);
    // ...
  } catch (error) {
    // ...
  }
}
```

　　別忘了移除在 `main()` 啟用的視角控制，免得玩家在遊戲過程中誤觸、改變視角；至此遊戲結束的文字看板也就完成囉！如圖 7-41：

▲ 圖 7-41　Game Over！遊戲結束文字看板

7-94

7-5 碰撞島嶼判定、結束遊戲

UI 的部份，為了讓遊戲結束更完整，來把向左、向右螢幕按鍵藏起來，並且方便玩家重玩，加入一個一開始藏起來的『重新整理』按鈕，對 HTML 加上所需的 CSS 與重新整理按鈕：

```html
<!DOCTYPE html>
<html lang="en">
<head>
  <!-- ... -->
  <style>
    /* html, body, #canvas, #loading, .btn #status, #options ... */
    #refresh {
      margin-right: 0.5rem;
      padding: 0.2rem 0.5rem;
    }
    #refresh.hidden {
      display: none;
    }
    #ui {
      /* ... */
      display: flex; gap: 1rem;
    }
    /* .sail-btn ... */
    #ui.hidden {
      display: none;
    }
  </style>
</head>
<body>
  <!-- ... -->
  <div id="options">
    <button class="btn hidden" id="refresh">重新整理</button>
    <label for="resolution-ratio">解析度</label>
    <select id="resolution-ratio" name="resolution-ratio">
      <!-- ... -->
    </select>
  </div>
  <div id="ui">
    <button class="sail-btn btn" id="sail-left">↖</button>
    <button class="sail-btn btn" id="sail-right">↗</button>
```

7 Catch The Wind 小遊戲

```html
  </div>
  <script type="module" src="main.js"></script>
</body>
</html>
```

重新整理按鈕身上有一個 `id="refresh"`，使用 DOM API 取得並且使之被按下時呼叫 `window.location.reload()`[11] 執行重新整理：

```javascript
function initGame(app) {
  app.state = { /* ... */ };

  // ...

  document.getElementById('refresh').addEventListener('click', () => {
    window.location.reload();
  });
}
```

最後在 `showGameOver()` 把方向按鈕的容器 `<div id="ui" />` 藏起來，並且把重新整理按鈕身上的 `hidden` CSS class 取消，使之顯示出來：

```javascript
function showGameOver(app) {
  app.input = listenToInputs(app.gl.canvas, app.state);
  document.getElementById('ui').classList.add('hidden');
  document.getElementById('refresh').classList.remove('hidden');
  const gameOverCanvas = createGameOverCanvas(app);
  app.textures.text = createTextTexture(gl, gameOverCanvas);
}
```

好的，這樣一來遊戲結束時的 UI 也調整完畢，整體如圖 7-42，重新整理按鈕放在右上角，如圖 7-43：

---

11 https://developer.mozilla.org/en-US/docs/Web/API/Location/reload

▲ 圖 7-42　遊戲結束時隱藏方向按鈕，使畫面更乾淨

▲ 圖 7-43　重新整理按鈕

⌘ Issue #1：`gl.readPixels()` 相容性

在實做碰撞判定讀取 framebuffer 回到 CPU 時，傳入 `gl.readPixels()` 的 `type`、`format` 兩個參數，事實上 WebGL spec[12] 唯一有定義的組合是 `format: gl.RGBA`、`type: gl.UNSIGNED_BYTE`，也就是一般顏色型別，而其他組合則是要看瀏覽器的實做，目前直接指定 `gl.RED_INTEGER`、`gl.INT` 會導致相容性不佳，筆者在 Chrome 測試沒問題，但是到了 Firefox 就不能用了，幸好 WebGL 有 API 可以直接詢問當前 framebuffer 使用的型別：

12　https://registry.khronos.org/webgl/specs/latest/1.0/#5.14.12

7 Catch The Wind 小遊戲

```
function checkSailboatCollision(app) {
  const { gl, framebuffers, state } = app;

  twgl.bindFramebufferInfo(gl, framebuffers.landMap);
  const format = gl.getParameter(gl.IMPLEMENTATION_COLOR_READ_FORMAT);
  const type = gl.getParameter(gl.IMPLEMENTATION_COLOR_READ_TYPE);

  // ...

  gl.readPixels(
    left, top, width, height,
    format, type,
    /* format: */ gl.RED_INTEGER,
    /* type: */ gl.INT,
    app.sailboatAltitudes,
  );

  // ...
}
```

`gl.getParameter()`[13] 可以用來詢問許多 WebGL 設定的參數、狀態，這邊詢問的值就如同字面上的意思：

- `gl.IMPLEMENTATION_COLOR_READ_FORMAT`：瀏覽器實做的讀取用 `format`
- `gl.IMPLEMENTATION_COLOR_READ_TYPE`：瀏覽器實做的讀取用 `type`

把詢問到的 `format`、`type` 傳入 `gl.readPixels()`：，但是改完之後會發現 Firefox 依然沒有辦法正確進行碰撞測試，在 Console 上看到這行錯誤：

```
WebGL warning: readPixels: buffer too small
```

經過檢查之後，發現 Firefox 在這邊的 `format` 是 `gl.RGBA_INTEGER`，並且在加大 `app.sailboatAltitudes` 的大小可以看到讀取完畢的 `app.`

---

[13] https://developer.mozilla.org/en-US/docs/Web/API/WebGLRenderingContext/getParameter

7-98

sailboatAltitudes 是真的 RGBA 四個元素一個 pixel：

```
>> app.sailboatAltitudes
<- ▼ Int32Array(2800) [ -65536, 0, 0, 1, -65536, 0, 0, 1,
    -65536, 0, … ]
    ▼ [0…99]
        0: -65536
        1: 0
        2: 0
        3: 1
        4: -65536
        5: 0
        6: 0
        7: 1
        8: -65536
        9: 0
        10: 0
        11: 1
        12: -65536
        13: 0
        14: 0
        15: 1
        16: -65536
        17: 0
        18: 0
        19: 1
```

▲ 圖 7-44　Firefox 上使用支援的 format gl.RGBA_INTEGER 讀取出來的結果

這樣前面 WebGL 的警告大小不足也就說的通了，在 `format` 為 `gl.RGBA_INTEGER` 的狀況下，建立 `app.sailboatAltitudes Int32Array` 時要把大小乘以 4，同時我們關注的資料是 index 為 0、4、8、12 的元素，因此定義一個 dataStep 表示『檢查時每次要跳的 index 數量』，根據 `format` 決定是 1 或是 4，並乘在長度相關的運算中：

```
function checkSailboatCollision(app) {
  // format, type ...
  const dataStep = format === gl.RGBA_INTEGER ? 4 : 1;
  if (!app.sailboatAltitudes) {
    app.sailboatAltitudes =
      new Int32Array(width * height * 2 * dataStep);
  }
  // gl.readPixels() ...
  for (
```

7 Catch The Wind 小遊戲

```
    let i = 0, len = width * height * dataStep;
    i < len; i += dataStep
) {
    if (app.sailboatAltitudes[i] > 0) return true;
}
return false;
}
```

⌘ Issue #2：沈船倒影

帆船沈下去之後，稍微轉一下畫面可以發現一個詭異的狀況：

▲ 圖 7-45　可以看到在水面下的帆船的倒影

7-100

為什麼會這樣呢？回想 CH5-2 製作鏡面效果時，我們直接切換到鏡面中的鏡頭來拍攝，意味著水面下的物件如果有被鏡面相機拍到，那麼繪製水面時取樣到對應的位置依然會出現水面下的物件；對於這個問題筆者使用一個簡單的方式解決：在 fragment shader 檢查發現該點在水面下時，`outColor` 的 alpha 值回傳 0 使之變成透明：

```glsl
// fragmentShaderSource ...
in mat3 v_normalMatrix;
in vec4 v_lightProjection;
in vec4 v_worldPosition;
// ...
void main() {
  // ...
  outColor = vec4(
    clamp(
      diffuse * diffuseLight +
      u_specular * specularBrightness +
      u_emissive,
      ambient, vec3(1, 1, 1)
    ),
    v_worldPosition.y < -0.1 ? 0 : 1
  );
}
```

為了得知是否在水面下，需要 `v_worldPosition` 這個 varying，表示該點在場景中的位置，稍微檢查 vertex shader 發現都有實做好了，因此直接加上 `in vec4 v_worldPosition`，並實做『水面下呈透明』邏輯。

可惜改完之後，結果還是不正確：

7-101

7　Catch The Wind 小遊戲

▲ 圖 7-46　水面下帆船船體形成了黑影

　　不知道有沒有讀者想到為什麼？這是半透明物件繪製順序的問題，在繪製鏡面世界時，先畫了帆船，水面下的部份呈現透明因而是黑色，但是深度圖被寫入了，因此在畫天空時比對發現前面有東西，因而留下船身在水面下的黑影；解決問題的方法是調整渲染順序，從『帆船、天空』改成『天空、帆船』，並且因為帆船、天空的 program 不同，連帶著 program 也需要調整切換順序：

```
function render(app) {
  // lightProjection ...

  gl.useProgram(programInfos.main.program);
  twgl.setUniforms(programInfos.main, globalUniforms);
```

7-5 碰撞島嶼判定、結束遊戲

```
{ // reflection
  twgl.bindFramebufferInfo(gl, framebuffers.reflection);

  gl.clear(gl.COLOR_BUFFER_BIT | gl.DEPTH_BUFFER_BIT);

  renderSailboat(app, reflectionViewMatrix, programInfos.main);

  gl.useProgram(programInfos.skybox.program);
  renderSkybox(app, projectionMatrix, inversedReflectionCameraMatrix);

  gl.useProgram(programInfos.main.program);
  twgl.setUniforms(programInfos.main, globalUniforms);

  renderSailboat(app, reflectionViewMatrix, programInfos.main);
}

gl.bindFramebuffer(gl.FRAMEBUFFER, null);
// ...
}
```

好的，這樣改完之後沈船倒影就正確囉：

▲ 圖 7-46　修正過後的沈船倒影，不再有水面下的成像出現

7-103

7 Catch The Wind 小遊戲

至此『Catch the Wind!』小遊戲就完成啦！完整程式碼可以在這邊找到：

</>	ch7/07/index.html
</>	ch7/07/main.js

也可以參考上線版本，透過瀏覽器打開此頁面：

https://webgl-book.pastleo.me/ch7/07/index.html

7-6 結語

▲ 圖 7-47 　『Catch the Wind!』標題畫面

不知道讀者是否有嘗試認真玩玩看『Catch the Wind!』呢？筆者可是有用遊戲性的思維去調整、設計遊戲的呢，希望可以讓玩家覺得好玩，筆者自己認真嘗試大約可以拿到 2500 分左右，不過當然有運氣的成份在啦，地圖產生是由遊戲開始時間經由偽隨機生成的。

▲ 圖 7-48　筆者認真嘗試大約會落在 2500 分

　　本書手把手的從一個三角形的繪製到整個『帆船與海』場景，最後製作成『Catch the Wind!』小遊戲，而且僅使用 TWGL 作為讓程式碼簡短的輔助，整個繪製流程還是使用底層的 WebGL API 來製作，希望本書有讓各位學習到東西，筆者其實在踏入業界很早期就知道有 WebGL 這個東西的存在，但是因為要做到有些成果需要非常多基石而一直沒有深入下去研究，就如同各位看到的，本書於 CH3 才在畫面上出現比較實際的 3D 畫面，不過也因此學到非常多東西，也更能了解 3D 渲染系統或是遊戲引擎的設計，畢竟 GPU 底層的運作方式與限制是大同小異的，例如說 3D 物件的 transform 之所以叫做 transform 是因為在 GPU、shader 的實做事實上是使用線性代數的矩陣進行 transform。

　　文章中製作的範例主要是示範該文主旨的概念，筆者為了讓這些成品比較有成品的感覺，有時候會直接在程式碼中出現一些調校好的數字，筆者在實做這些範例時當然不會是第一次就完美的，都是一改再改，改完覺得滿意了才用一個理想的順序去描述實做的過程作為文章內容，甚至最後是事先製作出『Catch the Wind!』完整的遊戲，再為讀者們拆解各個功能來介紹，實際開發應用程式肯定會有不少來來回回，在遊戲上更是如此，筆者在地圖產生的 procedural generation 就嘗試過許多版本，成果覺得不好玩就回去繼續改，最終採用讀者們所看到的島嶼產生方式。

7 Catch The Wind 小遊戲

　　同時，本書幾乎每個程式碼變動貼出來的手把手教學方式，是希望讓讀者們比較不容易失去方向，並藉由實做獲得成就感；但是若要真的掌握一項技術，那麼除了跟著範例製作之外，還需要有『產出』才是真的學會，例如改變 3D 場景的內容，使用不同的 obj 檔案，改變遊戲機制等等，通常筆者看到一項技術跟著範例做完之後，會嘗試沿著範例發想一些可以做的改變，然後自己實做出來，在這個過程中多多少少會踩到跟著範例沒有踩到的困難點，但是製作完成之時才是真的表示自己是或多或少能掌握這項技術的。

　　在最後筆者要感謝 @greggman[14] 撰寫的 WebGL2 Fundamentals[15]、WebGL Fundamentals[16] 與 Three.js Fundamentals[17]，有完整、深入的教學讓筆者可以有系統地學習，為了分享研讀的成果，寫下 2021 年鐵人系列文章『如何在網頁中繪製 3D 場景？從 WebGL 的基礎開始說起』[18]，並延伸改良文章內容出版本書以分享給各位讀者，感謝各位的閱讀！

14　https://greggman.com/
15　https://webgl2fundamentals.org/
16　https://weblglfundamentals.org/
17　https://threejsfundamentals.org/
18　https://ithelp.ithome.com.tw/2020-12th-ironman/articles/3929

Note

Note

Deepen Your Mind

Deepen Your Mind